# MODERN CLIMATOLOGY

# MODERN CLIMATOLOGY

**P.K. Saha**

**ALLIED PUBLISHERS PRIVATE LIMITED**

New Delhi • Mumbai • Kolkata • Chennai • Nagpur
Ahmedabad • Bangalore • Hyderabad • Lucknow

**ALLIED PUBLISHERS PRIVATE LIMITED**

1/13-14 Asaf Ali Road, New Delhi–110002
Ph.: 011-23239001 ● E-mail: delhi.books@alliedpublishers.com

47/9 Prag Narain Road, Near Kalyan Bhawan, Lucknow–226001
Ph.: 0522-2209942 ● E-mail: iko.hooms@alliedpublishers.com

17 Chittaranjan Avenue, Kolkata–700072
Ph.: 033-22129618 ● E-mail: cal.books@alliedpublishers.com

15 J.N. Heredia Marg, Ballard Estate, Mumbai–400001
Ph.: 022-42126969 ● E-mail: mumbai.books@alliedpublishers.com

60 Shiv Sunder Apartments (Ground Floor), Central Bazar Road
Bajaj Nagar, Nagpur–440010
Ph.: 0712-2234210 ● E-mail: ngp.books@alliedpublishers.com

F-1 Sun House (First Floor), C.G. Road, Navrangpura,
Ellisbridge P.O., Ahmedabad–380006
Ph.: 079-26465916 ● E-mail: ahmbd.books@alliedpublishers.com

751 Anna Salai, Chennai–600002
Ph.: 044-28523938 ● E-mail: chennai.books@alliedpublishers.com

5th Main Road, Gandhinagar, Bangalore–560009
Ph.: 080-22262081 ● E-mail: bngl.books@alliedpublishers.com

3-2-844/6 & 7 Kachiguda Station Road, Hyderabad–500027
Ph.: 040-24619079 ● E-mail: hyd.books@alliedpublishers.com

**Website:** www.alliedpublishers.com

© 2012, Allied Publishers Pvt. Ltd.

ISBN : 978-81-8424-756-5

Typeset at Business Port, 9 Ganesh Chandra Avenue, Kolkata 700013

Published by Sunil Sachdev and printed by Ravi Sachdev at Allied Publishers Private Limited
Printing Division, A-104 Mayapuri Phase II, New Delhi 110064

# PREFACE

Since time immemorial weather and climate in any part of the world shape the destiny of people. Not only the human, all the flora and fauna have the imprint of weather and climate on their living system. In other words the weather and climate become part of the natural environment. The understanding of weather and climate helps innate perception of sustenance of life on our planet.

There is now growing keen interest on weather and climate at all levels in the society. Study of climate has gained importance amongst the academics and been transforming to a *frontier science*. In the later half of the twentieth century there has been remarkable advance in science and technology. Innovative ideas and spectacular observations have resulted in new interpretation of climate science. Many new concepts have emerged breaking the barriers of traditional outlook.

In the first decade of the twenty-first century we are now concerned to critically assess the role of climatic processes on a pedestal of innovative science. Climatic anomalies are now interpreted to modulate the climate on a global scale. Climate change appears to be inevitable and has also become a global lively debate. Global warming has become a great concern to our ecosystem, whether anthropogenic and natural. Weather disturbances affect humans, sometime causing unprecedented loss of life and property. Interestingly climatic disturbances are on the rise. With our limited knowledge we fail to forecast and take appropriate action to mitigate sufferings.

This book contains twelve chapters explaining the basics of weather and climate, climatic elements and processes, climatic events and anomalies, global warming and climate change issues, relationship of climate with ecosystem, impact of climate on diseases and classification of world climates. Chapter I introduces *Understanding the Basics of Weather and Climate*. Chapter II, Chapter III and Chapter IV elaborate *Solar Energy, Motion in the Atmosphere* and *Moisture in the Atmosphere: Forms and Characteristics*. The role of *Airmass* has been discussed in Chapter V. Chapter VI and Chapter VII include discussion on *Weather Disturbances* and *Climatic Anomalies* respectively. Chapter VIII presents in depth studies on *Global Warming*. Chapter IX covers various aspects of *Climate Change*. Chapter X explains the interaction between *Climate and Ecosystem*. Chapter XI portrays the causal relationship of *Climate and Diseases*. Chapter XII approaches *Classification of Climates* based on varying aspects.

This treatise aims at understanding the climatic processes and delving into causal ambit of the problems with modern scientific parameters. It is a colossal task to arrange large variable data and information into a systematic study on climate. Literatures from various sources have been collected that include published text books, research articles, notes and sharing ideas through consultation. My experience for more than a three decade of teaching climatology has endowed me the insight and strength to write this book. There has been dearth of useful textbooks on climatology for the Indian students. After my retirement from the service I considered it my duty to contribute a little for the students and also others in different institutes pursuing studies and research in climate science. My endeavour will be fruitful if this book on climate science can benefit them.

I acknowledge the help of many persons in preparing the texts. Namita, my wife, all along encouraged me to write the book that may transcend the horizon of knowledge. She took all pains during the period for preparation of the text. Many of my students always look forward for the publication of this book. Write-up of the text is one thing and publication of the book a different one. It was my privilege to have a reputed publisher like Allied Publishers, agreed to publish the book. Mr. Jayant Manaktala of Allied Publishers in Kolkata took so much initiative in the publication of this book. I will be ever grateful to him for coordination and help. Thanks are also due to other colleagues of M/s Allied Publishers for printing and production of the book.

**Pijushkanti Saha**

# CONTENTS

# LIST OF TABLES

# LIST OF FIGURES

# Chapter 1

## UNDERSTANDING THE BASICS OF WEATHER AND CLIMATE

### Planet Earth

A gaseous envelope covers the planet earth. Essentially this cover has created an environment for sustenance of life for all species on the planet earth. Our blue planet possesses a green environment under this gaseous envelope. We call this cover the atmosphere of our earth. All weather processes are generated in the atmosphere and influence the biotic processes for living. This is perhaps unique in the universe. The atmosphere is constituted principally with gases like *nitrogen* and *oxygen*. Nitrogen is the source of amino acids that constitutes life. Oxygen is essentially required for respiration of all animal species and for burning. Other important gases are *carbon dioxide* and *ozone*. Carbon dioxide is required for photosynthesis by the plant species. Carbon dioxide also plays important role in moderating temperature conditions on earth's surface. Carbon dioxide in the atmosphere absorbs the radiating energy from the earth to keep the earth's surface warm for living. If there be no carbon dioxide in the atmosphere the temperature on earth's surface will drop to –18 degree Celsius from +15 degree Celsius. The earth's surface would then be the frozen surface inhibiting life systems. Ozone, though negligible in the atmosphere protects life on the earth's surface by absorbing the ultra violet rays. We shall not forget the importance of atmosphere and do not disturb the natural processes by our unwise action.

Long term weather phenomena generate the climatic conditions. Climate of any region on the earth's surface is defined as the expression of weather conditions based on temporal characteristics. Statistically the period may be thirty years where the temporal variations may be negligible. The term 'climate' has been derived from the Greek word *'klima'*. *'Klima'* means the slope of the earth. The Greek scholars considered the variation of regional climate due to variation of slope of the earth's surface. It resembles the present understanding of *latitudes.* As the axis of the earth inclines at an angle of sixty six degree, the insolation varies from the equator to the poles. The *'logos'* means the study. Hence the term 'climatology' emerges. Weather is the day-to-day state of the atmosphere, and is not in a stable state and characterised by dynamic processes. On the other hand, *climate*—the average state of weather—is fairly stable and predictable. Climate includes the average temperature, amount of precipitation, days of sunlight, and other variables that might be measured at any given site. However, there are also changes within the Earth's environment that can affect the climate.

**Nature and Natural Processes**

Our blue planet is unique in the universe. This is our mother earth. We live on the earth's surface. Not only the humans, but also all living species get their sustenance from mother earth. So earth is not only beautiful—it elegantly displays prerequisite conditions for living of all species. Our earth is a planet in the solar system. Sun in the solar system is the source of all energy. It is not exactly known whether any planet bears the sign of life. In that context the earth is unique and bountiful of living organisms as nature is endowed with favourable conditions for biotic regeneration.

For our better understanding of climate and ecology it is necessary to study nature and natural processes in brief. The important issues have been discussed in detail in the later chapters.

**Planet Earth: Origin and Evolution**

As mentioned the earth is a planet and a small part of the solar system. The solar system belongs to the universe. The universe is defined as the total of all the matter, energy and space. So in the universe we observe vast space along with many galaxies and constellations. The galaxies and constellations contain innumerable number of stars, planets and meteorites. A galaxy is an island of matter in space—where tremendous collection of gas, dust and many hundred millions of stars are present. The galaxy we live in is known as the Milky Way. It has a disc like shape and the earth is located halfway out from the centre. As we look toward or away from the centre or in any direction in the plane of that disc we find many stars and much dust and gas. For better understanding of the origin of earth as a planet it is desirable to have a glimpse of the celestial bodies.

**Evolution of Stars**

Stars initiate with the collapsing of gases of clouds in interstellar spaces and turn to be stars when nuclear activity starts in their interiors, fusing hydrogen into helium. In the type of nuclear fusion that occurs inside the sun and most other stars, four hydrogen nuclei fuse to become one helium nucleus. During their active period energy from the fusion creates a pressure that pushes outward, balancing the inward force of gravity. All stars have life cycles—they are born and die. Usually the hottest stars, much hotter than the sun have shorter lifetime as they spend their fuel at a furious pace, become cool or eventually die within 100,000 years. Stars similar to sun can survive for ten billion years. Stars, cooler than the sun, can live much longer, even 50 billion years or more.

When the stars spend all the hydrogen in their interiors, they can no longer fuse hydrogen into helium. As a result they cannot offset gravity and their interiors begin to contract. During the contraction energy is released and outer layers are pushed out. These outer layers become larger and cooler, and the stars become *Red Giants.* Finally the outer layers of some Red Giants are disconnected and float outward, having nearly 20 percent of the star's mass out into space. Such expanding shell of gas is known as *Planetary Nebula.* Nebulae are the clouds of gas and dust that look hazy to our view. A few nebulae represent shells of gas ejected by dying stars. The escaped nebula appears blue as it shows the inner layers of the star and is hot. Say after 50,000 years or so, the planetary nebula drifts away, and the central star cools. As a result the central star contracts until it is about the size of earth. If a mass less than 1.4 times the mass of the sun is left, the star continues at this stage indefinitely. This stage is called *White Dwarf.* The objects like Red Giants, nebulae and White Dwarfs are actually different stages in the life history of ordinary stars.

After a star becomes a Red Giant, it grows larger and turns to be super-giant. At that point it may explode totally, giving birth to a *Supernova.* A Supernova is the brightest object in the galaxy. The Supernova completely destroys the star, forming heavy elements and then ejecting these elements out into space. When our solar system was formed, the sun and the planets contained these heavy elements. It may be presumed that our solar system came into being when a Supernova made a cloud of gas and dust began to collapse. During a Supernova explosion, the core of the star can be extremely compressed and collapsed until it is quite small. Stars having 1.4–4 times mass of the sun collapse until they reduce to a quite small size. Most of the stars at that point contain gas of neutrons only, or of smaller particles called *Quarks* that make up neutron. The star is then known as *Neutron Star.*

In some cases a mass greater than about 4 times that of the sun exists even after a supernova explosion. Then the collapse of the star is inevitable. Because of its gravity, it becomes more and more compressed. The *Theory of Relativity* suggests that even light will be bent so much by the strong gravity that the light no longer will be able to escape. In fact nothing would escape. The star will turn to be a *Black Hole.* Black Holes are even fainter than the Neutron Stars.

**Sun in the Solar System**

Most of the stars are observed at night, the sun is only visible during day hours. Sun is the closest star to earth, 150 million kilometre away. The light from sun takes only 8.3 minutes to reach the earth's surface, whereas the nearest distance of any other star is more than 4 light years. The sun like all other stars is a ball of hot gas. It has a very hot core and relatively less hot atmosphere. The light and heat emitted from the surface of sun provide energy to the earth, which is of great importance both to climate and ecology. The sun is the sole determinant of existence of all species on earth.

*Energy*: The energy is created at the core of the sun's interior by the process of nuclear fusion in which hydrogen is converted to helium. Each helium atom that results from fusion has less mass (0.7 percent) than the sum of four hydrogen atoms. Albert Einstein's equation, $E = mc^2$, explains that the little bit of disappeared mass will be transformed on multiplication by $c^2$, (the square of the velocity of light) into a large amount of energy. The sun is a good example of a nuclear fusion reactor that humans have not yet been able to master for power generation.

*Photosphere*: The solar surface is called the *Photosphere*. The term has been derived from the Greek word *photo*, meaning light. The surface of the sun has, in general, uniform brightness, being slightly darker along its edges. Even there are few darker areas on its surface. These dark areas are relatively cooler regions of the solar surface and are called *Sunspots*. The Sunspot cycle significantly influences the global climate.

## Origin of the Atmosphere

Present atmosphere is constituted by nearly 4/5th of nitrogen and 1/5th of oxygen along with little amount of carbon dioxide, argon and many other gases in trace amounts. Since the inception of the earth the atmosphere had undergone many changes in different geological periods in the past. The atmosphere had been stabilised through various processes in the present form by the Cambrian period, about 580 million years ago.

It is observed that the elements which constitute the earth's atmosphere seem to be scarce in other planets of the solar system. Plausible explanation for its origin might be sought by understanding the composition of the earth's atmosphere. Presently dominant gases are: nitrogen (78.08%), oxygen (20.95%), argon (0.93%) and carbon dioxide (0.035%).

All these gases form 99.9 percent of the total volume of the atmosphere. Two lightest elements, i.e. hydrogen and helium are found in the outer reaches of the earth's atmosphere, but rarely found close to the earth's surface. This condition developed through eons of geologic time along with slow changes that shaped the present lithosphere and hydrosphere. At the initial stages of the inception of the earth from cosmic gases a weak gravitational field persisted and might have allowed the lightest gases like hydrogen and helium to wander close to the space. Moreover the present gases that prevail in the atmosphere are not direct residues of the earliest form of the planet. These gases probably have been transformed, constituted the second atmosphere, which evolved from volcanic eruptions, hot springs, chemical breakdown of solid matter, later moderated by the plant life. At the beginning there might have been an excess of water vapour (60–70%), carbon dioxide (10–15%) and nitrogen (8–10%), but no oxygen in the earth's atmosphere. The condition of the earth's surface was very hot and with cooling the resultant condensation yielded incessant rainfall continuing for long geological periods. The water from the rainfall filled in the cavities of the earth's surface to form the oceans, seas and lakes. Thereafter there was the evolution of plant life. The emergence of plant life on earth had tremendous effect upon the atmosphere. Slowly but steadily there was transformation of the atmosphere, plant life assimilating carbon dioxide from the atmosphere for photosynthesis and releasing oxygen to the atmosphere.

Even today the interaction of land, water, air and plant and animal life constantly uses and renews the atmosphere and a delicate balance is maintained for the principal constituent gases of the atmosphere. The decay of plants, weathering of rocks, burning of fuel and breathing of animals—all use oxygen and release carbon dioxide. On the other hand, the balance is maintained by the utilisation of carbon dioxide and release of oxygen by the plant life and also partly by the oceans. Even there has been exchange of carbon dioxide between the oceans and the atmosphere—an estimated 200 billion tons of carbon dioxide is exchanged per year. Nitrogen undergoes a complex cycle of transformation through bacterial activity in the soil, animal tissue, organic processes in decay, and released again to the atmosphere.

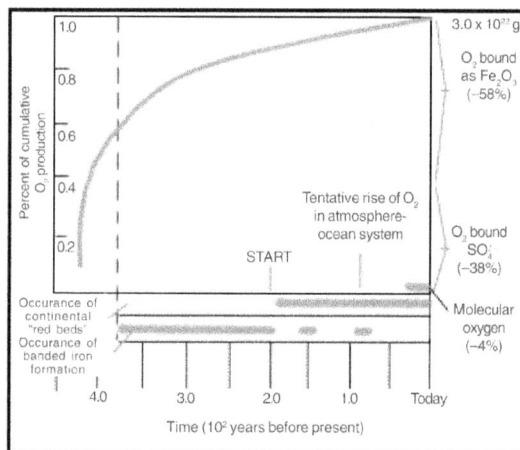

Figure 1.1: Cumulative Oxygen Production in the Earth-Atmosphere System

Such transformation of the status of the atmosphere is very important in understanding the environment of the past and also the present. We will also examine the changing pattern of climate in the past and present decades with particular reference to global warming in later chapter.

**Components of the Earth**

We can distinguish different components around us. These are: *Abiotic Components—* (i) Atmosphere, (ii) Lithosphere and (iii) Hydrosphere and *Biotic Components—* Biosphere.

**Atmosphere**

The atmosphere is one of the principal components of environment. The atmosphere means the gaseous envelope of air which clings to the surface of earth. It is as much a part of the earth as the land and the oceans but differs from them in many respects. The nature, characteristics and different processes in the atmosphere determine the status of environment. It is worthwhile to understand the nature, characteristics and dominating processes in the atmosphere.

The atmosphere in other word is called *air*. Air is the mixture of a number of gases, water vapour and a variety of fine particulate and suspended materials. The gaseous mantle, called air, has some unique characteristics.

*Nature of Air*

Pure air appears to be colourless, odorless, and tasteless. The existence of air cannot be felt except in motion. The other characteristics of air are its mobility, compressibility and expanding ability. All such characteristics are of immense importance, as we find density of air varying with altitude. The density of air decreases with increasing altitude, so the air close to the earth's surface exerts more pressure. It can also transmit compression waves such as sound waves. Air is transparent to many forms of radiation, but not to some *Green House Gases* (GHGs), otherwise the earth would have been turned into frozen land surface. Air also forms a canopy in offering resistance to the objects like meteors that invade the atmosphere from the outer space. The meteors are destroyed through friction before they can reach the earth's surface. Without the air there would have been no clouds, winds, storms, and better to no generation of weather systems. As a result there could not be any life-form on earth's surface. The envelope of air protects the life-forms as shield from the ultraviolet radiation of the sun. It is unique feature of our living planet.

The total mass of the atmosphere is approximately $5.2 \times 10^{15}$ tonnes* and nearly 50 percent of the total mass of the atmosphere (air) lie within 5500 metres (5.5 km.) and about 90 percent lie within 30 kilometres from the surface of earth.

*Composition of Air*

We have broadly accounted for the composition of the atmosphere in our discussion on the origin of atmosphere. The air in the atmosphere is a mixture of different constituent gases, but not a compound as all the constituent gases in atmosphere maintain their unique individual properties. Moreover it has been found that the composition of different gases in the lower reach of the atmosphere is nearly constant. Hence the zone (up to 100 km.) of this constant composition of constituent gases is termed *Homosphere*. Beyond this limit the composition of gases differs, and is called the *Heterosphere*.

The composition of different gases in the Homosphere is considered very important, as this is the most effective zone of weather-making, which exerts tremendous control over the

---

* Calculated on the basis of air pressure (weight) per unit area and total area of the earth's surface as 1.026 kg. per sq. cm. and $5.1 \times 10^{18}$ sq. cm. respectively.

environment. In the following table the composition of different gases in the lower atmosphere has been stated.

**Table 1.1:** Compositions of Different Gases in the Lower Atmosphere
(0–25 km.)

| Element | Symbol | Percent by volume (dry air) |
|---|---|---|
| Nitrogen | $N_2$ | 78.08 |
| Oxygen | $O_2$ | 20.98 |
| Argon | Ar | 0.93 |
| Carbon dioxide | $CO_2$ | 0.035 |
| Neon | Ne | 0.0018 |
| Helium | He | 0.0005 |
| Ozone | $O_3$ | 0.00006 |
| Hydrogen | H | 0.00005 |
| Krypton | Kr | Trace |
| Xenon | Xe | Trace |
| Methane | $CH_4$ | Trace |
| Water vapour | $H_2O$ | Small but variable (0–4 %) |

From Table 1.1 it appears that nitrogen, oxygen, argon and carbon dioxide constitute nearly 99.99 percent of the total volume of air in the lower atmosphere. Other constituent gases are neon, helium, ozone, hydrogen, krypton, xenon and methane. Of these, ozone is very important, though it occupies only 0.00006 percent volume of air and when compressed its thickness would be only 3 mm. in hundreds of kilometres of the atmosphere. As shown in Table 1.1, "dry air" is a mixture of nitrogen, oxygen, argon, carbon dioxide, etc., which are all, practically speaking, perfect gases, and so obey the gas laws of Boyle, Charles and Avogadro. The composition of dry air is nearly homogeneous below 20 km. except for water vapour and ozone, whose concentrations vary greatly.

**Properties of Principal Gases in the Atmosphere**

*Nitrogen*

Nitrogen is an inert gas. Nitrogen cannot combine directly with other gases. Nitrogen is present in different organic compounds particularly in proteins. Amino acids, essential for the creation of life, are derived from nitrogen. The atmosphere is the storehouse of free gaseous nitrogen, whereas nitrogenous compounds are found in the bodies of the organisms and soils. The living organisms, except some nitrogen fixing bacteria like *Azotobacter, Clostridium, Derxia, Rhizobium* etc., cannot utilise nitrogen directly from air. It is to be converted into nitrate to be utilised by the plants. The plants use them in the synthesis of amino acids and proteins and convert them into their tissues. Herbivores consume the plant proteins and convert them into their tissues. Carnivores get proteins from the herbivores and convert them into their body tissues. The dead plants and animal bodies and their excreta are decomposed by various decomposers and subsequently gaseous nitrogen is released to the atmosphere. The nitrogen cycle in the biosphere is very important for the sustenance of life. We will discuss the nature and significance of nitrogen cycle below.

*Nitrogen Cycle*

The *nitrogen cycle* in general resembles the carbon cycle, but differs in some respects. The nitrogen cycle is more complex than the carbon cycle. Nitrogen is the source of proteins and nucleic acids and in that sense it is essential for life. Atmosphere is the reservoir of free gaseous nitrogen as it constitutes nearly 80 percent of the atmosphere by volume. Nitrogen as compound also resides in the bodies of living organisms and in the soils. But free nitrogen cannot be directly utilised by the organisms, with the exception of few nitrogen-fixing bacteria and blue green algae.

As mentioned the nitrogen cycle being a complex one, gaseous free nitrogen in the atmosphere is converted into ammonia or oxidised to nitrate forms at different stages. Nitrogen-fixing bacteria and blue green algae play significant role in converting that gaseous nitrogen into organic compounds and finally to nitrates soluble in water. Nitrates are utilised by plants for the synthesis of amino acids and proteins. The plant proteins are consumed by the herbivores, which convert them into their tissues. When the plant and animals die or their remains are decomposed by various organisms, the nitrogen is again released to the atmosphere.

The steps in the nitrogen cycle are: Nitrogen fixation—Ammonification—Nitrification—Denitrification—Release of gaseous nitrogen.

Nitrogen fixation is done by (i) non-biological or electrical methods, and (ii) biological methods. Thunderstorm and lightning play a significant role in producing nitrates by non-biological or electrical methods. Blue green algae like *anabaena, nostoc* and *oscillatoria* fix nitrogen by biological methods. Many free-living bacteria can also fix nitrogen—both non-symbiotically and symbiotically. Of the non-symbiotic nitrogen-fixing bacteria *clostridium, azotobacter,* and *derxia* are important. *Rhizobium* living in root nodules of leguminous plants can fix nitrogen symbiotically.

Ammonification is the process in which nitrogen in organic matter of dead plants and animals is converted to ammonia and amino acids.

In the nitrification process ammonia is first converted into nitrites by a group of microorganisms like *nitrosomonas, nitrospira, nitrosogloea* and *nitrococcus*. Nitrites are then converted to nitrates mostly by *nitrobacter* and *nitrocystis*.

Denitrification is the process in which ammonia and oxides of nitrogen are reverted back to nitrogen by different forms of bacteria, namely *bacillium cereus, B. licheniformis* and *P. denitrificans.* The gaseous nitrogen is released to the atmosphere and the cycle continues.

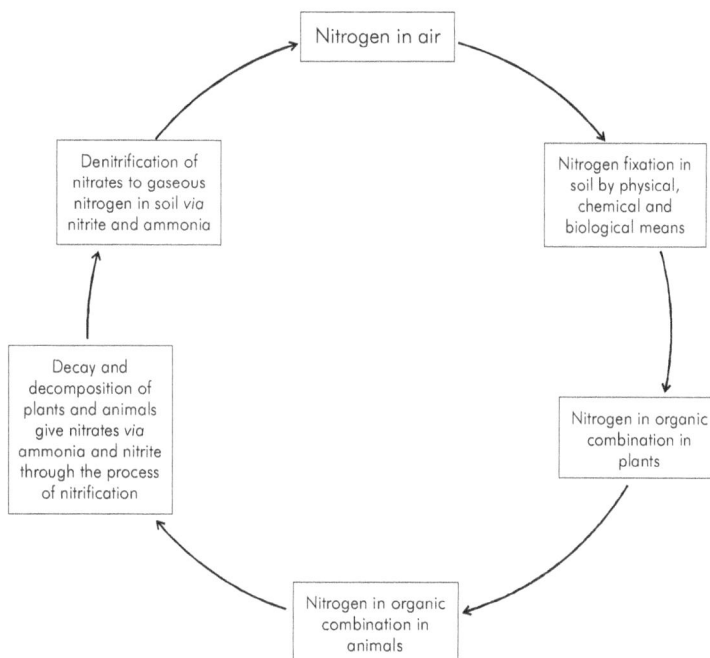

**Figure 1.2:** Nitrogen Cycle

## *Oxygen*

Oxygen is required for oxidation that includes combustion and burning of anything. It can readily combine with many other chemical elements. Oxidation is a very crucial process in the environment responsible for chemical weathering of rocks. Oxygen is considered a life-saving gas, as all organisms except anaerobic microbes require free oxygen for their respiration. Oxygen helps the animal metabolism in maintaining the energy and temperature level in the body. Oxygen is released into the atmosphere by the plants through the process of photosynthesis. For this reason the role of plants in maintaining the level of oxygen in air has been stressed upon in recent years by the scientists. The oxygen cycle, that maintains a balance between oxygen consumption and its regeneration, is also considered very important to the biosphere.

### *Oxygen Cycle*

The *Oxygen Cycle* is relatively a simple one. Oxygen constitutes nearly 21 percent of the atmosphere by volume. Oxygen is also present in water and rocks in combined form. Free oxygen in air is essential for living animals to breathe, whereas the aquatic animals get their sustenance from the dissolved oxygen (DO). The status of oxygen is exactly balanced in the atmosphere. The green plants assimilate carbon dioxide in photosynthesis and release oxygen from water to the atmosphere. So the air becomes fresh. The plants and animals use oxygen of the air in respiration and release carbon dioxide that is accurately balanced by the uptake of carbon dioxide and the release of oxygen by green plants in photosynthesis.

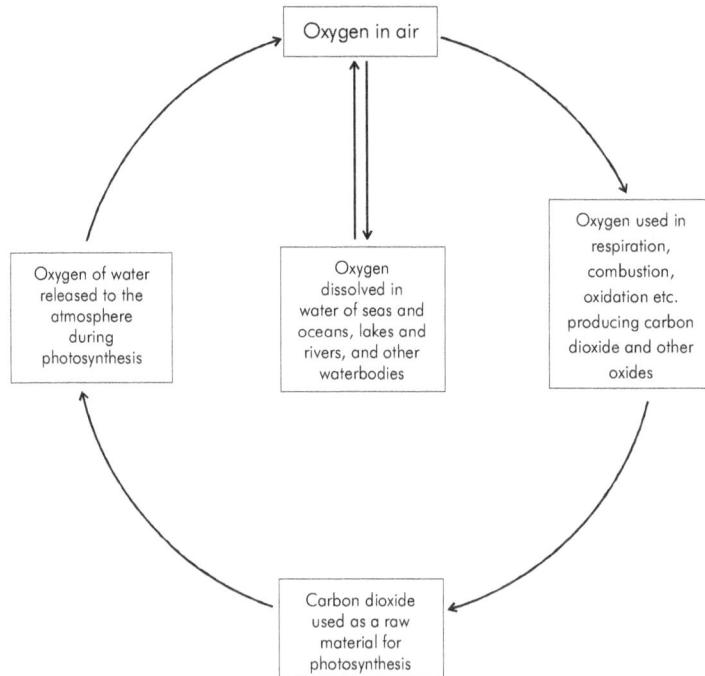

**Figure 1.3:** Oxygen Cycle

### *Carbon Dioxide*

Carbon dioxide is formed due to burning of organic carbon-containing matter. Due to respiration plants and animals also release carbon dioxide in the atmosphere. On the other hand plants consume carbon dioxide for their sustenance by the process of photosynthesis and in the process release oxygen and water vapour into the atmosphere. Carbon dioxide is a Green House Gas (GHG). Carbon dioxide absorbs infrared radiation radiated back from the earth in the long-wave length region of the spectrum. Such absorption of energy in the troposphere causes warming of the earth's surface. Actually by holding back the terrestrial radiation carbon dioxide plays significant role in making the earth habitable of all species. If there be no Green House Gas like Carbon dioxide in the earth's atmosphere, the temperature of earth's surface would have dropped to $-18$ degree Celsius, thus carbon dioxide contributes

+33 degree Celsius rise of temperature. So if there be increase in carbon dioxide content in the atmosphere, the global temperature will tend to rise. It has been observed that there had been a tremendous rise of carbon dioxide with increased burning of fossil fuels since 1860. Incidentally 1860 was the beginning of industrial age. In recent years, particularly after 1950 the rise has become spectacular. It is estimated that each year nearly 6 billion tons of carbon dioxide is transferred to the atmosphere. All such contributions are caused by human action, which cannot be offset by any natural processes. Hence results the accumulation of carbon dioxide in our atmosphere and we are the culprits. Industrial nations share the largest contribution. According to a 1989 estimate the concentration of carbon dioxide in the atmosphere was 353 ppm. (0.0353%). At the present rate of addition the level of carbon dioxide in 2050 A.D. will be ranging between 400 to 600 ppm. With such a rise of carbon dioxide in the atmosphere, the global temperature rise will be around 4 degree Celsius by 2050 A.D. At the end of present millennium the rise of carbon dioxide might be 15 times more contributing a rise of 22 degree Celsius temperature. The rise of carbon dioxide due to burning of fossil fuel has become a global issue of great concern. The Intergovernmental Panel on Climate Change (IPCC) had been set up by the United Nations to regulate the carbon dioxide emissions and suggest appropriate remedies.

*Carbon Cycle*

The *Carbon Cycle* is very important for conversion of radiant energy into chemical energy (C-C bonds of carbohydrate) in the plants by the process of photosynthesis. In the process carbon dioxide is absorbed by the plants from the atmosphere, which remains the storehouse of carbon dioxide. Carbon dioxide is released into the atmosphere by various means—by respiration of the plants and animals and burning of the fossil fuels. The respiration by plants and animals releases about 60 billion tons of carbon dioxide, which is exactly balanced through the process of photosynthesis by plants. The fossil rocks contain a vast store of carbon (40,000 billion tons). The burning of fossil fuels releases nearly 6 billion tons of carbon dioxide annually. An estimated 6 billion tons carbon dioxide is thus added to the atmosphere each year. The amount of carbon dioxide is only 0.035 percent by volume of the constituent gases in the atmosphere. The estimated total amount, however, comes to about 23 billion tons. Carbon dioxide is also released when some carbon compounds are decomposed by microorganisms.

Another important regulatory role is played by the oceans in the exchange of carbon dioxide. The oceans form the largest reservoir of carbon dioxide containing nearly 130,000 billion tons of carbon dioxide, i.e. about 50 times more than the atmosphere. There is an exchange of 200 billion tons of carbon dioxide between the oceans and the atmosphere each year. Carbon dioxide is dissolved in the ocean water to form carbonic acid, which converts carbonates into bicarbonates. The bicarbonates are dissociated during photosynthesis to precipitate carbonate.

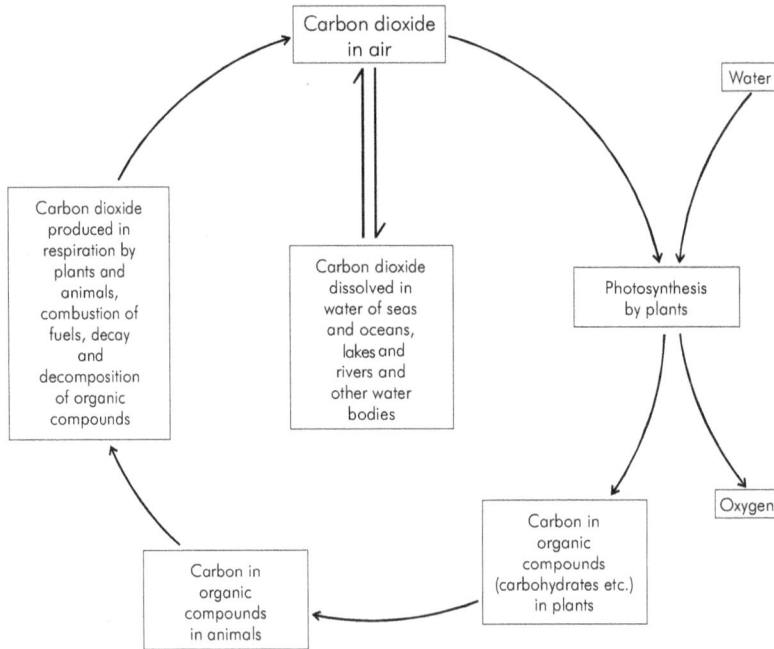

**Figure 1.4:** Carbon Cycle

The carbonate rocks in the lithosphere also have small reserve of carbon dioxide—amounting to only 0.1 billion tons of carbon dioxide. This amount of carbon dioxide (0.1 billion tons) is absorbed by the rocks through weathering process.

It can be seen from the following table that natural gain of carbon dioxide by the atmosphere from different sources is exactly balanced by the natural loss, except the additions contributed by human action through burning of fossil fuels and cultivation of soils.

**Table 1.2:** Gain and Loss of Carbon Dioxide by the Atmosphere

| Gain by the atmosphere (billion tons) | Process |
|---|---|
| 2 | Released from the soils by cultivation |
| 6 | Burning of the fossil fuels |
| 0.1 | Released from the hot sprigs and volcanoes |
| 60 | Respiration and biological decomposition |
| 100 | Released from the oceans |
| Loss from the atmosphere | Process |
| 0.1 | Weathering of rocks |
| 60 | Photosynthesis by the plants |
| 100 | Absorption by the oceans |

## *Ozone*

Ozone is a relatively simple molecule, consisting of three oxygen atoms bonded together. Ozone is present in the atmosphere as an unstable gas and in negligible amount. If compressed the thickness of ozone in the atmosphere will be only 3 mm. But its effects are considered important. Near the surface of the earth the concentration of ozone is toxic to living organisms, because it reacts strongly with other molecules. Because ozone reacts easily with biological tissues, it tends to be destructive to them. Its effect on human lungs resembles a slow burn. Medical researchers have discovered that breathing ozone over several months to years at levels that are now common causes irreversible damage to the lungs of mammals in laboratories. Being mammals ourselves, we can expect respiratory impairment from overexposure to ozone. Children, seniors and all adults who exercise regularly outdoors in summer are particularly vulnerable. Plant species and ecosystems suffer from ozone as well. Productivity drops significantly in several species of important agricultural crops when ozone concentrations reach levels that are now common.

Contrary its role in the upper layer (stratosphere and above) is beneficial. It turns to be a life saving gas in higher altitudes. It plays a significant role in absorbing ultraviolet rays and thus protects the life on earth's surface. Ozone is mostly (90 percent) concentrated in the stratosphere at an altitude of 15 to 35 km. Ozone is formed in the atmosphere by the breakdown of an oxygen molecule into two atoms and the subsequent combination of an atom with another oxygen molecule. The breakdown of the oxygen molecule into two atoms of oxygen is caused by the ultraviolet rays in higher atmosphere usually at an altitude of 80 to 100 km.

### *Formation of Ozone*

Ozone forms readily in the stratosphere as incoming ultraviolet radiation breaks molecular oxygen (two atoms) into atomic oxygen (a single atom). In that process, oxygen absorbs much of the ultraviolet radiation and prevents it from reaching the earth's surface where we live.

In the language of a simplified chemical formula,

$$O_2 + \text{sunlight} \rightarrow O + O \qquad (Reaction\ 1)$$

as one electrically excited free oxygen atom encounters an oxygen molecule, they may bond to form ozone.

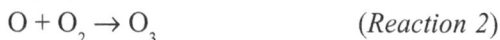

$$O + O_2 \rightarrow O_3 \qquad (Reaction\ 2)$$

Destruction of ozone in the stratosphere takes place as quickly as formation of ozone, because the chemical is so reactive. Sunlight can readily split ozone into an oxygen molecule and an individual oxygen atom.

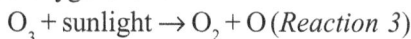

$$O_3 + \text{sunlight} \rightarrow O_2 + O \, (Reaction\ 3)$$

When an electrically excited oxygen atom encounters an ozone molecule, they may combine to form two molecules of oxygen.

$$O + O_3 \rightarrow O_2 + O_2 \qquad (Reaction\ 4)$$

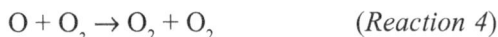

The ozone formation-destruction process in the stratosphere occurs rapidly and constantly, maintaining an ozone layer.

In the troposphere near the Earth's surface, ozone forms through the splitting of molecules by sunlight as it does in the stratosphere. However in the troposphere, nitrogen dioxide, not molecular oxygen, provides the primary source of the oxygen atoms required for ozone formation. Sunlight splits nitrogen dioxide into nitric oxide and an oxygen atom.

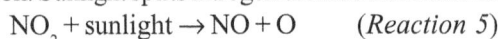

$$NO_2 + sunlight \rightarrow NO + O \qquad (Reaction\ 5)$$

A single oxygen atom then combines with an oxygen molecule to produce ozone.

$$O + O_2 \rightarrow O_3 \qquad (Reaction\ 6)$$

Ozone then reacts readily with nitric oxide to yield nitrogen dioxide and oxygen.

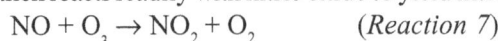

$$NO + O_3 \rightarrow NO_2 + O_2 \qquad (Reaction\ 7)$$

The process described above results in no net gain in ozone. Concentrations occur in higher amounts in the troposphere than these reactions alone account for. In the 1950s, chemists discovered that two additional chemical constituents of the troposphere contribute to ozone formation. These constituents are nitrogen oxides and volatile organic compounds, and they have both natural and industrial sources.

*Depletion of Ozone Layer*

Worldwide monitoring has indicated that stratospheric ozone layer declined for at least two decades, with losses of about 10 percent in winter and 5 percent in summer and autumn in various locations of the earth's surface, particularly over the Antarctic, and also in such diverse locations as Europe, North America and Australia. Researchers also observe depletion of ozone over the North Pole. The problem is becoming more and more critical each year. From the satellite imageries taken recently the depletion of ozone over the poles has been confirmed.

The major culprit for the depletion of ozone layer happens to be chloroflurocarbon ($CFC_{11}$ and $CFC_{12}$). F. Sherwood Rowland and Mario Molina who shared the Nobel Prize in chemistry in 1995 for their discovery of the role of CFCs in the destruction of ozone suggested that most of the chlorine atoms would combine with ozone and initiate a chain reaction (see the formation of ozone). As a result of chain reaction, a single chlorine atom can remove as many as 100,000 molecules of ozone. Rowland and Molina predicted that if industry continued to release a million tons of CFCs into the atmosphere each year, atmospheric ozone would eventually drop by 7 to 13 percent. With the depletion of ozone in the atmosphere, more ultraviolet radiation would reach the earth's surface. It is estimated that increased exposure would lead to higher incidence of skin cancer, cataracts, and damage to the immune system and to slower plant growth. As some CFCs would persist in the atmosphere for more than

100 years, these would last through the twenty-first century. Rowland and Molina called for the ban on further release of CFCs in the air. Alerted by the ozone depletion in the atmosphere, the United States of America, Canada, Norway and Sweden banned the use of CFCs in aerosol sprays in late nineties. Many other countries also followed their pathways.

**Nitrogen oxides ($NO_x$):** Nitric oxide and nitrogen dioxide are together known as $NO_x$ and often pronounced "nox." Sources of $NO_x$ include lightning, chemical processes in soils, forest fires, and the intentional burning of vegetation to make way for new crops (biomass burning). $NO_x$ also come from smokestack and tailpipe emissions as by-products of the combustion of fossil fuels (coal, oil, and natural gas) at high temperatures. Coal-fired power plants are the primary sources of $NO_x$ in the United States. Automobiles, diesel trucks and buses, and non-road engines (farming and construction equipment, boats, and trains) also produce $NO_x$.

**Water vapour:** Of the different constituent gases in air, water vapour is the most variable one and in a local atmosphere may vary from 0–4 percent; the concentration of major gases remains more or less constant over time. The level of water vapour in the air has however a significant effect on the climatic conditions. Rainfall, snowfall, dew, fog etc. are the resultant effects of condensation of water vapour in the air. If there is absence of water vapour in the air, evaporation will be extremely rapid and shallow ponds or similar surface water bodies will simply dry away in no time. Water vapour has also a role in maintaining the heat balance, as like several other green house gases it can trap radiated heat and raise ambient temperature.

### Evaporation/Sublimation

Whenever a water molecule leaves a surface, it is said to have evaporated. Individual water molecule which transitions between a more associated (liquid) and a less associated (vapour/gas) state does so through the absorption or release of kinetic energy. The aggregate measurement of this kinetic energy transfer is defined as thermal energy and occurs only when there is differential in the temperature of the water molecules. Liquid water that becomes water vapour takes a parcel of heat with it, in a process called 'Evaporative Cooling'. The amount of water vapour in the air determines how fast each molecule will return back to the surface. When a net evaporation occurs, the body of water will undergo a net cooling directly related to the loss of water.

In USA, the National Weather Service measures the actual rate of evaporation from a standard "pan" open water surface outdoors, at various locations nationwide. Others do likewise around the world. The US data is collected and compiled into an annual evaporation map. The measurements range from under 30 to over 120 inches per year.

Evaporative cooling is restricted by atmospheric conditions. Humidity is the amount of water vapour in the air. The vapour content of air is measured with devices known as hygrometer. The measurements are usually expressed as specific humidity or percent relative humidity. The temperatures of the atmosphere and the water surface determine the equilibrium vapour pressure; 100% relative humidity occurs when the partial pressure of water vapour is equal to the equilibrium vapour pressure. This condition is often referred to as complete saturation. Humidity ranges from 0 gram per cubic metre in dry air to 30 grams per cubic metre (0.03 ounce per cubic foot) when the vapour is saturated at 30° C. Another form of evaporation is sublimation, by which water molecules become gaseous directly from ice without first becoming liquid water. Sublimation accounts for the slow mid-winter disappearance of ice and snow at temperatures too low to cause melting.

### Condensation

Water vapour will only condense onto another surface when that surface is cooler than the temperature of the water vapour, or when the water vapour equilibrium in air has been exceeded. When water vapour condenses onto a surface, a net warming occurs on that surface.

The water molecule brings a parcel of heat with it. In turn, the temperature of the atmosphere drops slightly. Condensation produces clouds, fog and precipitation (usually only when facilitated by cloud condensation nuclei). The dew point of an air parcel is the temperature to which it must cool before water vapour in the air begins to condense.

Also, a net condensation of water vapour occurs on surfaces when the temperature of the surface is at or below the dew point temperature of the atmosphere. Deposition, the direct formation of ice from water vapour, is a type of condensation. Frost and snow are examples of deposition.

### Relation of Water Vapour with Pressure and Temperature

The amount of water vapour in an atmosphere is constrained by the restrictions of partial pressures and temperature. Dew point temperature and relative humidity act as guidelines for the process of water vapour in the water cycle. Energy input, such as sunlight, can trigger more evaporation on an ocean surface or more sublimation on a chunk of ice on top of a mountain. The *balance* between condensation and evaporation gives the quantity called vapour partial pressure.

The maximum partial pressure (*saturation pressure*) of water vapour in air varies with temperature of the air and water vapour mixture. A variety of empirical formulae exist for this quantity; the most used reference formula is the Goff-Gratch equation for the SVP over liquid water:

$$\log_{10}(p) = -7.90298\left(\frac{373.16}{T} - 1\right) + 5.02808\log_{10}\frac{373.16}{T}$$

$$-1.3816.10^{-7}\left(10^{11.344\left(1-\frac{T}{373.16}\right)} - 1\right)$$

$$+8.1328.10^{-3}\left(10^{-3.49149\left(\frac{373.16}{T}-1\right)} - 1\right)$$

$$+\log_{10}(1013.246)$$

Where $T$ is temperature of the moist air, given in units of kelvins, and $p$ is given in units of millibars (hectopascals).

The formula is valid from about $-50$ to $102°$ C; however there are a very limited number of measurements of the vapour pressure of water over super-cooled liquid water.

Under adverse conditions, such as when the boiling temperature of water is reached, a net evaporation will always occur during standard atmospheric conditions regardless of the percent of relative humidity. This immediate process will dispel massive amounts of water vapour into a cooler atmosphere.

Exhaled air is almost fully at equilibrium with water vapour at the body temperature. In the cold air the exhaled vapour quickly condenses, thus showing up as a fog or mist of water droplets and as condensation or frost on surfaces.

Controlling water vapour in air is a key concern in the heating, ventilating, and air-conditioning (HVAC) industry. Thermal comfort depends on the moist air conditions. Non-human comfort situations are called refrigeration, and also are affected by water vapour. For example many food stores, like supermarkets, utilise open chiller cabinets, or *food cases*, which can significantly lower the water vapour pressure (lowering humidity). This practice delivers several benefits as well as problems.

**Future Trends in the Atmospheric Compositions**

Atmospheric composition determines air quality and affects weather, climate, and critical constituents such as ozone. Exchanges with the atmosphere link terrestrial and oceanic pools within the carbon cycle and other biogeochemical cycle. Solar radiation affects atmospheric chemistry and is thus a critical factor in atmospheric composition. The ability of the atmosphere to integrate surface emissions globally on time scales from weeks to years couples several environmental issues including global ozone depletion and recovery and its impact on surface ultraviolet radiation, climate forcing by radiating active gases and aerosols, and global air quality. Thus, atmospheric chemistry and associated composition are a central aspect of Earth system dynamics. A few critical questions have been raised.

* Is atmospheric composition changing and how?
* Whether the change in atmospheric constituents and solar radiation would affect the global climate and to what extent?
* How do atmospheric trace constituents respond to and affect global environmental change?
* How the regional air quality would be affected by global atmospheric chemical and climate changes?

NASA expects to provide the necessary monitoring and evaluation tools to assess the effects of climate change on ozone recovery and future atmospheric composition; improved climate forecasts based on our understanding of the forcing of global environmental change; and air quality forecasts that take into account the feedbacks between regional air quality and global climate change.

We need to have quantitative understanding of:

* Changes in atmospheric composition and the timescales over which they occur,
* Forcing (anthropogenic and natural) that drive the changes,
* Response of atmospheric trace constituents to global environmental change and the subsequent effects on global climate, and effects of global atmospheric chemical and climate changes on regional air quality.

Drawing on global observations from space, augmented by suborbital and ground-based measurements, NASA is uniquely poised to address these issues. This integrated observational strategy is furthered via studies of atmospheric processes using unique sub-orbital platform-sensor combinations to investigate, for example: (1) the processes responsible for the emission, uptake, transport, and chemical transformation of ozone and precursor molecules associated with its production in the troposphere and its destruction in the stratosphere; and (2) the formation, properties, and transport of aerosols in the Earth's troposphere and stratosphere. The research strategy in the atmospheric composition encompasses an end-to-end approach for instrument design, data collection, analysis, interpretation, and prognostic studies.

Through our study over the past two decades, we have made significant progress in our current level of understanding of the variability of, forcing on, responses to, and consequences of changes in atmospheric composition. However, many questions stand. For example, halogen chemistry is known to be responsible for stratospheric $O_3$ loss, but the roles of chemistry vs. dynamics remain to be precisely quantified. The connection between climate change and ozone chemistry has been recognised, but uncertainties remain regarding the effects on the timing and extent of ozone recovery. In the troposphere, we have observed varying trends in ozone; however its geographical evolution and trends remain to be quantified. Similarly, the spatial and temporal variations in the oxidizing capacity require further characterisation.

Global observations show the transport of tropospheric ozone over large (hemispheric) distances. However, the extent to which regional pollution can be attributed to such long-range transport remains to be quantified. In the climate area, important changes due to effect of radiation in atmospheric water vapour have been observed, but these temporal variations are not quantitatively understood so that future changes can be predicted. Observational advances have yielded important information on the geographical and vertical distribution of atmospheric aerosols.

## Layering of the Atmosphere

A few horizontal layers are identified in the atmosphere. With the advancement of technology and application of new tools and methods like satellites, rockets, radio waves, radiosonde, balloons etc., we have now advanced knowledge on these layers.

Initially the atmosphere can be divided based on temperature variation. Three warm layers (1. near the earth's surface, 2. at an altitude of 50–60 km. and 3. at an altitude above 120 km.) and two cold layers (1. at an altitude of 10–30 km. and 2. at an altitude of 80 km.) could be recognised. These layers are interspersed with the following spheres:

In the Lower Atmosphere –
1. Troposphere,
2. Stratosphere,

In the Upper Atmosphere –
1. Mesosphere
2. Thermosphere (or called Ionosphere)
3. Exosphere and Magnetosphere

### *Lower Atmosphere*

#### *Troposphere*

Atmosphere, close to the earth's surface is known as *Troposphere*. Temperature decreases with the increasing height in the troposphere. The rate of change in temperature with altitude is called the *Environmental Lapse Rate of temperature (ELR)*. ELR varies from day-to-day at a place, and from place to place on any given day. Normally a decrease of 6.5° C temperature is recorded with a rise of 1000 metre altitude, which is the average value of ELR. The troposphere is the lowest layer of the atmosphere. From the surface it extends to a height, between 8 km (8,000 m) at the poles and 16 km (16,000 m) at the equator, with some variation due to weather factors. The isomorphic temperature condition prevails at the end of troposphere. This isomorphic layer determines the limit of troposphere and is known as *Tropopause*. So tropopause is a transition zone between the troposphere and the stratosphere.

The term troposphere has been coined from the Greek word *tropo* meaning mixing or turning. So troposphere is a layer where there is mixing or turning of air. This is the zone of turbulence in the atmosphere. Weather making processes are confined to this layer of the atmosphere. The tropopause acts as lid to restrict the movement of air further upwards. Three fourth of the gaseous matter and nearly all aerosols are found in the troposphere.

**Selected Properties
of Earth's Atmosphere**

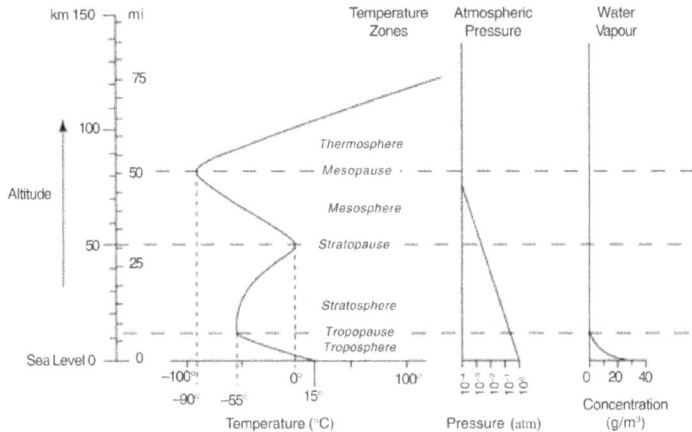

Figure 1.5: Nature of the Atmospheric Layers

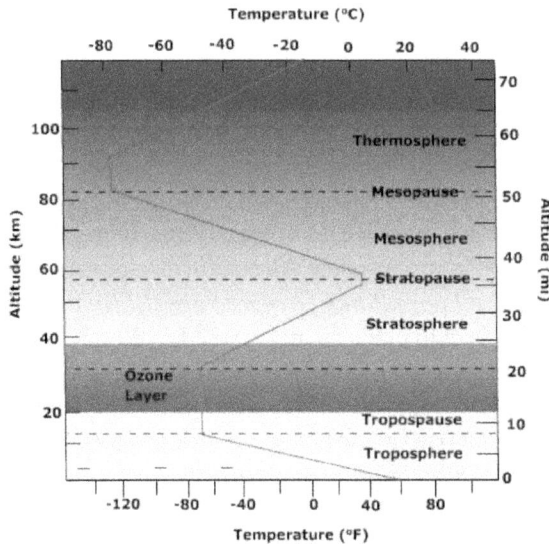

Figure 1.6: Structure of the Atmosphere

The height of the troposphere varies according to the latitudes. Over the equator the height stands at 16 km. (16000 metre) and over the poles at 8 km. (8000 metre). The variation of height plays important role in weather making processes. The tropopause appears to be an *inversion layer* and usually restricts movement between the troposphere and stratosphere. However some gaps are found in the tropopause, through which there is exchange of matter between the troposphere and stratosphere. Water vapour can penetrate to the stratosphere and the ozone gas from the stratosphere subsides into the troposphere. The gap is called the *ozone hole.*

*Stratosphere*

The second layer in the atmosphere is known as *Stratosphere.* It extends for about 50 km. above the tropopause. The thermal structure changes in the stratosphere. Contrary to the troposphere the temperature rises with increasing altitude in the stratosphere. Near the tropopause the temperature remains to be –80° C, but rises to be about 0° C at the outer limit of the stratosphere. The top of the stratosphere has a temperature of about 270 K (–3° C or 26.6° F), a little below the freezing point of water. This top is called the stratopause, above which temperature again decreases with height. The vertical stratification, with warmer layers above and cooler layers below, turns the stratosphere dynamically stable. Regular convection and associated turbulence are absent in this layer of the atmosphere. Stratosphere is almost free from clouds. Most of the long distance airliners fly through the stratosphere.

Ozone is usually concentrated in the stratosphere at an altitude of 25–35 km. Ozone in the atmosphere can absorb the ultraviolet (UV) rays of the sun. The absorption of ultraviolet rays causes heating of the stratosphere. Heating may be very fast, raising temperature from –80° C to –40° C within 48 hours. In northern hemispheric winter, sudden stratospheric warming is often be observed which are probably caused due to absorption of Rossby waves (see General Circulation) in the stratosphere.

Intensive interactions occur in the stratosphere among radiational, dynamical, and chemical processes, in which horizontal mixing of gaseous components proceeds much more rapidly than vertical mixing. A stratospheric circulation, known as Quasi-Biennial Oscillation (QBO) is evidenced over the tropical latitudes, which is driven by gravity waves. The gravity waves appear to be convectively generated in the troposphere. The QBO induces a secondary circulation which is important for the global stratospheric transport of tracers such as ozone or water vapour (as noted earlier). Many of the unresolved questions on global circulation may be resolved, if the nature of this global transport is wholly understood.

The stratosphere merges with the next layer Mesosphere, separated by a pause, known as the *stratopause.* This is also an isomorphic layer having uniform vertical temperature condition.

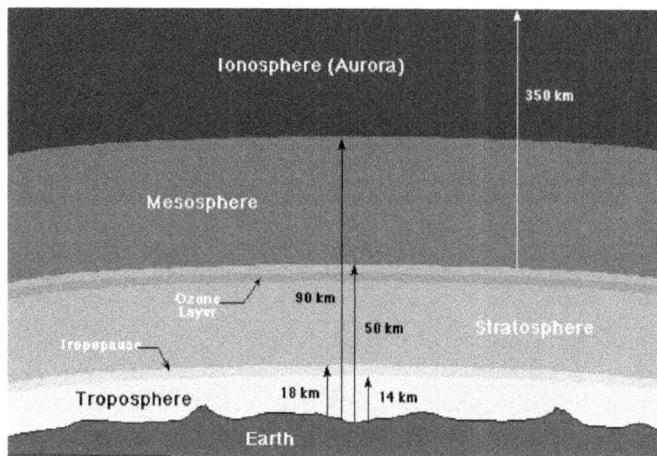

Figure 1.7: Stratosphere with Adjoining Troposphere Below and Mesosphere Above

## Upper Atmosphere

### Mesosphere

*Mesosphere* is the third atmospheric layer next to the stratosphere and just below the thermosphere or ionosphere. In this sphere again temperature decreases with the altitude above the stratopause. Temperature continues to drop up to a height of 80–90 km. and is recorded at about –100° C. The altitude ranges from 50 to 80 km. Many scientists prefer the zoning of this sphere from 20 to 80 km. The term *meso* in Greek means middle. So it is considered the middle layer of the atmosphere.

As the temperature drops with height and due to presence of water vapour, though negligible, the clouds are formed in this sphere. In high latitudes, particularly in summer, cloud formation is evidenced and known as *Noctilucent cloud.* Dust particles from the comets and upper convective current help the formation of noctilucent cloud. These noctilucent clouds appear white or pearly in color and can have a wavy, web-like structure. Atmosphere is rarified and air pressure is very low in this layer. Air pressure drops from 1 mb. (50 km.) to 0.01 mb. (80 km). At the base of the mesosphere the pressure is only 1/1000 of that at sea level and when the top is reached at 95 km it is a mere millionth. For practical purposes it is a vacuum. Millions of meteors used to burn up daily in the mesosphere as a result of collisions having the gas particles. This generates enough heat to vapourise almost all of the falling objects long before they reach the ground, resulting in a high concentration of ferrous and other metal atoms there.

With increasing distance from Earth's surface the chemical composition of air becomes strongly dependent on altitude and the atmosphere becomes enriched with lighter gases. At very high altitudes, the gases begin to form into layers according to molecular mass (weight), because the force of gravity is greater on the heavier molecules. It is in this layer that foreign bodies (such as meteors and spacecraft) entering the atmosphere start to warm up.

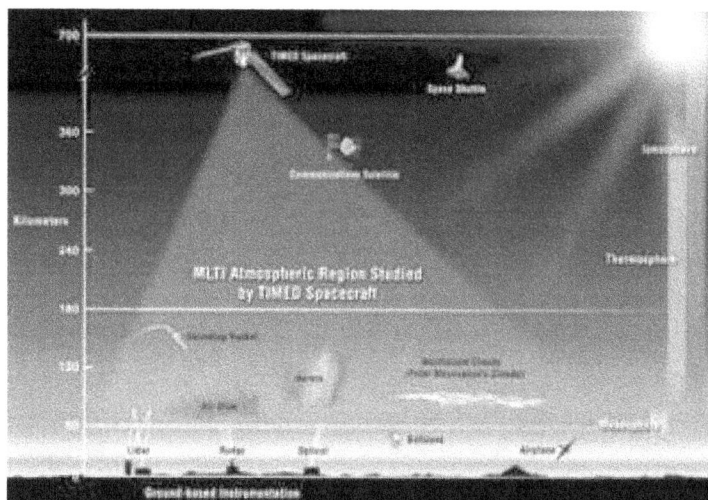

Figure 1.8: Timed Mission Study of Thermosphere, Ionosphere and Mesosphere Energetics
(Source: NASA)

## Thermosphere

After the mesopause temperature again continues to rise. The layer just above the mesopause is called the *Thermosphere*. This layer begins at 80 km. and extends for about 500 km. above the mesopause. Nitrogen molecules, oxygen molecules and oxygen atoms are predominantly found in the lower part of the thermosphere. Above 200 km. oxygen atoms predominate. Oxygen atoms can absorb the ultraviolet rays of the sun. So intense heating occurs in this layer raising the temperature very high. At an altitude of 350 km. 1200° K (Kelvin) temperature is observed. Temperatures are highly dependent on solar activity, and can rise to 15,000° C. As the atmosphere is nearly vacuum one would not feel warm in the thermosphere, because it is so near vacuum that there is not enough contact with the few atoms of gas to transfer much heat. A normal thermometer will record significantly below 0° C.

Above 100 km. ionisation process starts due to absorption of X-Rays and UV rays. Oxygen atoms and nitrogen molecules are converted to electrons by this process. Sometimes these electrons penetrate lower to an altitude from 300 km. to 80 km. *Aurora Borealis* and *Aurora Australis* could be observed over high latitudes from the earth's surface. These are unique phenomena observed in this layer. Due to active process of ionisation this layer is also known as *Ionosphere*. Electrons are heavily concentrated at altitude 100–300 km.

Radiation causes the atmospheric particles in this layer to become electrically charged enabling radio waves to bounce off and be received beyond the horizon. Three important

layers which help radio wave transmission are found in the ionosphere. These are—
1. Kennely-Heaveside Layer or E Layer and 2. Appleton Layer or F Layer and 3. D Layer.
Of these Kennely-Heaveside Layer is very important and lies at an altitude of 90–130 km.
This layer transmits radio wave to the earth's surface. Appleton Layer is not very specified
and less important.

*Exosphere and Magnetosphere*

The base of the *exosphere* lies at an altitude of 500–750 km. It is almost the highest layer
of the atmosphere, beyond which exists the magnetosphere that merges into the space.
This layer is composed of mainly hydrogen, with some helium, carbon dioxide, and atomic
oxygen near the *exobase*. From the exosphere the atmospheric gases, atoms, and molecules
can, to any appreciable extent, escape into space. However the hydrogen formed near
mesopause due to breaking of water vapour and methane fills the gap. Similarly helium is
regenerated due to interaction of cosmic rays with the nitrogen molecules. The lowest
altitude of the exosphere is known as *Exobase.* It is also called the *critical level* as the
height above which there are negligible atomic collisions between the particles and the
constituent atoms are on purely ballistic trajectories. It is also the designated layer where
the space shuttle orbits.

*Magnetosphere* lies 2000 km. above the exosphere. In this layer ionisation becomes
more dominant. Electron (positive) and proton (negative) exist in this layer. These are spaced
as belts at altitudes of 2000 km. and 4000 km. These belts are known as Van Allen Radiation
Belt. These belts were probably formed due to interaction of cosmic rays on concentrated
ions resulting from the solar wind and earth's magnetic field. Solar flare in this upper
atmosphere also influences the weather disturbances in the lower atmosphere.

At an altitude of 80000 km. the earth's atmosphere merges with the solar atmosphere.

## Hydrosphere

Hydrosphere is the other important part of the earth's surface. The extensive and vast
water-bodies are defined as oceans and seas and are distinctly different in their characteristics
from the land surface. The water-bodies on land surface are also found, but these have less
significance than oceans. The oceans and seas play very important role in weather-making
processes. The oceans and seas cover nearly 71 percent of the total area of earth's surface.
The distribution of land and water surface in two hemispheres appears to be just opposite or
antipodal—northern hemisphere dominated by the land surface and southern hemisphere
almost occupied by ocean bodies.

### Oceans and Seas

The term world-ocean refers to combined ocean bodies and seas of the globe. Hydrosphere on the globe covers nearly 361,059,200 square kilometres. Of these the Pacific Ocean, the Atlantic Ocean and the Indian Ocean are the principal water-bodies. These oceans in total include nearly 89 percent of the total area of hydrosphere. The area and volume of different oceans are presented in the following table.

**Table 1.3:** Oceans and Seas—Mean Areas and Volumes

*Oceans:*

| Name | Area (sq. km.) | Volume (cu. Km.) |
|------|----------------|------------------|
| Atlantic Ocean | 82,441,500 | 323,613,000 |
| Indian Ocean | 73,442,700 | 291,030,000 |
| Pacific Ocean | 165,246,200 | 707,555,000 |
| Total | 321,130,400 (88.94%) | 1322,198,000 (96.5%) |

*Intercontinental Seas:*

| Name | Area (sq. km.) | Volume (cu. Km.) |
|------|----------------|------------------|
| Arctic Sea | 14,090,100 | 16,980,000 |
| Malay Sea | 8,143,100 | 9,873,000 |
| Central American Sea | 4,319,500 | 9,573,000 |
| Mediterranean Sea | 2,965,900 | 4,238,000 |
| Total | 29,518,600 (8.2%) | 40,664,000 (2.96%) |

*Smaller Enclosed Seas:*

| Name | Area (sq. km.) | Volume (cu. Km.) |
|------|----------------|------------------|
| Baltic Sea | 422,300 | 23,000 |
| Hudson Bay | 1,232,300 | 158,000 |
| Red Sea | 437,900 | 215,000 |
| Persian Gulf | 238,000 | 6,000 |
| Total | 2,331,300 (0.64%) | 402,000 (0.03%) |

*Fringing Seas:*

| Name | Area (sq. km.) | Volume (cu. Km.) |
|------|----------------|------------------|
| Bering Sea | 2,268,200 | 3,259,000 |
| Okhotsk Sea | 1,527,600 | 1,279,000 |
| Japan Sea | 1,007,700 | 1,361,000 |
| East China Sea | 1,249,200 | 235,000 |
| Andaman Sea | 797,600 | 694,000 |
| Californea Sea | 162,200 | 132,000 |
| North Sea | 575,300 | 54,000 |
| English Channel & Irish Sea | 178,500 | 10,000 |
| Laurentian Sea | 237,800 | 30,000 |
| Bass Sea | 74,800 | 5,000 |
| Total | 8,078,900 (2.22%) | 7,059,000 (0.51%) |

*Hydrosphere – Total*

| 361,059,200 (sq. km.) | 137,032,500 (cu. Km.) |
|------------------------|------------------------|

It appears from the table of all the oceans, the Pacific Ocean is the largest in area and has the largest volume of water, followed by the Atlantic Ocean and the Indian Ocean. The inter-continental seas cover nearly 8 percent area and 3 percent volume of water in the hydrosphere. Next to it come the 'fringing seas', which occupy nearly 2 percent area and 0.5 percent volume of water in the hydrosphere. The 'smaller enclosed seas' occupy the least area (0.64 percent) and least volume of water (0.03 percent) in the hydrosphere, but these enclosed seas are of great significance.

*General Characteristics of the Oceans*

The depth of the oceans and seas vary considerably. The marginal seas along the land-front are less deep, where the continental shelf is extensive. The continental shelf is the extended part of the continents into the oceans. The depth of the continental shelf lies within 200 m. It can be seen that the Atlantic Ocean has the largest extension of the continental shelf (13.3 percent), compared to the Pacific Ocean (5.7 percent) and the Indian Ocean (4.2 percent). Next to continental shelf the relief of the ocean floor changes to a comparatively sloping gradient known as continental slope. The continental slope ultimately meets the nearly flat bottom surface of the oceans. The flat bottom areas of the oceans are called 'Deep Sea Plains', which account for the maximum area (nearly 83 percent) of the oceans (excluding the adjacent seas). The deep sea plains have depths in different oceans ranging 2000 m. to 5000 m. The Pacific Ocean, the Indian Ocean and the Atlantic Ocean occupy deep sea plain covering respectively 90.6 percent, 90.6 percent and 85.8 percent of the total area. Many parts of the oceans record depths more than 5000 m., but occupy not more than 1.2 percent of the total area of the oceans. These are called 'ocean deeps'. Most of the ocean deeps are found in the Pacific Ocean and a fewer in the Atlantic and Indian Ocean. Of different ocean deeps the Mariana or the Challenger Deep in the North Pacific Ocean has the maximum (5269 m.) depth in the world. Next is the Tonga Deep, located in the Central South Pacific Ocean (5022 m.). Other important ocean deeps are the Philippine Deep (4767 m.), the Japan Deep (4655 m.), the Murray Deep (3540 m.) in the Pacific Ocean, the Puertorico Deep (4662 m.), the Romanche Deep (4030 m.) in the Atlantic Ocean and the Sunda Deep (3828 m.) in the Indian Ocean.

Distribution of temperature and extent of salinity in the oceans vary latitude-wise. Relatively high temperature condition prevails over the tropical oceans. The warm equatorial currents originate close to the equator and flow towards the poles as surface current. On the other hand, cold conditions dominate uniformly over the polar regions, which become the source of cold currents that move towards the lower latitudes. Temperature in the oceans also varies according to the depth. Usually three structured layers of temperatures are observed—near the surface up to a depth of 500 m. warm surface temperature, sharp decline of temperature in the zone of 500 m. to 1000 m. known as thermocline and uniform cold temperature below the depth of 1000 m. The vertical temperature distribution in the ocean is very important to study the environment for life. The upper layers where the sunlight can penetrate are known as *'limnetic zone'*. In this zone active photosynthesis and growth occur resulting in abundance of oxygen and rapid consumption of nutrients. So the limnetic zone is usually rich in aquatic species. Next to limnetic zone are the *'profundal zone'* and *'benthic zone'*. These zones are dark, where decomposition, mineralisation and nutrient accumulation occur. In these dark zones dissolved oxygen is deficient and anaerobic microbial activity predominates. Salinity of sea water appears to be high under arid condition. The salinity increases sharply in the enclosed seas where the mixing of water is nearly absent. The Red Sea and the Mediterranean Sea record very high salinity. The salinity in the low latitudes also varies with depth, increasing steadily to a certain depth (–100 to –200 m.) and then decreasing sharply to a minimum at a depth of 800 m.

## Significance of Hydrosphere

The hydrosphere is of immense importance to mankind. Actually it seems to be the lifeline of all living organisms on earth's surface. Through the process of evaporation the oceans supply water vapour to the atmosphere. The water vapour being condensed in the atmosphere causes precipitation on the earth's surface. The running water flows as river and finally meets the sea. A part of the precipitation is collected through seepage as underground water and forms the water table. The plants absorb water from the underground sources and the water is again released to the atmosphere through the process of evapotranspiration. The process continues and the oceans and the atmosphere actually determine the status of water and moisture on our living earth. The process is called the 'Hydrologic cycle'.

*Other Characteristics of Hydrosphere*

The oceans are a great reservoir of carbon dioxide. They can absorb more carbon dioxide than the atmosphere. The stored carbon dioxide amounts to about 130,000 billion tons, nearly 50 folds more capacity than the atmosphere. Moreover the oceans exchange 200 billion tons of carbon dioxide with the atmosphere each year. The reserve of carbon dioxide in the oceans acts, within certain limits, as a regulator to retain a balance. When there is excess carbon dioxide in the atmosphere the oceans absorb, but where there is deficit the oceans replenish the atmosphere.

The oceans are the storehouse of vast resources—supplying water, salts, minerals and food. The resources of the oceans have not yet been wholly tapped and will survive many millenniums, if there be judicious utilisation. The oceans also play important role as receptacle of pollutants and waste materials from the land surface. But this role of the oceans must be controlled and kept within limits through less generation of hazardous wastes and pollutants on land surface. The oceans since the dawn of civilisation play the significant role in transportation.

The role of the hydrosphere appears to be critical when we consider the sustenance of life on earth's surface. It is known that the first emergence of life occurred under marine conditions. The hydrologic cycle illustrates the importance of the oceans supplying fresh water to the land surface otherwise the existence of life on land surface would not have been possible. We must realise that this water resource on land surface is very limited and scarce.

---

## Further Readings

Barry, R.G. and R.G. Chorley (1968) *Atmosphere, weather and climate,* Methuen & Co., London.

Court, Arnold (1957) 'Climatology: Complex, Dynamic and Synoptic', *Annals of Association of American Geographers,* Vol. 47, **2**.

Craig, R.A. (1965) *The Upper Atmosphere: Meteorology and Physics,* Academic Press, Newyork.

Critchfield, H.J. (1975) *General Climatology,* Prentice Hall of India Pvt. Ltd., New Delhi.

Goody, R.M. and James C.G. Walker (1972) *Atmosphere,* Prentice Hall Inc., Engelwood Cliffs, N.J.

Hare, F.K. (1962) The Stratosphere, *Geographical Review,* **52**.

Landsberg, H.E. (1953) 'Origin of the Atmosphere', *Scientific American,* **189**, **2**.

Rasool, S.I. (ed.), *Chemistry of the Lower Atmosphere,* Plenum Press, New York.

Ratcliffe, J.A. (ed.), (1960) *Physics of the Upper Atmosphere,* Academic Press, New York & London.

Saha, P.K. (2000) 'Nature and Natural Processes', *Environment,* University of Calcutta, Calcutta.

Saha, P.K. and P.K. Bhattacharya (1994) *Adhunik Jalavayu Vidya,* West Bengal State Book Board, Calcutta.

Walker, J.C. and N.W. Spencer (1968) 'Temperature of the Earth's Upper Atmosphere', *Science,* **162**, **3861**.

# Chapter 2

## SOLAR ENERGY

### Basic Points

I. Sun is the source of energy to all planets in the solar system. Our earth belongs to the solar system. We have discussed the unique position of the earth in its relation to the sun. Sun is a big star and nuclear reactor supplying energy through the process of burning helium. The energy contributed by sun comes in the form electromagnetic wave and supplies heat and light to our earth. Heat energy reaches the earth's surface through the passage of earth's atmosphere for a few hundred kilometers. Heat energy from the solar source plays the most significant role in weather-making. This energy drives the climate and weather and supports virtually all life on earth.

II. A sunray emitted from the Sun takes about 9 minutes to reach the earth.

III. Sun is a fusion reactor emitting 3,800 million, million, million, million watts of energy each second and the earth receives only 1/2,00,000,000,000 portion of this amount equal to $1.3 \times 10^{17}$ w/h. which is 20,000 times the energy requirement of the world.

IV. The temperature of the Sun at the centre is 15 million °C and at the surface it is about 6000° K.

V. About 50% of the energy received outside earth's atmosphere actually can reach the earth and is about 1 cal/sq.cm./min at sea-level.

### Energy from the Sun

Earth receives 174 petawatts (Petawatt = $10^{15}$ watt) of incoming solar radiation (insolation) at the upper atmosphere at any given time. When it meets the atmosphere, 6 percent of the insolation is reflected and 16 percent is absorbed. Average atmospheric conditions (clouds, dust and pollutants) further reduce insolation traveling through the atmosphere by 20 percent due to reflection and 3 percent *via* absorption. These atmospheric conditions not only reduce the quantity of energy reaching the earth's surface, but also diffuse approximately 20 percent of the incoming light and filter portions of its spectrum. After passing through the atmosphere, approximately half the insolation is in the visible electromagnetic spectrum with the other half mostly in the infrared spectrum (a small part is ultraviolet radiation).

The absorption of solar energy by atmospheric convection (sensible heat transport) and evaporation and condensation of water vapour (latent heat transport) affects the winds and the water cycle. Upon reaching the surface, sunlight is absorbed by the oceans, land masses and plants. The energy captured in the oceans drives the thermohaline cycle. As such, solar energy is ultimately responsible for temperature-driven ocean currents such as the thermohaline cycle and wind-driven currents such as the Gulf Stream. The energy absorbed by the earth, in conjunction with that recycled by the greenhouse effect, warms the surface to an average temperature of approximately 14° C. The small portion of solar energy captured by plants and other phototrophs is converted to chemical energy via photosynthesis. All the food we eat, wood we build with, and fossil fuels we use are products of photosynthesis. The flows and stores of solar energy in the environment are vast in comparison to human energy needs.

- The total solar energy available to the earth is approximately 3850 zettajoules ($ZJ$ = Zettajoule = $10^{21}$ Joules ) per year.
- Oceans absorb approximately 2850 ZJ of solar energy per year.
- Winds can theoretically supply 6 ZJ of energy per year.
- Biomass captures approximately 1.8 ZJ of solar energy per year.
- Worldwide energy consumption was 0.471 ZJ in 2004.

The solar radiation at the top of the earth's atmosphere varies with latitude, while the annual average ground-level insolation varies temporally. For example, in North America, the average insolation at ground level over an entire year (including nights and periods of cloudy weather) lies between 125 and 375 W/m² (3 to 9 kWh/m²/day). At present, photovoltaic panels typically convert about 15 percent of incident sunlight into electricity; therefore, a solar panel in the contiguous United States, on average, delivers 19 to 56 W/m² or 0.45–1.35 kWh/m²/day.

**Heating of the Earth's Surface**

Solar energy is essentially required for heating of the earth's surface. Heating of the earth's surface is not uniform and depends on many factors. Differential heating of the earth's surface causes dynamic weather systems and generates motion in the atmosphere. We must understand the basic reasons for differential heating of the earth's surface.

A. Earth revolves the sun inclined in 62 degree 30 minute angle. Earth stands as a globe to incoming solar rays. The central part of the globe has the vertical interface to incoming solar rays. On the other hand poles and surrounding areas stand tilted to incoming solar rays. Energy that falls vertically on small unit area will be more effective in heating the areas than the slanted rays covering comparatively wide areas. As a result latitudes act as a control for heating the earth's surface. Earth as a globe has a median plane, known to be the Equator (0 Degree), dividing the globe into two hemispheres. Poles are the extreme points in the globe on two sides of the hemispheres, stand tangential to incoming solar rays. The inclination increases moving away from the Equator, specified in terms of latitudes. So higher the latitudes the solar rays are more inclined and receive lower and lower energy on the earth's surface.

B.  Earth's surface has different covers for absorption of solar rays. Solar rays come in the short wave lengths and are reflected by the earth's surface to the space. Areas with dark cover absorb more heat energy than snow cover. This phenomenon is known as *albedo*. Albedo is known as the surface reflectivity of solar radiation. The term has its origins from a Latin word *albus*, meaning 'white'. An ideal white body has an albedo of 100 percent and an ideal black body, 0 percent. The typical amounts of solar radiation reflected from various objects are shown in Table 2.1. Mean annual albedo values differ appreciably between the equator and the poles, largely due to the presence of snow and ice-covered surfaces in higher latitudes. Atmospheric reflectance also varies with dust concentration, the zenith angle of the Sun, and the type and/or amount of cloud cover. Well-developed convective clouds reflect up to 90% of incident solar energy, making thick clouds appear bright from space. The characteristics of a surface change from one season to another, even affects the reflectance properties. This fact is most evident throughout the high latitudes where snow cover and ice extent reach maximum values during the cold seasons, significantly increasing the surface reflectance values. But in the spring and short summer when ice melts it helps reduction of reflectance value and results in more absorption of heat energy.

**Table 2.1:** Reflectivity Values (albedo) of Various Surfaces

| Surface | Details | Albedo |
|---|---|---|
| Soils Sands | Dark and Wet/Light and Dry | 0.15 – 0.45 |
| Grass | Long/Short | 0.16 – 0.26 |
| Agricultural crops | – | 0.18 – 0.25 |
| Tundra | – | 0.18 – 0.25 |
| Forests | Deciduous | 0.15 – 0.20 |
|  | Coniferous | 0.05 – 0.15 |
| Water | Small zenith angle | 0.03 – 0.10 |
|  | Large zenith angle | 0.10 – 1.00 |
| Snow | Old/fresh | 0.40 – 0.95 |
| Ice | Sea | 0.30 – 0.45 |
|  | Glaciers | 0.20 – 0.40 |
| Clouds | Thick | 0.60 – 0.90 |
|  | Thin | 0.30 – 0.50 |

*Source:* Oke, 1998; Ahrens, 2001.

C.   Heat energy on the earth's surface is expressed by temperature as weather condition. Temperature always refers to the thermal status of the atmosphere. Earth's ground surface and atmosphere close to it get more heat energy and record higher temperature. Atmospheric space away form the earth gets lesser and lesser heat energy with increasing altitudes. So temperature on the earth's surface decreases with increasing altitudes.

D.   Differential heating of land and water causes temperature variation over land and sea. The albedo over the landmass is higher than the watermass. Under uniform conditions the watermass gets more solar energy. However, the watermass becomes heated slowly as the specific heat of the landmass is higher. The landmass during day-hours becomes heated at a faster rate and also cools rapidly during night-hours. On the contrary the water retains the heat energy received during the day-hours favoured by low cooling rate during night hours. Such contrast of temperature conditions over land and sea is of great significance in weather-making process.

### *Electromagnetic Waves*

Any body above the absolute zero can emit radiant energy. A hot body like sun emits heat in the form of radiant energy that can be absorbed by a colder body, such as the earth, thus raising its temperature. Radiant energy has a wave like form. Maxwell defined this as electromagnetic waves. The electromagnetic waves travel in all directions like an ever expanding sphere. The radiant energy can travel through space or a vacuum and gases. It must be noted that the only way the energy can travel in the vacuum is as electromagnetic waves. The wavelength is defined as distance between adjoining wave crests. The wave's frequency is defined as the number of wave crests (or troughs) passing through a fixed point in one second. The velocity is determined by multiplying the frequency with the wavelength. The wave's frequency is inversely proportional to its wavelength. Much of the radiation is distributed in a relatively narrow band around the wavelength indicating the maximum amount of radiation. The wavelength of this maximum radiation is inversely proportional to the temperature of the hot body, i.e. hotter the body shorter the wavelength and *vice-versa*.

The electromagnetic spectrum indicates the range of all possible wavelengths. It contains short wavelengths, long wavelength radiation as well as visible light. Human eye can detect the emission of radiant energy in the wavelengths ranging between 0.4 to 0.7 microns. This is known as VIBGYOR (violet, indigo, blue, green, yellow, orange and red). Red is the longest and violet is the shortest in this range. Radiation in the wavelength beyond this range, i.e. higher than 0.7 microns and lesser than 0.4 microns, is not visible. The radiation in the wavelength higher than 0.7 microns is called infrared radiation. The terrestrial radiation is returned to the space as infrared radiation. The infrared radiation also includes heat from a stove, radio waves, and radiation from radar and microwave ovens. The radiation in the wavelength less than 0.4 microns is termed short wavelength and includes ultraviolet (UV)

**Table 2.2:** Monthly Insolation for Specific Latitudes for the Year 2000

Year (AD) = 2000
Eccentricity of the Earth's eliptical orbit = .016704
Obliquity (degrees) = 23.4398
Spatial angle from Vernal Equinox to Perihelion (radians) = 4.93746
Solar Constant (W/m^2) = 4*(global annual insolation) = 1367.00

| Lat | Jan | Feb | Mar | Apr | May | Jun | Jul | Aug | Sep | Oct | Nov | Dec | Annual |
|---|---|---|---|---|---|---|---|---|---|---|---|---|---|
| -90 | 500 | 308 | 57 | 0 | 0 | 0 | 0 | 0 | 11 | 216 | 445 | 552 | 173.9 |
| -85 | 498 | 307 | 69 | 0 | 0 | 0 | 0 | 0 | 23 | 215 | 443 | 550 | 175.3 |
| -80 | 492 | 304 | 100 | 6 | 0 | 0 | 0 | 0 | 52 | 222 | 438 | 544 | 179.7 |
| -75 | 483 | 308 | 135 | 24 | 0 | 0 | 0 | 8 | 86 | 240 | 429 | 533 | 187.1 |
| -70 | 471 | 325 | 170 | 54 | 4 | 0 | 0 | 29 | 122 | 265 | 423 | 519 | 198.3 |
| -65 | 468 | 346 | 204 | 87 | 23 | 4 | 12 | 59 | 156 | 292 | 429 | 505 | 215.2 |
| -60 | 474 | 368 | 237 | 121 | 50 | 25 | 37 | 91 | 190 | 319 | 441 | 505 | 237.9 |
| -55 | 482 | 389 | 268 | 156 | 81 | 52 | 66 | 125 | 223 | 343 | 453 | 508 | 262.0 |
| -50 | 490 | 408 | 297 | 189 | 114 | 83 | 98 | 158 | 254 | 366 | 464 | 512 | 285.8 |
| -45 | 496 | 424 | 324 | 222 | 148 | 116 | 131 | 191 | 283 | 386 | 472 | 514 | 308.7 |
| -40 | 499 | 438 | 348 | 253 | 181 | 149 | 165 | 223 | 310 | 403 | 479 | 514 | 330.1 |

| Lat | Jan | Feb | Mar | Apr | May | Jun | Jul | Aug | Sep | Oct | Nov | Dec | Annual |
|---|---|---|---|---|---|---|---|---|---|---|---|---|---|
| -35 | 500 | 449 | 370 | 283 | 214 | 183 | 198 | 254 | 335 | 418 | 482 | 512 | 349.7 |
| -30 | 498 | 457 | 389 | 311 | 246 | 216 | 230 | 284 | 357 | 430 | 483 | 507 | 367.1 |
| -25 | 492 | 461 | 405 | 337 | 277 | 249 | 262 | 311 | 377 | 438 | 481 | 499 | 382.2 |
| -20 | 484 | 462 | 418 | 360 | 306 | 280 | 292 | 336 | 393 | 443 | 475 | 487 | 394.7 |
| -15 | 473 | 460 | 428 | 380 | 334 | 310 | 320 | 359 | 407 | 446 | 466 | 473 | 404.6 |
| -10 | 458 | 455 | 435 | 398 | 359 | 338 | 347 | 380 | 418 | 444 | 454 | 456 | 411.7 |
| -5 | 441 | 446 | 438 | 413 | 382 | 364 | 371 | 398 | 425 | 440 | 439 | 436 | 416.0 |
| 0 | 420 | 434 | 439 | 425 | 402 | 388 | 393 | 413 | 429 | 432 | 421 | 413 | 417.4 |
| 5 | 397 | 420 | 435 | 434 | 420 | 409 | 413 | 425 | 430 | 421 | 401 | 388 | 415.9 |
| 10 | 371 | 402 | 429 | 439 | 435 | 428 | 429 | 433 | 428 | 407 | 377 | 360 | 411.5 |
| 15 | 343 | 381 | 419 | 441 | 446 | 444 | 443 | 439 | 423 | 390 | 351 | 330 | 404.3 |
| 20 | 313 | 357 | 406 | 441 | 455 | 458 | 454 | 442 | 414 | 370 | 323 | 299 | 394.3 |
| 25 | 281 | 331 | 390 | 436 | 461 | 468 | 462 | 442 | 402 | 347 | 293 | 265 | 381.7 |
| 30 | 248 | 303 | 371 | 429 | 464 | 476 | 468 | 438 | 387 | 322 | 261 | 231 | 366.5 |
| 35 | 213 | 272 | 349 | 419 | 464 | 481 | 470 | 431 | 370 | 294 | 227 | 195 | 349.0 |
| 40 | 178 | 240 | 325 | 405 | 461 | 483 | 470 | 422 | 349 | 265 | 193 | 160 | 329.4 |
| 45 | 142 | 207 | 298 | 389 | 456 | 483 | 467 | 410 | 326 | 234 | 158 | 124 | 307.9 |

| Lat | Jan | Feb | Mar | Apr | May | Jun | Jul | Aug | Sep | Oct | Nov | Dec | Annual |
|---|---|---|---|---|---|---|---|---|---|---|---|---|---|
| 50 | 107 | 172 | 269 | 370 | 448 | 481 | 462 | 395 | 300 | 201 | 123 | 89 | 285.0 |
| 55 | 73 | 137 | 238 | 349 | 438 | 478 | 455 | 378 | 272 | 167 | 89 | 56 | 261.0 |
| 60 | 41 | 102 | 205 | 326 | 427 | 475 | 448 | 359 | 243 | 132 | 56 | 27 | 236.9 |
| 65 | 15 | 67 | 171 | 301 | 417 | 475 | 443 | 339 | 211 | 97 | 26 | 4 | 214.2 |
| 70 | 1 | 36 | 135 | 276 | 413 | 488 | 447 | 320 | 178 | 63 | 6 | 0 | 197.3 |
| 75 | 0 | 11 | 99 | 252 | 420 | 502 | 459 | 306 | 145 | 31 | 0 | 0 | 186.0 |
| 80 | 0 | 1 | 64 | 238 | 429 | 512 | 468 | 304 | 111 | 9 | 0 | 0 | 178.5 |
| 85 | 0 | 0 | 33 | 235 | 434 | 518 | 473 | 308 | 83 | 0 | 0 | 0 | 174.2 |
| 90 | 0 | 0 | 21 | 236 | 435 | 520 | 475 | 309 | 71 | 0 | 0 | 0 | 172.8 |
| Globe | 353 | 350 | 345 | 339 | 334 | 331 | 331 | 333 | 338 | 344 | 349 | 353 | 341.8 |

*Source: Atmosphere—Ocean Model, NASA, GISS.*

radiation, X-rays and gamma rays. Long exposures to short wave radiation may cause undesired health hazards. Ultraviolet radiation is now considered the cause of skin cancer and a few other diseases. X-rays and gamma rays can destroy all living cells.

## Heat Budget of the Earth

Heat budget of the earth relates to the radiational energy balance of earth. We know that the solar energy is the source of heat energy on our earth's surface. It is the radiation budget that endows the heat energy of the earth. The *radiation budget* represents the balance between incoming energy from the Sun and outgoing thermal (long wave) and reflected (short wave) energy from the earth.

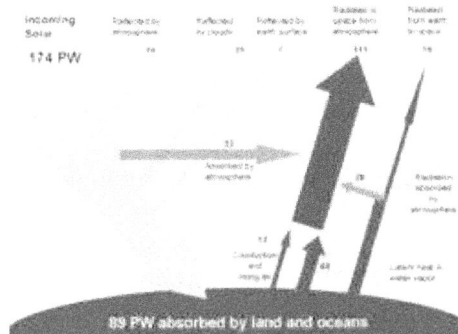

**Figure 2.1:** About Half the Incoming Energy from the Sun is Absorbed by Water and Land Masses, while the Rest is Reradiated Back into Space (Values are in PW = $10^{15}$ W)

On a global scale the budget is balanced. The incoming energy is expended to maintain a balance. Otherwise the temperature would rise constantly. On a regional and local scale the budget appears to be not balanced. We find that the tropical areas get more energy than they spend, while at higher latitudes of the winter hemisphere more is spent than gained. The terrestrial balance is maintained by energy transfers in currents of ocean and atmosphere. For such transfers, radiation energy is transformed to latent heat, heat or even motion (kinetic energy). Finally, it all becomes thermal energy and radiates out of the earth-atmosphere system.

### *Surface Radiation Budget*

The Surface Radiation Budget includes downward shortwave radiation, reflected shortwave radiation, downward longwave radiation, upward longwave radiation, net radiation.

The *downward shortwave radiation* may be reflected to the space absorbed in the atmosphere and absorbed at the ground.

$$Esun = AEsun + Eatm + (1–Asfc)Esfc$$

where *Esun* is the incident solar radiation, A is albedo of TOA (Top of Atmosphere), *Eatm* is the energy of absorbed in atmosphere, *Esfc* is downward irradiance at surface and *Asfc* is surface albedo. This equation implicitly includes scattering and multiple reflectors between the surface and clouds. Usually the incident solar radiation and TOA albedo is measured by satellite measurement.

The *reflected solar radiation* is the product of surface albedo and the downward solar radiation, the surface albedo should be determined. One method to estimate surface albedo is the minimum albedo technique. Because few locations are likely to be cloud-covered for an entire month, the minimum albedo is likely to indicate the clear-sky albedo. It can be calculated from narrow band AVHRR (Advanced Very High Resolution Radiometer) observations.

The *downward longwave radiation* is mostly radiated from the atmosphere. It depends on the temperature and moisture of the atmosphere. The water vapour and other gases, aerosols absorb some solar energy and emit some longwave radiation energy. The computation of downward longwave radiation from the atmosphere is difficult, even when the distributions of water vapour, carbon dioxide, cloudiness, and temperature are measured. Some satellite measurements estimate downward longwave radiation.

A small amount of *longwave radiation* is reflected by the surface. Natural surface emission is dominant. It is also difficult to measure and define the surface temperature especially of the vegetation surface. Agglomerating the above four components the calculation of *net radiation* at the surface may be derived. However, the measure is not accurate because of errors in each accumulate.

### Heat Balance of the Earth

During its travel through the space the solar energy is depleted. Even the available solar income is reduced and variations result due to many factors like latitudes, seasons and local conditions. The depletion occurs as the solar rays are scattered, reflected and absorbed in varying degrees by atmospheric gases and aerosols in the air. Scattering results due to deflections of some of the wavelengths from the direct beam by air molecules and fine dust particles including smoke haze as solar rays pass through a relatively transparent medium like air. True scattering develops only when the diameters of the obscuring particles are smaller than the wavelengths of the radiation. In case the diameters of the obscuring particles are larger than the wavelengths of radiation, the diffuse reflection will result. This is particularly caused by cloud droplets and larger dust and salt particles in the air. In general the atmospheric gases are transparent to most of the wavelengths in solar rays. However a few gases absorb and are selective in their action, absorbing more of some wavelengths than others. Water vapour in the atmosphere plays most significant role in absorption, absorbing six times as much solar radiation as all the other gases combined.

The amount of depletion of solar rays by scattering, reflection and absorption in the atmosphere depends on two factors: (a) the length of the passage through air and (b) the transparency of air. The length of the passage can be determined mathematically. But it is difficult to assess the transparency status of air, as it varies spatially and periodically and is controlled by the amount of cloudiness and turbidity of the atmosphere. As a result it appears to be difficult in computing the amount of depletion of solar radiation traveling through the atmosphere. There are different estimates of such depletion and solar radiation available on earth's surface. Let us consider the estimates made by Houghton *et al.* and Budyko separately.

*Estimate by Houghton and Others*

A simple estimate for heat balance of the earth has been made by Houghton and others. They considered the solar radiation in terms of 100 units (or 100 percent) as noted below:

**Table 2.3:** Heat Balance of the Earth (after Houghton and Others)

*(in units)*

| | | |
|---|---|---|
| *Reflected to space* | | |
| By clouds | 25 | |
| By the earth's land & sea surface | 2 | |
| Scattered by molecules of air & fine dust | 7 | 34 units |
| *Absorbed in the atmosphere* | | |
| By the atmosphere | 17 | |
| By clouds | 2 | 19 units |
| *Transmitted through the atmosphere and absorbed at the land-sea surface* | | |
| As direct sunlight | 19 | |
| As diffused radiation through clouds | 23 | |
| As scattered radiation | 5 | 47 units |
| | | 100 units |

From Table 2.3 it appears that 34 units or percent of solar rays reaching the outer limit of the earth's surface is returned to space in its original short wavelength. This radiation is not available in heating the earth-atmosphere system. Only 19 units or percent of solar radiation is absorbed by the atmosphere and this absorbed radiation results in heating the atmosphere directly. The remaining 47 units or percent of solar radiation can reach the earth's surface and is absorbed by land-sea surface. This results in heating the land-sea surface.

Thus we find that 34 percent of solar radiation is returned to space and not available for heating the earth-atmosphere system. Only 66 percent is available for heating the earth-atmosphere system. This 66 percent is called the *effective solar radiation*. Moreover the atmosphere directly absorbs only 19 percent for heating. The atmosphere receives most of its heat (47 percent) indirectly by the transfer of heat from earth's land-sea surface. Such transfer is made through long wave terrestrial infrared radiation, which the Green House Gases in the atmosphere can absorb. *"It is of utmost significance that the atmosphere is fuelled mainly from below* (Trewartha, 1968)".

Interestingly we observe that in general the annual temperature on earth's surface neither increases nor decreases. It is assumed that the 66 percent of solar radiation gained is exactly balanced by an equal amount of energy radiated back to space as long wave terrestrial radiation. This is known as *terrestrial heat balance.*

Let us examine the heat budget of the earth. We presume that out of 120 units of terrestrial radiation 114 units are absorbed by the atmosphere and remaining 6 units lost to space. The atmosphere also gets 10 units and 23 units respectively from (a) heat transport through turbulence and convection and (b) heat transfer as latent heat of condensation through evaporated moisture from ground to atmosphere. The atmosphere gains directly 19 units from solar short wave radiation. Thus total 166 units (19 + 114 + 10 + 23) are available to the atmosphere. Of this, 106 units are re-radiated back to the earth's surface and 60 units returned to space. The earth also radiates back 6 units directly to space. As a result total 66 units (60 + 6) of earth's radiation are returned to space. This equals to 66 units of solar radiation gained by the earth-atmosphere system. This equation of terrestrial heat budget is known as *Heat Balance of the Earth.*

*Budyko's Estimate for Heat Budget of the Earth*

Budyko estimated the solar income of the earth per year at 250 kilo calories per square centimetre. The *albedo* or reflecting power of the earth is still debated. However, Budyko accepted an amount of 14 percent for the earth's surface and 26 percent for cloud cover as albedo. Thus nearly 40 percent of the income is lost and the earth's *effective solar income* appears to be 150 kilo calories/square centimetre/year. With this opening amount of solar energy Budyko prepared the heat budget or heat balance of the earth as noted below.

As per Budyko's estimate the effective total radiation is 150 kilo calories per square centimetre per year. Of this 111 and 39 are received by the earth and atmosphere respectively as shortwave absorption. The net long-wave back-radiation from earth is 43 kg cal/cm$^2$/year and returned to space. By evaporation and condensation 56 kg cal/cm$^2$/year is transferred to the atmosphere. Similarly net turbulent transfer from earth to atmosphere is 12 kg cal/cm$^2$/year. On the other hand net long-wave back radiation from atmosphere is 107 kg cal/cm$^2$/year and returned to space. Thus we find that incoming 111 kg cal/cm$^2$/year is exactly balanced by outgoing effective radiation. The incoming 107 (39 + 56 + 12) kg cal/cm$^2$/year is expended as outgoing effective radiation. The effective total radiation of 150 kg cal/cm$^2$/year to earth is offset by return to space. In this way radiational heat balance

**Table 2.4:** Heat Balance of the Earth (after Budyko)

*in kilo calories/square centimeter/year*

| Process | Earth | | Atmosphere | | Space | |
|---|---|---|---|---|---|---|
| | In | Out | In | Out | Returned to earth | |
| Total radiative income | | | | | 250 | |
| Reflected | | 35 | | 65 | 100 | |
| Effective total radiation | | | | | 150 | |
| Shortwave absorption | 111 | | 39 | | | |
| Net longwave back radiation from earth | | 43 | | | | 43 |
| Evaporation and condensation | | 56 | 56 | | | |
| Net turbulent transfer | | 12 | 12 | | | |
| Net longwave back radiation from atmosphere | | | | | 107 | 107 |
| Total effective radiation | +111 | −111 | +107 | −107 | +150 | −150 |
| *(in percent of total radiative income)* | | | | | | |
| Total radiative income | | | | | 100 | |
| Reflected | | 14 | | 26 | | 40 |
| Effective total radiation | | | | | 60 | |
| Shortwave absorption | 44.4 | | 15.6 | | | |
| Net longwave back radiation from earth | | 17.2 | | | | 17.2 |
| Evaporation and condensation | | 22.4 | 22.4 | | | |
| Net turbulent transfer | | 4.8 | 4.8 | | | |
| Net longwave back radiation from atmosphere | | | | | 42.8 | 42.8 |
| Total effective radiation | +44.4 | −44.4 | +42.8 | −42.8 | +60 | −60 |

of the earth-atmosphere system is maintained and the earth in turn neither gains nor loses heat.

*Latitudinal Heat Balance*

Though the gains in solar energy by earth are balanced by equal losses of earth's radiation to space, this is not observed in all latitudes. There are variations in the gains and losses of heat in different latitudes. In low latitudes ranging from 0 to 35/40 degree in both hemispheres the earth and atmosphere both gain more heat from solar radiation than they lose to space by earth's long wave back-radiation to space. The reverse situation is observed in high and middle latitudes (40 to 90 degree) in both hemispheres, where there is a net loss of energy than its gain. If this continues, the tropics would have been progressively warmer and the middle and high latitudes colder. But we find a balance in the net gain and loss of energy at different latitudes of earth's surface. It has become possible due to steady and continuous transfer of heat from low to middle and high latitudes by atmospheric and oceanic circulations.

From Table 2.5 it can be observed that in maintaining a balance the following amount of heat must be carried pole-ward. Such heat transfer takes place throughout the year. The rate of transfer is, however, high in winter months than in summer. Moreover in general the winter hemisphere has a net energy loss and the summer hemisphere possesses a net gain. The net summer-time gain is stored mainly in the surface layers of oceans. The variations in heat gain and loss on a latitudinal scale give rise to atmospheric and oceanic circulations. From the table it can also be observed that largest transfer of heat is required in the middle latitudes. In this region there is a maximum of advection and air-mass movement resulting in largest turbulence and storminess.

**Table 2.5:** Latitudinal Transfer of Heat

| Latitude, north in degree | $10^{19}$ calorie per day |
|---|---|
| 0 | 0 |
| 10 | 4.05 |
| 20 | 7.68 |
| 30 | 10.46 |
| 40 | 11.12 |
| 50 | 9.61 |
| 60 | 6.68 |
| 70 | 3.41 |
| 80 | 0.94 |
| 90 | 0 |

**Energy Budget of the Earth**

It is necessary to understand the Energy Budget of the earth. The dynamics of the atmosphere is closely associated with the energy budget.

## Incoming Energy

The total flux of energy entering the earth's atmosphere is estimated at 174 petawatts. This flux consists of:

- Solar radiation (99.978%, or nearly 174 petawatts; or about 340 W m$^{-2}$). This is equal to the product of the solar constant, about 1366 watts per square metre, and the area of the Earth's disc as seen from the Sun, about $1.28 \times 10^{14}$ square metre, averaged over the Earth's surface, which is four times larger. The solar flux averaged over just the sunlit half of the earth's surface is about 680 W m$^{-2}$. Note that the solar constant varies (by approximately 0.1% over a solar cycle); and is not known absolutely to within better than about one watt per square metre. Hence the geothermal and tidal contributions are less than the uncertainty in the solar power.
- Geothermal energy (0.013%, or about 23 terawatts (Terawatt $=10^{12}$ watts); or about 0.045 W m$^{-2}$). This is produced by stored heat and heat produced by radioactive decay leaking out of the Earth's interior.
- Tidal energy (0.002%, or about 3 terawatts; or about 0.0059 W m$^{-2}$). This is produced by the interaction of the Earth's mass with the gravitational fields of other bodies such as the Moon and Sun.
- Waste burning heat from fossil fuel consumption (about 0.007%, or about 13 terawatts; or about 0.025 Wm$^{-2}$).

## Outgoing Energy

The average albedo (reflectivity) of the Earth is about 0.3, which means that 30% of the incident solar energy is reflected back into space, while 70% is absorbed by the Earth and re-radiated as infrared. The planet's albedo varies from month to month, but 0.3 is the average figure. It also varies very strongly spatially: ice sheets have a high albedo, oceans low. The contributions from geothermal and tidal power sources are so small that they are omitted from the following calculations.

So 30% of the incident energy is reflected, consisting of:

- 6% reflected from the atmosphere.
- 20% reflected from clouds.
- 4% reflected from the ground (including land, water and ice).

The remaining 70% of the incident energy is absorbed:

- 51% absorbed by land and water, then emerging in the following ways:
  - 23% transferred back into the atmosphere as latent heat by the evaporation of water,
  - 7% transferred back into the atmosphere by heated rising air,
  - 6% radiated directly into space,
  - 15% transferred into the atmosphere by radiation, then reradiated into space.

- 19% absorbed by the atmosphere and clouds, including:
  - 16% reradiated back into space,
  - 3% transferred to clouds, from where it is radiated back into space,

When the Earth is at thermal equilibrium, the same 70% that is absorbed is reradiated:

- 64% by the clouds and atmosphere.
- 6% by the ground.

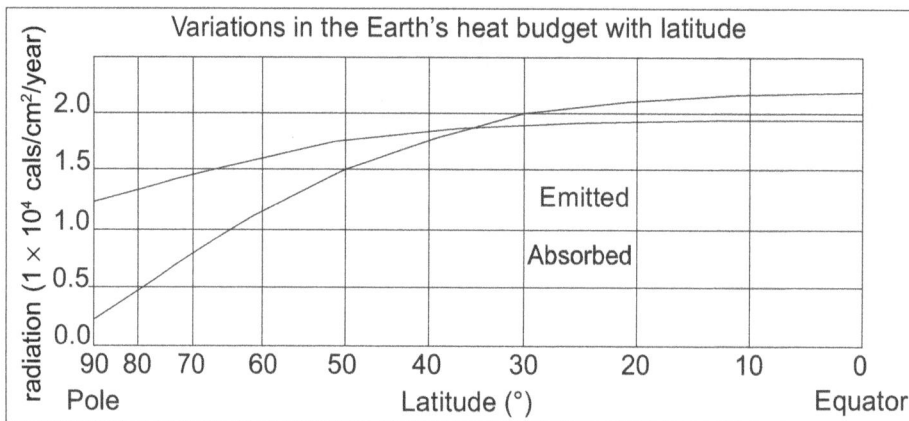

**Figure 2.2:** Energy Deficit and Energy Surplus

The area on the left (as shown in Fig. 2.2) is the energy deficit at the poles. Here, more energy is emitted than is absorbed. The area on the right is the energy surplus at the equator. Here, more heat is absorbed than is emitted due to the large amount of cloud acting as an insulator. As there is no overall temperature change there must be an overall transfer of energy between low and high altitudes. This occurs by *convection*, or the movement of air (wind) due to pressure systems.

*Anthropogenic Interference*

Green House Gas (GHG) emissions and other factors such as land-use changes, modify the energy budget slightly but significantly. The IPCC provides an estimate of this forcing, affecting solar input. The largest and best-known well-mixed greenhouse gases ($CO_2$, $CH_4$, halocarbons, etc.) contribute such change (2.4 W m$^{-2}$ compared to 1950). This is less than 1% of the solar input, but contributes to the observed increase in atmospheric and oceanic temperature.

**Feedbacks on Energy Balance**

Feedbacks in any system maintain the steady state. Information on earth's climate system will help understanding how rising GHG levels may affect Earth's energy balance. Feedbacks are interactions between climate variables such as temperature, precipitation, and vegetation and elements that control the greenhouse effect, such as clouds and albedo. Positive feedbacks amplify temperature change by making the greenhouse effect stronger or by reducing albedo, so they make the climate system more sensitive to the properties that induce them. Negative feedbacks have a dampening effect on temperature change, making the climate system less sensitive to the factors that trigger them.

Feedbacks can be very complex processes and may take place over short or long time spans. Important feedbacks in Earth's atmosphere include:

- *Water vapour feedback on temperature change (positive):* The atmosphere can hold increasing amounts of water vapour as the temperature rises, because the pressure of water vapour in equilibrium with liquid water increases exponentially with temperature. The presence of more water vapour as temperature rises increases the greenhouse effect, as well as the absorption of solar radiation, which further raises temperature. It is the fact that warm air can hold more water vapour than cool air.

- *Cloud feedback on terrestrial radiation (positive):* Because warmer temperatures increase water vapour amounts, they can increase cloudiness and further enhance temperature. It is hard to know whether or how much cloudiness will increase as temperature does, because cloudiness depends more on upward air motion than on temperature or water vapour levels directly.

- *Cloud feedback on solar radiation (negative):* As temperature and atmospheric water vapour levels rise, cloudiness may increase. Greater the cloudiness more will be the Earth's albedo, reflecting an increasing fraction of solar radiation back into space and decreasing temperature, although some cloud types are more reflective than others. It is, however, difficult to know whether or how much cloudiness will increase with temperature. Also, as noted earlier, clouds can also absorb infrared radiation, raising temperatures.

- *Vegetation feedback on solar radiation (negative):* As temperatures rise, deserts may extend, increasing Earth's albedo and decreasing temperature. This is a very complex feedback. It is uncertain whether deserts will extend, or conversely, whether higher $CO_2$ levels might stimulate higher plant growth levels and increase vegetation instead of reducing it.

- *Ice-albedo feedback on solar radiation (positive):* Rising temperatures cause polar glaciers and floating ice sheets to melt, decreasing earth's albedo and raising temperatures. As there is relatively little polar ice on land today this feedback is not likely to play a major role in near-term climate change. However, temperature increases large enough to melt most or all of the floating ice in the Arctic could sharply accelerate global climate change, because ocean water absorbs almost all of the incident solar radiation whereas ice reflects most sunlight.

Feedbacks cause much of the uncertainty in today's climate change models, and more research is required to understand how these relationships work. A 2003 National Research

Council study called for better measurement of many factors that affect climate feedbacks, including temperature, humidity, the distribution and properties of clouds, the extent of snow cover and sea ice, and atmospheric GHG concentrations.

---

## Further Readings

Budyko, M.I. (1962) 'Heat balance of the Surface of the Earth', *Soviet Geography*, Moscow.

Byers, H.R. (1959) *General Meteorology,* McGraw Hill Book Co., New York.

Haurwitz, B. and J.M. Austin (1944) *Climatology,* McGraw Hill Book Co., New York.

Kendrew, W.G. (1957) *Climatology,* Oxford University Press, Fairlawn, N.J.

Landsberg, H. (1958) *Physical Climatology,* Gray Printing Co., Du Bois, Paris.

Oliver, J.E. and John J. Hidore (2002) *Climatology,* Pearson Education, New Delhi.

Pettersen, Sverre (1958) *Introduction to Meteorology,* McGraw Hill Book Co., New York.

Ransom, W.H. (1973) 'Solar Radiation and Temperature', *Weather,* **8.**

Saha, P.K. and P.K. Bhattacharya (1994) *Adhunik Jalavayu Vidya,* West Bengal State Book Board, Calcutta.

Stone, R. (1955) 'Solar Heating of Land and Sea', *Geography,* **40.**

Trewartha, G.W. (1968) *An Introduction to Climate,* McGraw Hill Kogakusha Ltd., Tokyo.

Willett, H.C. and Frederic Sandera (1959) *Descriptive Meteorology,* Academic Press, New York.

# Chapter 3

## MOTION IN THE ATMOSPHERE

### Characteristics of Air

In general, air in the atmosphere is motionless. Air is in motion when differential pressure exists between two locations having comparatively high and low pressure.

Usually air moves from high pressure area to low pressure area on earth's surface. Difference in atmospheric pressure may be both—horizontal and vertical. Air rises up in the atmosphere and also sinks due to vertical change of pressure. But surface motion is generated on horizontal scale. *In situ* difference of surface pressure leads the air in motion.

Air pressure varies from place to place depending on the nature of air. Dry air exerts more pressure than the moist air. Dry air is principally constituted by oxygen and nitrogen only, whereas the moist air contains water (moisture) constituting hydrogen and oxygen molecules. The molecules of hydrogen and oxygen are lighter and so air with moisture turns to be lighter than the dry air.

We do not perceive the difference through our senses. But such difference occurs in the nature with the mixing of moisture with air. So we find that the dry areas on the earth's surface display high pressure condition than the areas with moist air. Air moves from the high pressure area to a low pressure area. Air moves from the high pressure area in clockwise direction in the northern hemisphere. But around the low pressure area air converges with anti-clockwise rotation. The opposite pattern is observed in the southern hemisphere. This is manifested by *Buys Ballot Law*. If any one stands back to the air in the northern hemisphere the low pressure area will exist to his left and *vice versa* in the southern hemisphere

### *Weight of Air Pressure*

Air pressure on the earth's surface is measured by a *barometer*. Barometer displays the air pressure by the column of mercury in the instrument. Normally air pressure is equal to a height of 76 cm. mercury column. So the air pressure is:

P = h$\rho$g = 76 $\times$ 13.50 $\times$ 980 dyne = 1013,000 dyne.

($10_6$ = 1 bar. So $10_3$ = 1 milibar)

So, normal air pressure is about 1013 milibar. Normal air pressure varies spatially and also temporally. In dry season, particularly in winter air pressure rises to 1020–1025 mb. But in rainy season (monsoon) the pressure drops to 990 mb. Even lower pressure may be associated with cyclones and storm conditions.

## Laws of Motion

Movement of any object follows the Laws of Motion as testified by Sir Isaac Newton. Newton's three laws state the 'Law of Inertia' as the first, the 'Law of Acceleration' as the second and the 'Law of Reciprocal Action' as the third. Newton's first and second laws of motion are of significance understanding the motion in the atmosphere.

### *Newton's First Law: Law of Inertia*

Lex I: *Corpus omne perseverare in statu suo quiescendi vel movendi uniformiter in directum, nisi quatenus a viribus impressis cogitur statum illum mutare.*

'Every body perseveres in its state of being at rest or of moving uniformly straight forward, except insofar as it is compelled to change its state by force impressed'.

The net force on an object is the vector sum of all the forces acting on the object. Newton's first law says that if this sum is zero, the state of motion of the object does not change. Essentially, it makes the following two points:

- An object that is not moving will not move until a net force acts upon it.
- An object that is in motion will not change its velocity (accelerate) until a net force acts upon it.

It states that a body will change its velocity of motion only if acted upon by an unbalanced force. That means if any object is moving will keep moving until a force modifies its motion.

There are numerous forces acting on the air that cause changes in its motion. Factors that affect the large-scale features of atmospheric circulation include the spherical shape and gravitational attraction of the Earth, its rotation, the tilt of the Earth's spin axis with respect to the plane of the Earth's orbit around the Sun, and the energy from the Sun. The principal factors that influence atmospheric circulation on a global scale are due to thermal processes that gives rise to *pressure gradient force* and rotation of the earth attributing *Coriolis force*.

### *Pressure Gradient Force*

A pressure gradient is the difference in pressure divided by the distance measured in the direction from high to low pressure. Therefore, the pressure gradient is responsible for exerting force acting on the air in the direction from high to low pressure. This is sometimes called the *pressure gradient force*. The pressure gradient force (PGF) is expressed by the equation:

$$fx/m = -1/\rho \ X \ dp/dx$$

The term $F/m$ is equal to the acceleration $dv/dt$ because this is an expression of Newton's law, $F = ma$. $dp/dx$ is the component of the pressure gradient along the x-axis. $\rho$ is density and $(1/\rho)$ shows that as the density increases, the acceleration due to the pressure gradient becomes smaller.

The pressure gradient force acts at right angles to isobars in the direction from high to low pressure. The greater the pressure difference over a given horizontal distance, the greater the force and hence the stronger the wind.

The diagram below shows how the pressure gradient force would be the difference in the pressure at constant height at between $C$ and $A$, divided by the horizontal distance in $x$ between $C$ and $A$. In the case of that diagram, the pressure is lower at $C$ than at $A$ and the pressure gradient force would be positive and trying to force air in the +x direction.

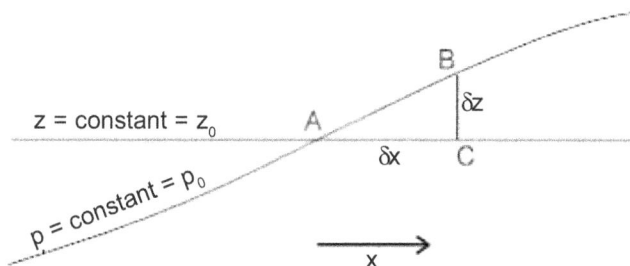

**Figure 3.1**: Slope of the Pressure Gradient Force

It may also be noted that the pressure gradient force in the N-S direction increases with height in the atmosphere because, as the slopes of the constant pressure surfaces are steeper higher up in the atmosphere. A steeper slope means that the $dp/dy$ has a greater magnitude and also, the density at the higher altitude will be lower, making the pressure gradient force stronger.

The magnitude of the PGF is particularly affected by several successive periods where the strength of the atmospheric flow was lower (negative trend) or higher (positive trend) than normal. The atmospheric flow appears stronger than before (positive trend), except in summer. As the direction of the PGF can be related to the trajectory of the flow *via* the Buys-Ballot law, its anomalies are directly related to anomalies in the advection of air masses. No definite trend can be detected for this variable. However, it appears that the northerly flow becomes rare in winter.

### *Coriolis Force*

The Coriolis force indicates an apparent force that is caused due to the rotation of the earth:

- Acts to the *left* in the southern hemisphere

- Acts to the *right* in the northern hemisphere
- Coriolis force is zero at the equator
- Coriolis force is proportional to the velocity of the air.

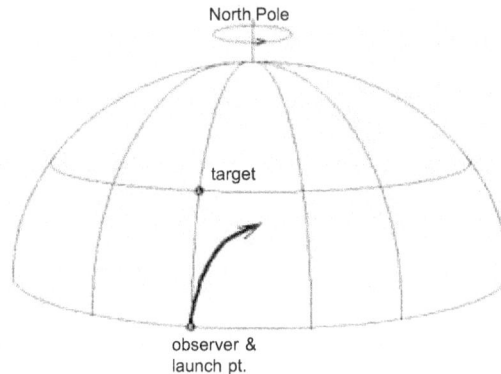

**Figure 3.2:** Coriolis Force
*Source:* Steven C. Wofsy

To understand the effect of Coriolis force, imagine standing on the top of a spinning globe (north pole). Say, a stone is thrown due south, parallel to the surface and the stone will veer to the thrower's *right* (west). It results as the platform is rotating anti-clockwise.

While standing at the south pole then the earth would be rotating in a clockwise sense. Therefore, the stone would veer to the *left* (west). The consequences of the Coriolis force due to the earth's rotation will be manifested for winds from any direction:

*Winds in the southern hemisphere will always be deflected to the left.*
*Winds in the northern hemisphere will always be deflected to the right.*

The magnitude of the Coriolis force $F$ depends upon the density of the air $r$, the wind speed $v$, the angular speed of the Earth's rotation $w$ ($w = 2\pi / T$ where $T$ is the period of the Earth's rotation: $T = 24$ h) and the latitude $\phi$ as noted in the equation below:

$$F = 2\ v\ \rho\ \omega\ Sin(\phi)$$

The Coriolis force is maximum at the poles ($Sin90° = 1$) and zero at the equator ($Sin0° = 0$). There are many principal prevailing surface wind patterns that are caused in part due to Coriolis forces. One of these is the trade winds, the prevailing east-to-west winds within the tropics. North of the equator, the wind is from the northeast, and south, it is from the southeast. The energy source for these winds is the constant warming of the air near the equator. This warmed air expands and rises, and cooler, denser air from north and south of the equator moves toward the equator near the surface. In the process, this moving air is forced westward in the northern hemisphere and also westward in the southern hemisphere by the Coriolis forces. Near the equator due to convergence no appreciable motion could be felt. This region is known as the equatorial doldrums. The north/south meeting point of the winds is not precisely the equator. The warming is highest in the summer hemisphere, which tends to shift the doldrums toward the summer hemisphere. Also, because there is more land in the northern hemisphere, heating of the air there is greater, and this shifts the median latitude of the doldrums slightly north of the equator.

Storms centre develops on regions of *low* pressure. Unstable weather conditions occur where the air moves towards the centre and then rises. If the air is moist then as it rises the air cools leading to condensation and rain. Because of the Coriolis Effect, the winds will move in a clockwise winds sense around such low pressure centres in the southern hemisphere (anticlockwise in northern hemisphere).

Clear and dry weather will usually be associated with the centres of high pressure. Surface winds flow away from such centres. In the southern hemisphere this outward flowing air veers to the left and forms the characteristic anticlockwise flow around high-pressure centres (opposite direction in the northern hemisphere).

## *Visualisation of Coriolis Force*

Assume that a person on a merry-go-round (a flat rotating platform) sees two points, $X$ and $Y$ as marked on the platform. He is at the point $X$ and intends to slide a frictionless puck (a rubber disc) from $X$ to $Y$. If the platform is stationary the puck would slide along a straight line path from $X$ to $Y$ with a constant speed.

When the platform is rotating anticlockwise the situation will be different.

Suppose he pushes the puck horizontally with a velocity $vx$. Since the platform is rotating, at the instant he releases it the puck will have the same vertical velocity $vy$ as of him. Therefore, the velocity $v$ of the puck with respect to the ground will have two components $v_x$ and $v_y$. This time the puck will not reach the point $Y$ but instead veers off to the right that is the puck now appears to travel in a curved path with constant speed from his point of view (non-inertial frame of reference). The puck has an *angular momentum L*. So,

$$L = m\, v_y\, R$$

Where $m$ is its mass, $R$ is the distance from the axis of rotation and $v_y$ is the *tangential* component of its velocity vector at any instant.

An important physical law states that the angular momentum of an object does not change unless acted upon by a *torque* (force times perpendicular distance). This is known as the *Law of Conservation of Angular Momentum* (think of ice skaters spinning). For the sliding puck as there are no torques acting on it, therefore the product is:

$$v_y R = \text{constant}$$

Therefore, when the radius $r$ increases, the velocity $v_y$ shall decrease. The puck must deflect to the right "as it is left behind" as it now has a smaller tangential speed than a point below it on the rotating platform. Also, a puck pushed toward the centre will also veer to the

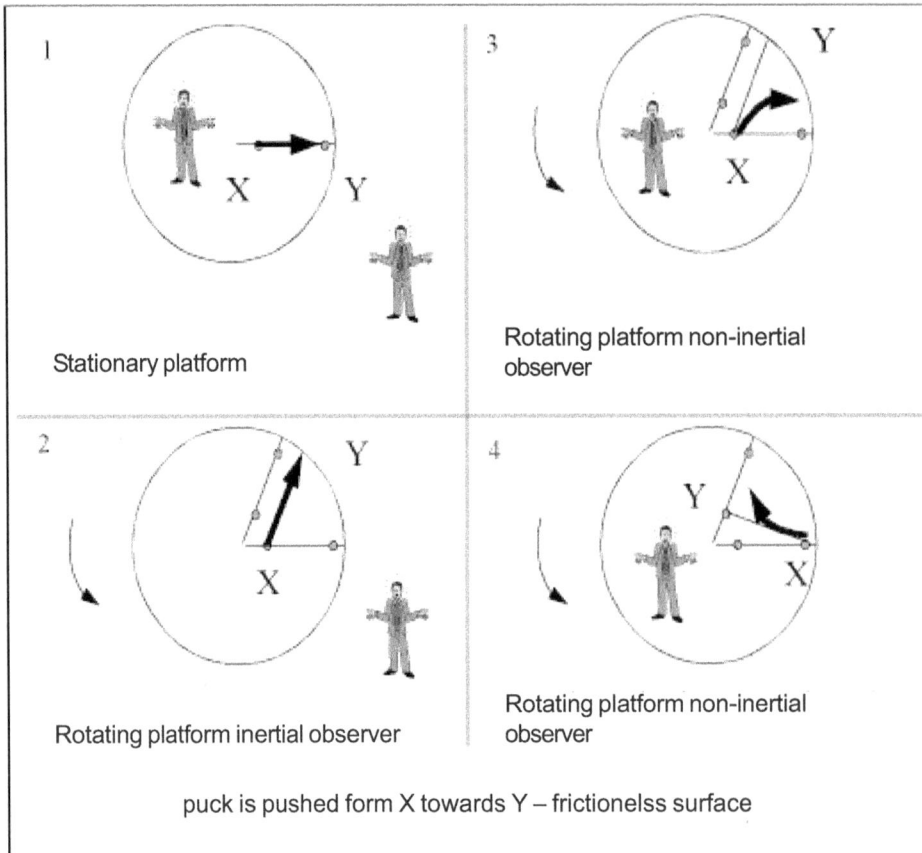

**Figure 3.3:** Explanation of Coriolis Force

right. In this case the radius decreases and the velocity of the puck increases, therefore, it moves faster than the rotating platform and is deflected to the right. To an observer in an inertial frame of reference, viewing the rotating platform from above always sees the puck slide in a straight line path with constant speed while the platform rotates under the moving puck. For the observer in the inertial frame of reference, the acceleration of the puck is zero. So, the total horizontal force acting on the puck must also be zero. For the observer in the non-inertial frame of reference, the puck must accelerate because it moves in a curved path and therefore the puck must be acted upon by some horizontal force. This force is called the *Coriolis force*. It is not a 'real' force, it is only an artifact of an observation made in a non-inertial frame of reference. The relatively slow rotation of the earth makes its effects very small in situations such as throwing stones or walking. However, many of the atmospheric and oceanic characteristics that we take for granted are due to the effects of the Coriolis force.

### Resulting Wind Circulation

*Geostrophic and Cyclostrophic Winds*

Winds blow because of horizontal and vertical pressure gradients and the atmospheric motion can be deduced from isobaric surface charts etc. If the *horizontal pressure gradient force* is exactly balanced in magnitude by Coriolis effect accelerations of the air will be relatively small and a *geostrophic wind (Gk. geo = earth, strophe = turning )* will flow horizontally at a constant speed proportional to the isobaric spacing gradient, perpendicular to the two opposing forces and parallel to straight isobars. Air will be accelerated to the extent that these forces are unbalanced. Transitory disturbances and vertical movement create imbalance. When vertical motion is present the horizontal wind can not be exactly geostrophic.

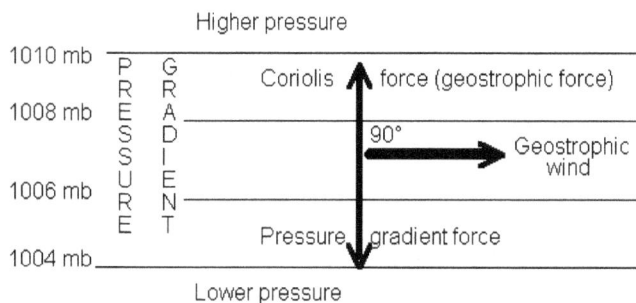

**Figure 3.4:** Geostrophic Winds

Geostrophic flow is predominant above the friction layer in very large scale weather systems where the pressure gradient force and the Coriolis force are nearly equal and opposite, e.g. the southern ocean west wind belt. Between 15°S and 15°N latitudes there is little geostrophic flow due to weak Coriolis, the latter being zero at the equator and winds tend to flow across the isobars, in which case it is more suitable to show wind flow as streamlines on upper air charts. *(A streamline arrow shows the direction of flow, whereas an isotach is a line along which the speed of flow is constant.)*

At the other end of the scale in a short span mesoscale systems Coriolis force has insufficient time to take effect, or is relatively weak compared to other forces, thus geostrophic balance is not present and air accelerations can be quite large.

If atmospheric circulation was always in perfect balance between geostrophic forces and pressure gradient forces geostrophic winds would flow and there would be no change in pressure systems. In reality the pressure distribution takes the form of curved isobars resulting in a third force. The *centripetal acceleration* pushes the flow inward of the curve.

The *gradient wind* is the equilibrium wind for the three forces—centripetal acceleration, pressure gradient and Coriolis force (or geostrophic). The vector difference between the geostrophic and the gradient winds causes the *ageostrophic wind*. Thus ageostrophic movement is large for small scale systems and small for large scale systems.

When the centripetal acceleration becomes the major control of the gradient wind there is an extremely strong curvature of the airflow and the winds are called *cyclostrophic (Gk = circle–turning)*. Examples are the *tropical cyclones* and *tornadoes*. When a body is moving in a curved path centripetal force is the radial inward force that constrains the body to move in that curved path and, even at constant speed, there is an inward acceleration resulting from the body's continually changing velocity. The equal and opposite centrifugal force that appears to act outward on a body moving in a curved path is a fictitious force but convenient to show the equilibrium forces for air moving in a cyclonically curved path e.g. around a surface low pressure system.

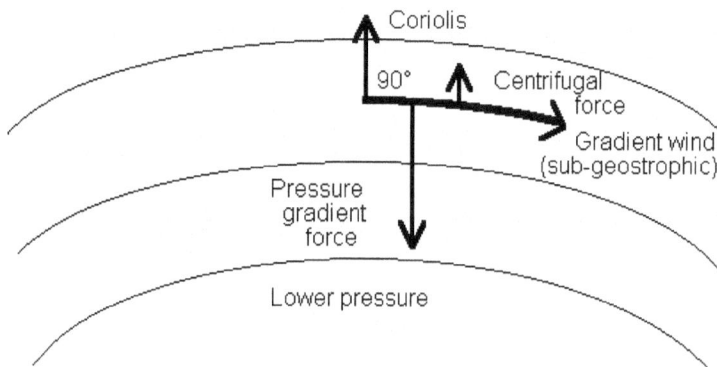

**Figure 3.5:** Generation of Gradient Wind—Sub Geostrophic (Balance between Centrifugal Force and Pressure Gradient Force)

For the gradient wind to follow cyclonically curved isobars the pressure gradient force must be slightly stronger than Coriolis force to provide the centripetal force. As the magnitude of the Coriolis force is directly dependent on wind speed it follows that the wind speed around a low is less than would be expected from the pressure gradient force and the gradient wind is called the *sub-geostrophic* wind.

For air moving in an anticyclonically curved path e.g. around a high, the opposite occurs, the Coriolis providing the centripetal force.

For the three forces to be in equilibrium the Coriolis force must exceed the pressure gradient force and consequently the gradient wind speed must be greater than would be expected from the pressure gradient force and is thus called the *super-geostrophic* wind.

Air moving within a pressure pattern possesses momentum. If the air moves into a different pressure pattern and gradient it will tend to maintain its speed and Coriolis force for some time, even though the pressure gradient force has changed. The resultant imbalance will temporarily deflect the airflow across the isobars in the direction of the stronger force—Coriolis or the pressure gradient force.

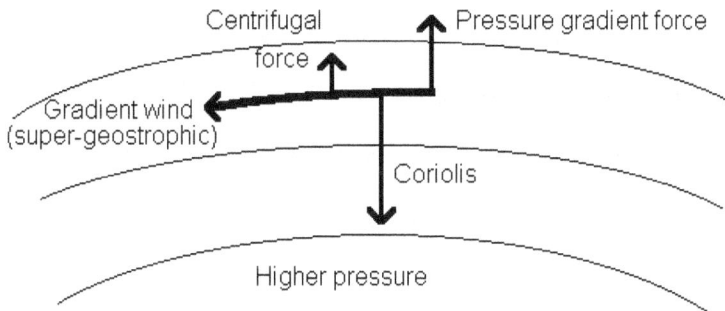

**Figure 3.6:** Generation of Gradient Wind—Super Geostrophic (Balance between Centrifugal + Pressure Gradient Force and Coriolis Force)

*The Ekman Spiral*

The earth's surface displays frictional interaction with air, its effect decreasing with altitude until between 1500 and 3000 feet asl (the boundary layer or friction layer), where the real wind is the sum of the geostrophic and ageostrophic components.

The *Ekman spiral* indicates, in vector form, how wind vector velocities change with altitude, flowing along the isobars above the boundary layer then decreasing in speed and veering as height decreases and friction effects increase, until the surface vector lies across the isobars at low speed. Thus in the presence of surface friction, wind force always acts opposite to wind direction, the veering low level air spirals in toward a low (clockwise rotation) and out from a high (anticlockwise rotation) resulting in surface convergence and divergence. However, frictional convergence has little effect on cloud development within a depression outside the tropics.

The terms *veering* and *backing* originally are referred to the shift of surface wind direction with time. Meteorologists now use the term when referring to the shift in wind direction with height. Winds shifting anti-clockwise around the compass are *'backing'*, those shifting clockwise are *'veering'*.

### Velocity Change between Surface and Gradient Wind

Over land the surface wind speed may be only 30 percent to 50 percent of the gradient wind speed. Wind blows across the isobars in the direction of the gradient force (i.e. towards the lower pressure) in the boundary layer. The stability of the boundary layer affects the strength of the friction force. A very stable layer suppresses turbulence and friction is weak except

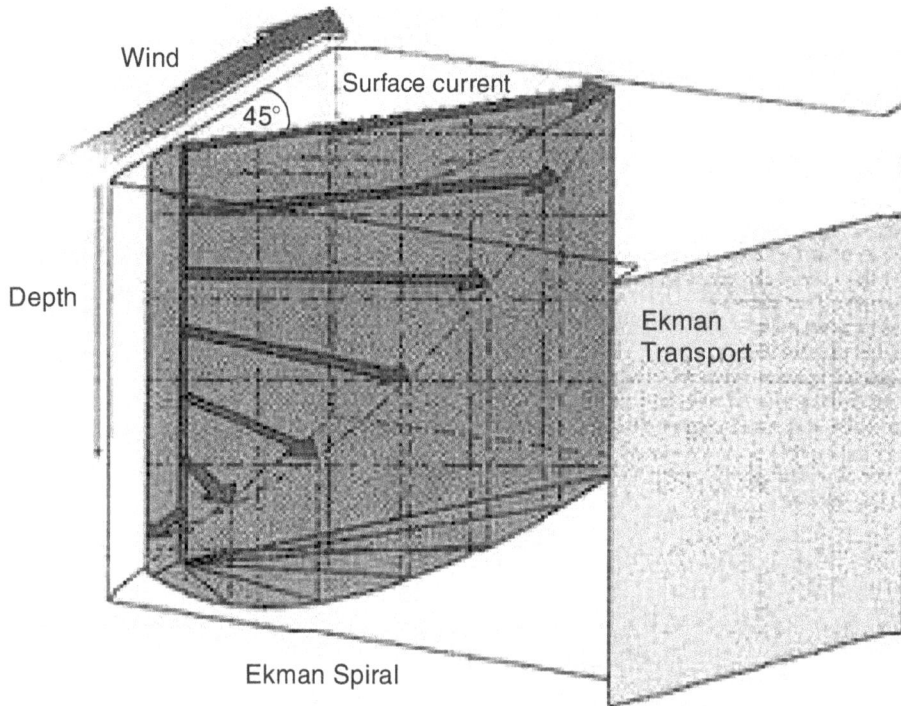

**Figure 3.7:** Ekman Spiral

near the surface. In a super-adiabatic layer convective turbulence is strong and the friction force will be strong. The following table is for a typically neutrally stable layer and shows the average angle change in wind direction for an average wind profile over various terrains beneath a moderately strong gradient wind of about 30 knots.

**Table 3.1:** Vertical Wind Profile

| Height-feet | Flat country | Rolling country | Hilly country | Wind speed–knots |
|---|---|---|---|---|
| Below 500 | +30° | +36° | +43° | 12 |
| 500–1000 | +22° | +30° | +36° | 20 |
| 1000–2000 | +10° | +17° | +25° | 25 |
| 2000 – 3000 | +2° | +5° | +10° | 28 |
| The wind is backing with height increase. The change in direction in the first 300 feet is negligible in strong winds but greatest in light winds (less than 10 knots) and may be as great as 15°–20° if the surface wind is less than 5 knots. The greatest change in wind speed occurs at night and early morning. | | | | |

**Table 3.2:** Low Level Geostrophic Wind Speed

**Calculating Low Level Geostrophic Wind Speed**

The geostrophic wind calculated from the isobar spacing on a surface [mean sea level] synoptic chart is usually a reasonable approximation of the wind speed at about the 3000 feet level.

Geostrophic wind speed (knots) = 3832 GT / P sine L
where G = horizontal pressure gradient in hPa/km
T = air temperature in kelvin units
P = msl pressure in hPa
L = the latitude in degrees

Because the proportion T/P normally does not vary greatly at msl the equation can be simplified to:

Geostrophic wind speed (knots) = 2175/D sine L
where D = the distance in km between the 2 hPa isobars on the chart.

*The sine of an angle less than 60° can be easily estimated without reference to tables by using the 1-in-60 rule of thumb i.e. the sine of an angle is roughly degrees x 0.0167 [or 1/60] e.g. sine 36°S = 36 x 0.0167= 0.601, or 36/60 = 0.6*

The following table is derived from the preceding and shows the estimated wind speed in knots for spacing between the 2 hPa isobars, from 40 to 600 km. If the surface chart shows 4 hPa spacing then just halve the distance between the isobars and use the table below.

Estimated wind speed from 2 hPa isobar spacings of 40 to 600 km

| Latitude | 40 km | 60 km | 80 km | 100 km | 120 km | 160 km | 200 km | 400 km | 600 km |
|----------|-------|-------|-------|--------|--------|--------|--------|--------|--------|
| 10°S | 300 | 210 | 160 | 130 | 110 | 80 | 60 | 30 | 20 |
| 20°S | 160 | 110 | 80 | 65 | 55 | 40 | 30 | 16 | 10 |
| 30°S | 110 | 75 | 60 | 45 | 35 | 30 | 25 | 12 | 8 |
| 40°S | 90 | 60 | 45 | 35 | 30 | 25 | 18 | 10 | 6 |

## Table 3.3: The Beaufort Wind Speed Scale

| No. | Wind speed | Gust speed | Meteorological classification | Terms used in general forecast | Wind effect on land |
|---|---|---|---|---|---|
| 0 | <1 knot | | Calm | Calm | Smoke rises vertically |
| 1 | 1–3 | | Light air | Light winds | Smoke drifts |
| 2 | 4–6 | | Light breeze | Light winds | Leaves rustle, water ripples; '15 knot' dry windsock tail drooping 45° or so |
| 3 | 7–10 | | Gentle breeze | Light winds | Wind felt, leaves in constant motion, smooth wavelets form on farm dams and small lakes, smoke rises at an angle above 30°; '15 knot' dry windsock tail 15° or so below horizontal |
| 4 | 11–16 | | Moderate breeze | Moderate wind | Small branches move, dust blown into air, crested wavelets form |
| 5 | 17–21 | | Fresh breeze | Fresh wind | Small trees sway, smoke from small fires blown horizontally; '15 knot' dry windsock horizontal |
| 6 | 22–27 | | Strong breeze | Strong wind | Large branches sway, whistling in wires |
| 7 | 28–33 | | Near gale | Strong wind | Whole trees in motion |
| 8 | 34–40 | 43–51 | Fresh gale | Gale wind | Twigs break off, difficulty in walking |
| 9 | 41–47 | 52–60 | Strong gale | Severe gale | Some building damage |
| 10 | 48–56 | 61–68 | Whole gale | Storm | Trees down |
| 11 | 57–62 | 69–77 | Storm | Violent storm | Widespread damage |
| 12 | 63+ | 78+ | Tropical cyclone | Tropical cyclone | Severe extensive damage |

*Beaufort scale number:*
0 – Sea is mirror-like
1 – Ripples present but without foam crests
2 – Small wavelets with glassy appearance do not break
3 – Large wavelets, crests begin to break with scattered white horses
4 – Small waves becoming longer, fairly frequent white horses
5 – Moderate waves, many white horses with chance of spray
6 – Large waves are forming with extensive white foam crests, spray probable
7 – The sea heaps up, white foam from breaking waves is blown in streaks
8 – The edges of crests break into spindrift with well marked foam streak lines
9 – High waves with tumbling crests and spray, dense foam streaks
10 – Very high waves with overhanging crests, surface appearance white, visibility affected
11 – Chaotic sea, large parts of waves blown into spume with foam everywhere
12 – Air filled with foam and spray, visibility severely impaired.

The following table shows the state of seas, with likely maximum wave height in metres.

**Table 3.4:** State of Seas Classification

| Calm | Zero | No waves |
|---|---|---|
| Rippled | 0.1 m. | No waves breaking on beach |
| Smooth | 0.5 m. | Small breaking waves on beach |
| Slight | 1.3 m. | Waves rock buoys and small boats |
| Moderate | 2.5 m. | Sea becoming furrowed |
| Rough | 4 m. | Sea deeply furrowed |
| Very rough | 6 m. | Disturbed sea with steep faced roller |
| High | 9 m. | Very disturbed sea with steep faced roller |

*The Compass and Wind Rose*

There are 32 compass 'points', each division being 11.25°. Winds shifting anti-clockwise around the wind rose are *'backing'*, those shifting clockwise are *'veering'*. The compass points and the associated compass direction (in degrees) have been shown in the table below:

**Table 3.5:** Compass and Wind-rose Points

| 11.25 | North by (one point) East | 191.25 | South by (one point) west |
|---|---|---|---|
| 22.50 | North-Northeast | 202.50 | South-Southwest |
| 33.75 | Northeast-North | 213.75 | Southwest-South |
| 45.00 | Northeast | 225.00 | Southeast |
| 56.25 | Northeast-East | 236.25 | Southwest-West |
| 67.50 | East-Northeast | 247.50 | West-Southwest |
| 78.75 | East-North | 258.75 | West-South |
| 90.00 | East | 270.00 | West |
| 101.25 | East-South | 281.25 | West-North |
| 112.50 | East-Southeast | 292.50 | West-Northwest |
| 123.75 | Southeast-East | 303.75 | Northwest-West |
| 135.00 | Southeast | 315 | Northwest |
| 146.25 | Southeast-South | 326.25 | Northwest-North |
| 157.50 | South-Southeast | 337.50 | Northwest-North |
| 168.75 | South-East | 348.75 | North-West |
| 180.00 | South | 360.00 | North |

## General Circulation of the Atmosphere

We know that the low latitude region gets more solar energy and is more heated than the higher latitude region. In contrast the polar region gets the least solar energy and remains to be perma-frost. In principle there is transfer of heat and energy from the lower latitude region towards the polar region to maintain a balance known as the *Latitudinal Heat Balance*. A thermal convection cell is thus formed between low latitude region and the polar region. The Hadley cell explains the simple General Circulation Model (GCM) attributing the transfer of heat from the *hot* Equator to the *cold* Pole. The model never suggests the role of eddies and break in the transport of air for many reasons.

A simple model for the latitudinal transport of heat and energy from the low latitude region towards the polar region may be conceived with the constituent three cells helping the circulation.

### *Three Cells Concept*

### *Hadley Cell*

The intense incoming solar radiation in the equatorial region generates rising air. The rising air cools at higher level for condensation and forms a region of intense clouds and heavy precipitation. The easterly winds converge in this region and called the Inter-Tropical Convergence Zone (ITCZ). The ITCZ is not stationary and moves north and south following the sun's declination during the year. Due to the stability of the stratosphere the rising air that reaches the tropopause can not rise further upward and moves towards the polar region. By the time the air moving northward reaches about 30° N it turns to be a westerly wind (it is moving to the east) due to the Coriolis force. To maintain the conservation of angular momentum, the pole-ward moving air increases speed. The increased speed and the Coriolis force are responsible for the formation of *subtropical Jet*. This pole-ward moving air piles up forming an area of high pressure at the surface—the *subtropical highs*. Some of the air sinks toward the surface. Subsidence acts as a lid to the rising air for convection and inhibits cloud formation in this region. For this reason many large deserts are located near 30° N and 30° S. Once the sinking air reaches the ground, some blows to the equator turning west (in the northern hemisphere) as controlled by the Coriolis force. This easterly surface air is known as the *Trade winds*, blows steadily from the northeast in the northern hemisphere and southeast in the southern hemisphere.

### *Ferrel Cell*

Ferrel conceived the idea that in the circulation model a secondary circulation feature exists, dependent for its existence upon the Hadley cell and the Polar cell. It behaves much as an atmospheric ball bearing between the Hadley cell and the Polar cell, and appears to be the eddy circulations (the high and low pressure areas) of the midlatitudes. On the southern margin (in the Northern hemisphere), it overrides the Hadley cell, and at its northern front, it overlaps the Polar cell. Some of the divergent air at the surface near 30° N moves towards the pole and is deflected to the east by the Coriolis force generating the prevailing westerly winds at the surface. At about 60° N the air tends to rise, cools and condenses. The clouds are formed and precipitation occurs due to cooling and condensation. This region is known to be the polar front. Some of this rising air also turns towards the equator as controlled by the global thermal convection.

*Polar Cell*

The area of heat sink is known as *polar cell*. Sinking air in the polar areas (above 60° latitude) settles and results in the formation of high pressure over the poles. At the surface, the pole-ward moving air turns to the right by the Coriolis force (in the northern hemisphere) forming the *Polar Easterly winds*. The cold polar air converges with the warm subtropical air moving towards the pole. Due to convergence of two contrasting air masses a boundary between these two air masses emerges, known as the *polar front*. The warm air from the subtropics rides over the cold equator-ward moving polar air. This polar front being very unstable appears to be the source of much of the changing weather in the high latitude region, particularly in fall, winter and spring. The large temperature contrast is evidenced in the polar front Jet Stream in the vicinity of the polar front.

The Three Cells Model is characterised by rising air at the Equator and 60° N, and by sinking air at 30° N and over the poles.

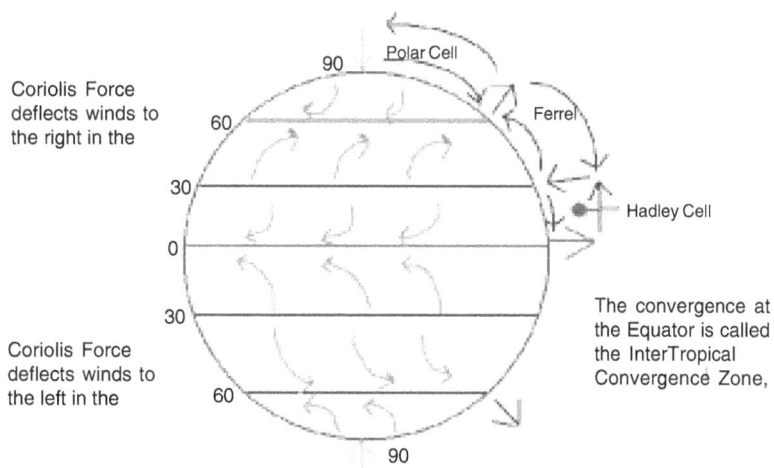

**Figure 3.8:** Three Cells Model

## Semi-Permanent Pressure Cells

Instead of cohesive pressure belts circling the earth as suggested by the three-cells model, semi-permanent pressure cells exist. These cells, which are either thermally or dynamically produced, fluctuate in strength and position on a seasonal basis. Semi-permanent pressure cells in the northern hemisphere include the Aleutian and Icelandic lows, and the Siberian, Hawaiian, and Bermuda-Azores highs. The oceanic lows gain maximum strength during the winter months while the oceanic highs peak in intensity during the summer. The thermally driven continental highs peak in winter while the continental lows have maximum strength during summer. The summer time Tibetan heat source is an important contributor to the development of the east-Asian monsoon. P. Koteswaram in his Heat Engine Concept highlighted the role of Tibetan Plateau drawing monsoon circulation to Indian landmass. Sinking atmospheric motions associated with the subtropical highs promotes desert conditions across affected latitudes. Seasonal fluctuations in the pressure belts relate to the migrating vertical ray of the Sun. The ITCZ lags slightly behind the vertical solar ray into the summer hemisphere. This causes a poleward migration of the subtropical highs in addition to a weakening of the higher latitude oceanic lows. In the winter hemisphere, opposite conditions result as the oceanic lows strengthen and the subtropical highs weaken and equator-ward shifting results. Such migrations greatly influence temperature and precipitation regimes across the globe. This is well understood in the tropics where seasonal precipitation is closely tied to variations of the subtropical highs and the ITCZ.

In fine, zonal surface winds on earth's surface are easterly (westward) in low latitudes and westerly (eastward) in mid-latitudes. In high latitudes winds are either easterly or nearly absent. The strength of the mean zonal surface wind varies seasonally. However the pattern of alternating easterlies and westerlies prevails throughout the year, with slight seasonal shifts of the latitudes at which the mean zonal surface wind changes sign. The mean meridional surface wind is weaker than the mean zonal surface wind. It moves towards the poles in regions of surface westerlies and towards the equator in regions of surface easterlies. In boreal summer, the monsoons of the northern hemisphere lead to a mean northward surface wind across the equator, which typically has a westerly component in monsoon regions. The mean surface winds of easterlies in the low latitude region had been recognised by the navigators in the past, who called it trade winds, a term we now use more restrictively to denote the tropical easterly winds.

## Global Surface Circulation

The visualised surface circulation on earth could be evidenced from 39 years of record. The circulation patterns as evidenced vary somewhat from the three cells model as described earlier. These differences are caused primarily by two factors. First, the earth's surface is not composed of uniform materials. The earth has two components—water and land. The water and land surface behave differently in terms of heating and cooling affecting latitudinal pressure zones to be less uniform. The second factor that influences actual circulation patterns is elevation. Elevation tends to cause pressure centres to become intensified when altitude is increased. This is especially true of high pressure systems.

On modified pattern one can visualise the *Intertropical Convergence Zone* (ITCZ), *Subtropical High Pressure Zone*, and the *Subpolar Lows.* The intertropical convergence zone is identified on the figures (January and July) by a bold line. The formation of this band of low pressure emerges as a result of solar heating and the convergence of the trade winds. In January, the inter-tropical convergence zone is found south of the equator. During this time of the calendar year, the Southern Hemisphere is tilted towards the Sun and gets higher inputs of solar radiation. The line denoting the inter-tropical convergence zone is not straight and parallel to the lines of latitude. The bends (rise and low) in the line occur because of the different heating properties of land and water. Over the continents of Africa, South America, and Australia, these bends are toward the south pole. This phenomenon results because heating of land surface is faster than the ocean.

During July, the intertropical convergence zone (ITCZ) is generally shifted north of the equator (Figure 3.10). This shift in position occurs as the altitude of the Sun rises in the northern hemisphere. The greatest spatial shift in the ITCZ, from January to July, occurs over the North Africa and Asian landmass. This shift is about 40° of latitude in some places. Due to intense heating in July the land areas of Northern Africa and Asia rapidly turn warmer forming the *Asiatic Low* which becomes part of the ITCZ. In winter months, the intertropical convergence zone moves south by the development of an intense high pressure system over central Asia (compare Figures 3.9 and 3.10). The extreme shift of the ITCZ is related to the development of a regional winds system called the *Asian Monsoon.*

The *Subtropical High Pressure Zone* is not a continuous area of high pressure stretching around the global surface. In reality, the system consists of several localised *anticyclonic* cells of high pressure. These systems are located roughly at about 20° to 30° of latitude and are shown with the letter **H** in Figures 3.9 and 3.10. The subtropical high pressure systems develop because of sinking air currents from the *Hadley cell.* These systems intensify over the ocean during the summer or high Sun season. During this season, the air over the ocean bodies remains relatively cooler because of the slower heating of water relative to land surfaces. Over land, the conditions differ in the winter months. At this time, land quickly becomes cooler, relative to ocean, forming large cold continental air masses.

The *Subpolar Lows* develop a continuous zone of low pressure in both hemispheres in latitude between 50° and 70°. The intensity of the subpolar lows varies with season. This zone is most intense during summer in the southern hemisphere. At this time, greater differences in temperature are between air masses, found on either side of this zone. North of subpolar low belt, summer heating warms subtropical air masses. South of the zone, the ice covered surface of Antarctica reflects much of the *incoming solar radiation* back to space. As a result, air masses above Antarctica remain cold because very little heating of

Sea-Level Pressure and Surface Winds                              Jan

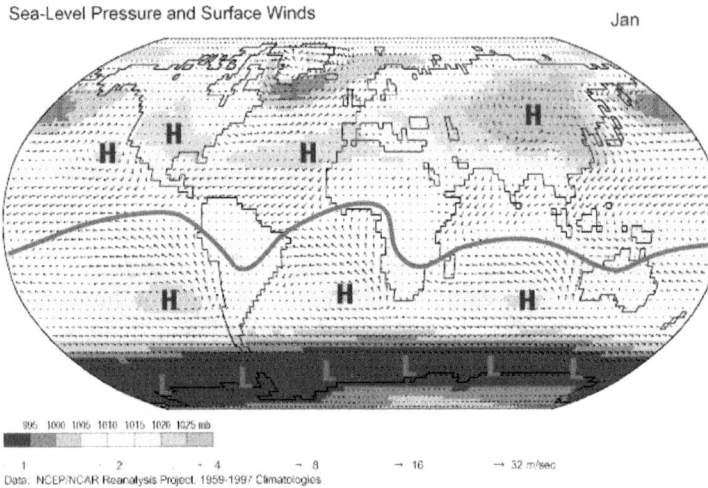

**Figure 3.9:** Mean January prevailing surface winds and centres of atmospheric pressure, 1959–1997. The bold line on this image represents the intertropical convergence zone (ITCZ). Centres of high and low pressure have also been labeled.

*Source of Original Modified Image:* Climate Lab Section of the Environmental Change Research Group, Department of Geography, University of Oregon—Global Climate Animations.

Sea-Level Pressure and Surface Winds                              Jul

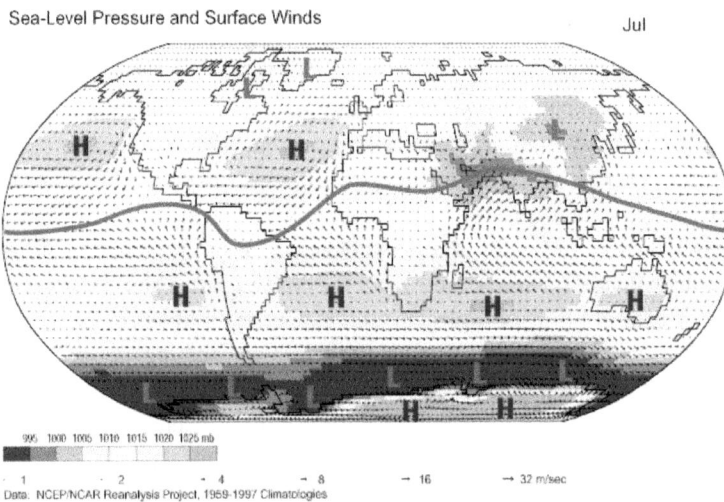

**Figure 3.10:** Mean July prevailing surface winds and centres of atmospheric pressure, 1959–1997. The bold line on this image represents the Intertropical Convergence Zone (ITCZ). Centres of high and low pressure have also been shown.

*(Source of Original Modified Image:* Climate Lab Section of the Environmental Change Research Group, Department of Geography, University of Oregon—Global Climate Animations).

the ground surface takes place. The meeting of the warm subtropical and cold polar air masses at the subpolar low zone enhances *frontal uplift* and the formation of intense low pressure systems.

In the Northern Hemisphere, the subpolar lows do not stretch as a continuous belt circling the globe. Instead, they remain as localised *cyclonic cells* (centres of low pressure). In the northern hemisphere winter, these pressure centres are intense and located over the oceans just to the south of Greenland and the Aleutin Islands (Figure 3.10). These areas of low pressure are responsible forming many *mid-latitude cyclones*. The development of the subpolar lows in summer only occurs (Figure 3.10) over Greenland and Baffin Island, Canada, but weak unlike the Southern Hemisphere. This is due to considerable heating of the earth's surface from 60° to 90°N latitudes. As a result, cold polar air masses generally do not form in summer.

*Impact of Rotation on Circulation*

We noted that the Coriolis force plays significant role in the general circulation in the atmosphere. Let us have a schematic explanation in the matter. Due to rotation the equatorial area is spinning 465 metre per second. Assume that the rising air will move towards 60° latitude from west to east then the circulation will attain a speed of 698 metre per second. However, due to frictional forces it is not possible to attain so high speed.

Let us explain it with the models. In Model A due to rotation wind will flow from the east in all latitudes. In some latitudes where the friction and direction of the wind remain opposite the momentum will be reduced, elsewhere the momentum will rise. Gradually stability will be attained. The wind systems will be divided in three cycles as shown in the Models B, C and D.

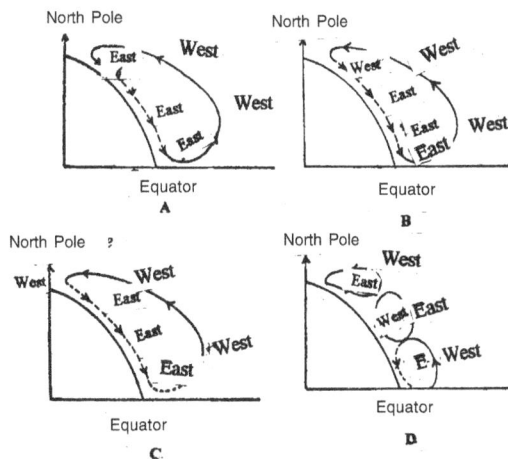

**Figure 3.11 (A, B, C and D):** Impact of Earth's Rotation on General Circulation

It is observed that near the 60 degree latitude low pressure (sub-polar low) is formed, one system will rise above to descend over the pole (polar high) and the other will rise to sink over the 30 degree latitude (sub-tropical high). So the one-cellular Hadley cell turns to be the tri-cellular pattern of global wind circulation (Model D).

### *Basic Pattern of Circulation*

We now have the schematic pattern, seven surface components of basic pattern of circulation from pole to equator; replicated north and south of equator:
1. Polar high,
2. Polar easterlies,
3. Subpolar low,
4. Westerlies,
5. Subtropical high,
6. Trade winds and
7. Intertropical convergence zone

Five major ocean basins of the subtropical latitudes serve as the source of major surface winds.

We will discuss below the seven basic components of surface pattern of circulation. It will be convenient to discuss first the Subtropical High, being the source region of two major global surface circulations.

### *Subtropical Highs*

Large semi-permanent *high-pressure cells* stay over each ocean basin at ~30° N/S. Two general high pressure ridges extend around globe at these latitudes. These cells are broken up over continents, especially in summer. Over the subtropical high general subsidence of air occurs from the upper air. Normally clear, warm and calm weather are observed over this region. Most of the principal deserts are located here. Due to calm weather in the past over this region the sailing ships used to dump horses overboard in the sea. So the latitude of this region has been defined as *Horse Latitudes.* Air circulation pattern remain anticyclonic over this zone. Air moves as divergent air flow from the subtropical high towards the equator and converges over the ITCZ. The air moves clockwise in northern hemisphere and anticlockwise in southern hemisphere. Wind flow is more pronounced on northern and southern sides than on eastern and western because of high pressure ridges in those latitudes. The subtropical high pressure cells remain to be the source of two (Trade Winds—equator ward and Westerlies—pole ward) of the three major global wind systems.

### *Trade Winds*

The Trade Winds are the major winds in the tropics. The trade winds are the prevailing winds in the tropics, blowing from the high-pressure area in the horse latitudes towards the low-pressure area around the equator. Trade winds blow within 25 degree north and 25 degree south. Trade winds blow from east to west and are known as *Easterlies*. In the northern hemisphere Trade winds blow from the northeast and are so called the *North Eastern Trades*. The direction of the trade winds in the southern hemisphere is just reverse blowing from the south east and so designated as the *South Eastern Trades*. Trade winds are considered to be the most reliable wind system blowing uniformly throughout the year maintaining nearly constant magnitude and direction. However the location of the source shifts related to the declination of sun. In the northern summer it is located further north and in the southern summer located further south. Trade winds are so reliable that the sailors used to navigate along the trade winds route and the winds were defined as trade Winds. Trade winds originate from the subtropical high region on the western side of the continent and usually blow over the dry continent. Around 30° N and S, the poleward flowing air begins to descend toward the surface in subtropical high-pressure belts. The sinking air is relatively dry because its moisture has already been released near the equator above the tropical rain forests. Near the centre of this high-pressure zone air descends and is called the horse latitudes. The moisture content appears to be low and precipitation is nearly absent.

The surface air that flows from these subtropical high-pressure belts toward the equator is deflected toward the west in both hemispheres by the Coriolis effect. Because winds are named for the direction from which the wind is blowing, these winds are called the northeast trade winds in the northern hemisphere and the southeast trade winds in the southern hemisphere. The trade winds meet at the doldrums. Doldrums happen to be a calm zone, where no surface wind is cognisable. As the trade winds converge over the equatorial low, vertical movement predominates than the horizontal movement. The area of convergence is known as Inter Tropical Convergence Zone (ITCZ).

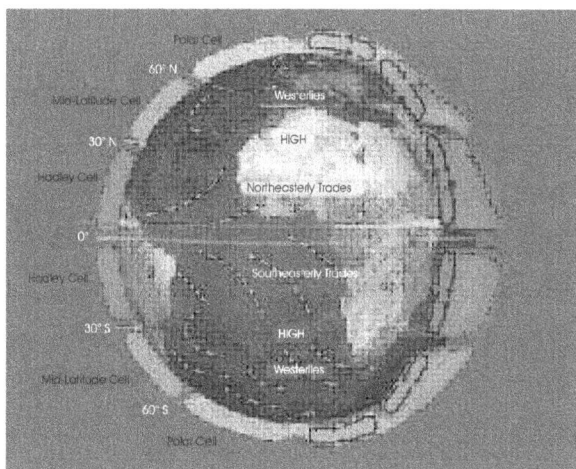

**Figure 3.12:** Schematic Diagram for General Circulation

Among the most well known trade winds is the *alizé* (sometimes *alize*), a steady, mild northeasterly wind which blows across Central Africa and the Caribbean. It brings cool temperatures between November and February. Trade winds are known to blow across Madagascar and other regions in the area. They are usually strongest in April to October but they do blow all year long.

*Intertropical Convergence Zone (ITCZ)*

The *Intertropical Convergence Zone (ITCZ)* is a belt of low pressure encircling globe at the equator. ITCZ is also referred as the *Inter-tropical Front, Monsoon trough, Doldrums* or the *Equatorial Convergence Zone.* It is caused due to the vertical ascent of warm and moist air from the latitudes above and below the equator. Northeast and Southeast Trade wind converge here, just north and south of the equator. The location of the intertropical convergence zone varies temporally. Over land, it moves back and forth across the equator following the sun's zenith point. Over the oceans, where the convergence zone is better defined, the seasonal cycle is more subtle, as the convection is regulated by the distribution of ocean temperatures.

Sometimes, a double ITCZ may form, with one located north and another south of the equator. Then a narrow ridge of high pressure may form between the two convergence zones, one of which is usually stronger than the other. Variation in the location of the intertropical convergence zone drastically affects rainfall in many equatorial countries, resulting in the wet and dry seasons of the tropics rather than the cold and warm seasons of higher latitudes. Longer term changes in the intertropical convergence zone can result in severe droughts or flooding in nearby areas.

In some cases, the ITCZ may become so narrow it may be interpreted as a front along the advancing edge of the equatorial air. Within the ITCZ the average winds are feeble

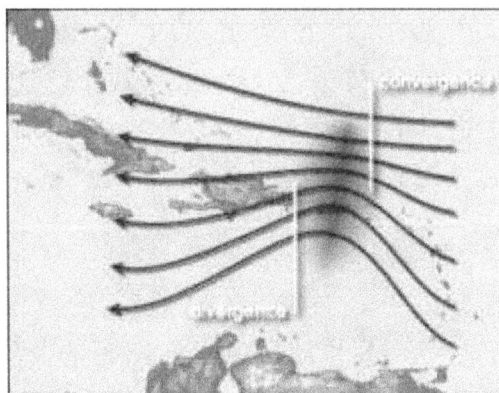

**Figure 3.13:** Location of ITCZ over the Tropical Seas

and erratic unlike the zones north and south of the equator where the trade winds blow. Early navigators in the tropical seas named this belt of calm seas the *doldrums* because of the stranded ships and stagnation due to no prevailing wind for many days. A hot and muggy climate could be experienced in the seas that might form a death trap for the sailors.

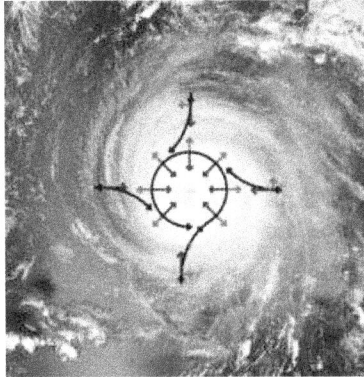

**Figure 3.14:** ITCZ—Convergence at the Centre of Low and Air Movement

ITCZ happens to be the source region of tropical cyclones. *Tropical cyclogenesis* requires low-level vorticity and the ITCZ/monsoon trough plays this role being a zone of wind change and speed, i.e. horizontal wind shear. When the ITCZ migrates more than 500 km from the equator during the respective summer hemisphere, the *Coriolis force* increases away from the equator. So it helps the formation of tropical cyclones within this zone north and south of the equator, but not exactly over the equator. In the north Atlantic and the northeastern Pacific oceans, tropical waves move along the axis of the ITCZ causing intensity in thunderstorm activity, and under weak vertical wind shear, these clusters of thunderstorms may turn to be the tropical cyclones. The convective cells thus formed help in the transfer of heat toward the poles.

*The Westerlies*

From the Subtropical High Pressure Zone winds blow towards the pole and meet the polar winds around the Sub-polar Lows. These winds domain the mid-latitude region lying between 30 degree and 60 degree N/S latitudes. The winds moving towards the pole bend sharply due to strong Coriolis force in the high latitudes and almost blow from west to east. So these surface winds are known as the westerlies and found over the both hemispheres. However, the surface westerlies are less consistent and persistent than the trade winds.

The winds are predominantly from the southwest in the northern hemisphere and from the northwest in the southern hemisphere. The westerlies are particularly strong, particularly in the southern hemisphere, where there is less land in the middle latitudes to cause friction and slow the winds down. The strongest westerly winds which blow in the middle latitudes of southern hemisphere between 40 and 50 degrees latitude are referred as the *Roaring Forties*. At latitude 50 and 60 degree south in the vast expanse of oceans the winds turn to have variable intensity and are referred as *Furious Fifties* and *Shrieking Sixties* according to the varying degrees of latitude.

But due to expanse of continents and orographic barriers causing friction in the northern hemisphere the westerlies are relatively weaker than in the southern hemisphere. But the westerlies in the northern hemisphere play significant role in modulating the weather conditions, where the warm moist maritime air from the subtropics meet polar air to form front. The zone of convergence, or polar front, is most strongly developed in winter, when the contrast in temperature and humidity of two air-masses appears to be strong. The formation of such fronts over the land leads to cyclogenesis in the mid-latitude region.

## *Jet Streams*

During the second world war the pilots of the bombing aircrafts while flying towards west from the east observed a strong wind blowing at an altitude of 9–12 km. Over Japan the wind speed was recorded to be about 100 to 300 knot. After the end of war the meteorologists conducted many studies on the dominating strong winds in the upper troposphere. It was observed that strong wind prevails above the subtropical highs blowing both equatorward and poleward. This strong wind along a narrow corridor in the upper air is known as *Jet Streams*. Over the Indian sky the aircrafts also experience this strong wind, particularly in the winter season, blowing from west to east, say from Delhi to Guwahati. In India the speed of this wind is generally upto 150 knot, but further east towards Japan the speed increases to about 300 knot. There are usually two jet streams in the northern hemisphere, both westerly. The jet stream above the suptropical high at about 13 km is the suptropical jet. The one near the polar front at about 10 km is the polar jet. The location of the jet stream deviates from the average. The polar jet meanders into broad loops that can sweep north and south and can split into two forks. Loops in the jet can steer cold polar air south or direct warm air poleward, thus just play a major role in the global transfer of heat.

Strong westerly winds prevail over most of the upper troposphere just below the tropopause. These are essentially thermal winds caused due to strong meridional gradient near-surface temperature in the mid-latitude region. The westerlies are strongest near the polar front, where the temperature gradient is the steepest. It is also observed that during the winter months, Arctic and tropical air masses form a stronger surface temperature contrast resulting in a strong jet stream. However, during the summer months, when the surface temperature variation is less dramatic, the winds of the jet are weaker. The region that separates warm tropical air from cold polar air is referred to as the polar front. Air pressure decreases more rapidly with altitude in the cold polar air mass than in the warmer tropical air. A pressure gradient force is thus generated between cold and warm masses of air, with relative lower pressure over the cold air. The difference between the air pressure in the warm and cold air increases with altitude above the surface. The horizontal pressure difference generates a pressure gradient force that accelerates the air from the region of higher pressure (the warm air) towards the lower pressure (the cold air). The Coriolis force eventually balances the horizontal pressure gradient force resulting in a strong stream of air that flows from the west towards the east in the upper troposphere in the vicinity of the polar front. This jet stream of air is referred to as the polar front jet. The polar jet forms at the polar front, from a strong pressure gradient resulting from the temperature and therefore pressure difference at the boundary between the polar and subtropical air masses. The jet is strongest in winter when the temperature difference is most pronounced and it extends further south in winter as the leading edge of the cold air extends into subtropical regions. In summer the polar jet is weaker and found over more northerly latitudes.

The jet stream is a current of fast moving air found in the upper levels of the atmosphere. This rapid current is typically thousands of kilometres long, a few hundred kilometres wide, and only a few kilometres thick. Jet streams are usually found somewhere between 10–15 km (6–9 miles) above the earth's surface. The existence of the polar front jet streams is tied to the presence of horizontal temperature gradients. If temperature gradients exist through a deep layer of the troposphere, a pressure gradient force increases with height throughout the layer, and therefore so does the wind. The position of this upper-level jet stream denotes the location of the strongest surface temperature contrast (as in the diagram below).

**Figure 3.15:** Jet Stream Blowing between Cold and Warm Air Mass

The principal jet streams are westerly winds (flowing west to east) in the northern hemisphere, although in the summer, easterly jets can form in tropical regions. The path of the jet typically appears to be meandering and these meanders themselves propagate east, at lower speeds than that of the actual wind within the flow. The theory of Rossby waves provides the accepted explanation for propagation of the meanders; Rossby waves propagate westward with respect to the flow in which they are embedded, but relative to the ground, they migrate eastward across the globe.

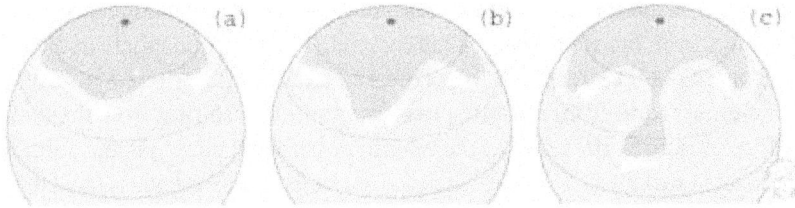

**Figure 3.16:** Meanders of Northern Hemisphere's Jet Stream Developing (a, b) and Finally Detaching a "Drop" of Cold Air (c)

In general, winds are strongest just under the tropopause (except during tornadoes, hurricanes or other anomalous situations). If two air masses of contrasting temperatures converge, the resulting pressure difference (which causes wind) will be the highest along the interface. The wind does not blow directly from the hot to the cold area, but is deflected by the *Coriolis effect* and blows along the boundary of the two air masses. So the jet streams flow along the boundary of two contrasting airmasses in the upper air.

The explanations may be derived from the consequences of the thermal wind relation. The balance of forces on an atmospheric parcel in the vertical direction is primarily between the pressure gradient and the force of gravity, which is known as *hydrostatic balance*. Horizontally the dominant balance outside of the tropics exists between the Coriolis effect and the pressure gradient. It is known as geostrophic balance. Given both hydrostatic and geostrophic balance, one can derive the thermal wind relation: the vertical derivative of the horizontal wind is proportional to the horizontal temperature gradient. In other sense temperatures decreasing poleward implies that winds develop a larger eastward component as one moving upwards. So the strong eastward moving jet streams are in part a simple consequence of the fact that the equator is warmer than the north and south poles.

The thermal wind relation, however, can not explain for why the winds are organised in tight jets, rather than distributed more broadly over the hemisphere. There are two reasons contributing the sharpness of the jets. One is the tendency for generating cyclonic perturbations in the midlatitude region to generate fronts. A front displays a sharp localised gradient in temperature. The polar front jet stream originates as a result of frontogenesis in midlatitude region, as the storms concentrate the north-south temperature contrast into a relatively narrow zone.

An alternative explanation has been suggested for the subtropical jet, which forms at the poleward limit of the tropical Hadley cell. This circulation appears to be symmetric with respect to longitude. Rings of air encircling the earth move poleward below the tropopause from the equator into the subtropics. When they move they tend to conserve their angular momentum. They are also moving closer to the axis of rotation, so they must spin faster in the direction of rotation, implying an increased eastward component of the winds.

The polar front and subtropical jets meet at some locations and times, while at other times they are distinctly separate. It was originally perceived that the polar front was a structure that had an existence independent of the cyclonic eddies. Now it is considered that the cyclonic eddies are growing from the potential energy in the broad north-south temperature gradient by a process known as the *baroclinic instability.* The resulting extratropical cyclones then turn the gradient into a front, thereby generating the polar front jet stream.

We now know that the path of the jet stream steers the cyclonic storm systems at lower levels in the atmosphere. So understanding the jet streams will help in weather analysis and weather forecasting.

## *Rossby Waves*

*Rossby* (or *planetary) waves* are large meanders in high-altitude winds that have a major influence on weather condition. They are caused due to the variation in Coriolis effect related to latitude. The waves were first identified in 1939 by Carl-Gustaf Arvid Rossby who went on to explain their motion. Rossby waves are a subset of inertial waves. Nature of the jet streams may be explained in terms of Rossby waves in the atmosphere.

Rossby waves in the atmosphere could be observed as large-scale meanders (usually 4–6) of Jet Stream. When these loops become very pronounced, they detach the masses of cold, or warm, air to be the cyclones and anticyclones and are responsible for day-to-day weather patterns in mid-latitude region. Rossby waves are usually large north-south undulation of the upper air westerlies. These help mixing the polar air with the tropical air, and cause severe weather changes in the mid-latitude region. Actually redistributions of cold air toward the equator and warm air toward the pole occur.

Oceanic Rossby waves are considered to result climatic changes due to variability in forcing, due to wind and buoyancy. Both barotropic and baroclinic waves play the role in variations of the sea surface height. It was difficult to detect the length of the waves earlier until the application of satellite altimetry. Baroclinic waves also generate significant displacements of the oceanic thermocline, often of tens of metres. Satellite observations have revealed the steady progression of Rossby waves across all the ocean basins, particularly at low and middle latitudes. These waves can take months or even years to cross a basin like the Pacific.

## *Polar Highs*

Due to intense cold condition over the polar region high pressure cells develop. High pressure cells are located over both poles. So these are thermally induced high pressure cells. Polar high is the result of less receipt of solar rays and is controlled by thermal condition. The polar surface is covered by ice almost throughout the year and known as permafrost region. Poleward wind descends over this region. Due to subsidence in the upper layer divergence is observed on the surface. So anticylonic circulation is generated and descending air diverges horizontally at the surface. The circulation moves clockwise in the northern hemisphere and anticlockwise in the southern hemisphere.

*Polar Easterlies*

The strong thermal deficit in the polar region causes the formation of polar high and drives the polar air masses to move equator ward as polar easterlies. The polar easterlies affect atmospheric characteristics and particularly air temperature on its route. The *Polar Easterlies* are the dry, cold prevailing winds that blow from the high-pressure areas of the polar highs at the north and south poles towards the low-pressure areas of the polar fronts between 60 and 90 degrees north and south. Cold air sinks at the pole creating the high pressure, forcing a southerly (northward in the southern hemisphere) outflow of air towards the equator; that outflow is then deflected eastward by the Coriolis effect. Unlike the westerlies in the middle latitudes, the polar easterlies are often weak and irregular. These prevailing winds blow from the east to the west. The polar easterlies are developed when the atmosphere over the poles cools. This cool air then subsides and spreads over the surface. As the air flows away from the poles, it is turned to the west by the Coriolis effect. As these winds moves from the east, they are called *Easterlies*.

The polar easterlies are usually weak in the summer, and very strong in the winter. In winter the polar wind can reach far into the mid-latitude region. When the easterlies encounter the westerlies they can become anticyclones or cyclones. In the northern polar region, where water and land are interspersed, the polar easterlies cause variable winds in summer. Due to the presence of major landmasses, notably in the Northern Hemisphere, there is significant variation in the polar easterlies. In addition, the wind systems and the associated climate are seasonally dependent. During the short summer season, the wind systems of the polar latitudes are greatly weakened. Close to the Antarctic continent the wind stress shows a reversal from eastward to westward, indicating the presence of polar easterlies along the coast.

The polar highs represent dense and coherent air masses with a mean diameter around 2,500 km, where the centre has the maximum atmospheric pressure. They circulate in the lower troposphere adopting zonal trajectories with a meridional component and their mean velocity is around 30 km/h.

*Sub-Polar Lows*

The Sub-Polar Lows represent the semi-permanent cyclone where surface mid-latitude south-westerlies converge with the polar north-easterlies. Hemispheric differences are observed in the sub-polar lows. The Sub-polar lows appear to be continuous throughout the year in the southern hemisphere over the cold seas around Antarctica, but discontinuous and poorly developed particularly in summer due to presence of landmass. The sub-polar lows are well formed in the southern hemisphere in the vicinity of Antarctic. In northern hemisphere two prominent Sub-polar lows are identified, the *Aleutian low* over the North Pacific Ocean and the *Icelandic low* over the North Atlantic. These are well developed in winter, but weaken or disappear in summer. In the southern hemisphere, the sub-polar low forms a continuous trough that completely encircles the globe. In winter the sub-polar lows deepen. This is due to the increase in equator to pole temperature gradient during winter. This gradient acts as the "thermal engine" for the development of sub-polar cyclones. In summer the continents warm up and the anticyclone dissipates. In many locations thermal lows (cyclones) appear (especially in arid and semi-arid areas). At the same time the equator to pole temperature gradient gets weaker and there is less energy to form sub-polar cyclones on the polar front. The Aleutian low vanishes in summer.

The origin of Sub-Polar Lows (SPLO) is yet to be understood exactly. They are generated at the surface as the seas are warmer than land (at the same latitude) in the winter. The temperature differences create pressure differences as well. Lows develop over the warmer water. The temperature differences appear to be less in the summer so the Sub-polar Lows disappear, thus they become semi-permanent features. The *Icelandic low* and *Aleutian low* are important winter-time weather makers. The north Atlantic is known for winter storms contributing to a miserable ocean voyage between Europe and the US. The *polar easterly winds* connect the Sub-polar Lows with the Polar Highs. The Sub-polar Lows are characterised with rising air and weather conditions like cloudiness, precipitation, unstable/ stormy weather.

### Local and Periodic Winds

*Slope and Valley Winds*

Valleys tend to develop their own air circulation, somewhat independent of the ambient wind overflow, and having a tendency to flow up or down the valley regardless of the prevailing wind direction. This circulation is modified by solar heating of the valley slopes.

*Anabatic* winds develop during the day hours when hillside slopes are heated more than the valley floor. The differential heating of contact air causes air to flow upslope. Wind speeds of 10 knots plus may be achieved. *Katabatic* winds normally form in the evening, the result of re-radiative cooling of upper slopes lowering the temperature of air in contact with it. Consequently the colder and denser air sinks rapidly down-slope. In specific circumstances katabatic winds can grow to strong breeze force during the night hours but cease with morning warming. Anabatic and katabatic winds are usually confined to a layer less than

500 feet deep. Katabatic winds are density or gravity currents and can also occur in the tropics. In some cases katabatic winds can persist for days; an extreme example being the large scale diurnal katabatic winds flowing from the dome of intensely cold, dense air over the Antarctic ice plateau (average elevation 6500 feet). These winds can achieve sustained speeds exceeding 80 knots but sustained speeds of 160 knots have been recorded at Commonwealth Bay, the windiest place on earth.

*Katabatic Winds*

A *Katabatic* wind originates in cold upland areas and cascade toward lower elevations under influence of gravity. *Katabatic* is a Greek word *katabatikos* meaning "going downhill". It is the technical name for a drainage wind, a wind that carries high density air from a higher elevation down a slope under the force of gravity. Such winds are sometimes also called *fall winds*.

It results from the cooling by radiation of air atop a plateau, a mountain, glacier, or even a hill. Since the density of air increases with lower temperature, the air will flow downwards, warming adiabatically as it descends. The temperature of the wind depends on the temperature in the source region and the amount of descent. In the case of the Santa Ana, for example, the wind can (but not always) become hot by the time it reaches sea level. In the case of Antarctica, by contrast, the wind is intensely cold. Katabatic winds are most commonly found blowing out from the large and elevated ice sheets of Antarctica and Greenland. The buildup of high density cold air over the ice sheets and the elevation of the ice sheets brings into play enormous gravitational energy, propelling the winds well over hurricane force. In Greenland these winds are called *Pitaraq* and are most intense whenever a low pressure area approaches the coast. In the Fuegian Archipelago (or Tierra del Fuego) in South America as well as in Alaska, a wind known as a *Williwaw* is a particular danger to harbouring vessels. It originates in the snow and ice fields of the coastal mountains. Williwaws commonly blow as high as 100 knots, and 200 knot Williwaws have been reported. *Mistral, Bora* and *Taku* are the other varieties of Katabatic winds and found respectively in Rhone Valley (France), Adriatic region and Southeastern Alaska.

*Squalls and Gusts*

*Squalls* or "squally winds" are a sudden onset of strong wind, but declining gradually, lasting several minutes and reaching speeds of 25–50 knots possibly gusting to 70–90 knots. Squalls may be associated with a thunderstorm (*rain squall, snow squall*), with a squall line (*line squall*), with a dry outflow from a thunderstorm in the interior (*dust squall*) or with an intense cyclone where the squall reinforces the strong wind. *Gusts* or "gusty winds" are the phenomena of increased wind speed exceeding the mean wind speed by at least 30 percent, but lasting only a few seconds and complemented by matching *lulls*. Squalls and gusts are associated with the development of thunderstorms. We will discuss the conditions for the development of thunderstorms later.

*Land and Sea Breeze*

Land and sea breeze blow over the littoral areas. The ancient Greeks had the knowledge of this wind system blowing over the Adriatic Coast. Aristotle in *Problemata* and Theophratus in *On the Winds* attempted to state the genesis and nature of the land and sea breezes. Differential heating of land and water surface causes the formation of land and sea breeze. This wind system has diurnal variation and is very much localised. During day hours the land surface becomes heated at a faster rate. So due to heating low pressure prevails over the land surface. The water surface of the sea remains comparatively cooler and high pressure dominates over the sea. So breeze blows from sea to land during day hours. Duration and intensity of the breeze are controlled by the declination of the sun. Land and sea breeze maintain hourly periodicity meticulously. Many studies have been made to show the diurnal variation of land and sea breeze in the coastal areas of varying latitude remaining nearly the same.

*Sea Breeze*

As the day dawns, coastal skies are cloudless or nearly cloudless, and the wind induced by large-scale weather patterns is light. With the sun rise increased solar energy heats the surface of the earth which, in turn, heats the lowest layers of the atmosphere. At sea, however, the radiant energy received is rapidly dispersed by a combination of turbulent mixing due to winds, waves, currents and the capacity of the water to absorb great quantities of heat with only slight alteration of its temperature. To the contrary air over land warms faster than that over the sea surface. Since warmer air is lighter the pressure over land becomes less than that over water, the average value of this difference being, about 1 millibar. So during day hours the sea breeze regime dominates over the coastal areas.

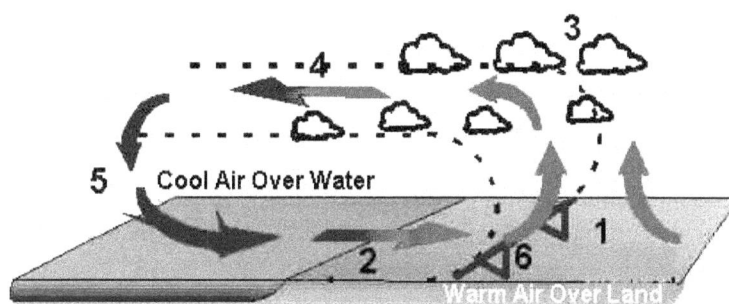

**Figure 3.17:** Sea Breeze Circulation

A few hours after sunrise, the pressure gradient becomes stronger allowing the sea breeze to move inland. As the sea breeze moves inland, the cooler sea air advances like a cold front characterised by a sudden wind shift, a drop in temperature and a rise in relative humidity. A temperature drop of 2° to 10° C within 15 to 30 minutes commonly occurs as the *sea breeze front* advances. The wind becomes stronger in the afternoon, sometimes blowing with a speed of 40 knot/hour. In the late afternoon the wind speed diminishes.

*Land Breeze*

After the sun set cooling starts both over the land and sea surface. Like daytime heating, cooling occurs at differential rates over water and land. The rapidly cooling land soon develops a higher air pressure over it relative to that over the sea, and the air begins to blow seaward. This is known as *land breeze*. The morphology of the coastline, strength of the large-scale winds, and coastal configuration influence the pattern of land breeze. Unlike the sea breeze, the land breeze is usually weaker in velocity and less common. The land breeze is often dominant for only a few hours and its direction is more variable.

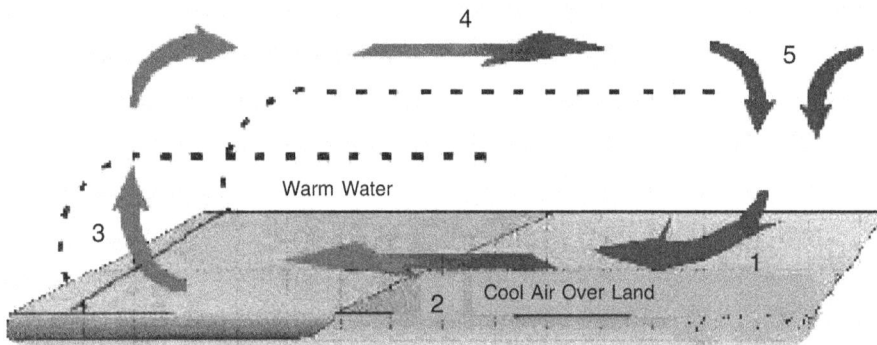

**Figure 3.18:** Land Breeze Circulation

*Monsoons*

Monsoon circulation is periodic in its nature, but not so much localised as in the case of land and sea breeze. Monsoon winds blow over the tropical region in summer, particularly during the apex of sun in the northern hemisphere in the months of June to September. Monsoon winds are westerly in direction, flow from the seas in the southwest to the lands in the northeast. Monsoon circulation replaces the northeastern trades in the summer months in northern hemisphere. In winter again the northeastern trades prevail over the monsoon region. In that sense the monsoon circulation appears to be periodic winds like the land and sea breeze, but on a seasonal scale.

The term monsoon has been derived from the Arabic word *'mausin'* (Sanskrit 'mausam'— meaning season). Monsoon circulation over the Arabian Sea was known to the sailors and traders long past. There was reference of wind reversal in summer and winter in the old Greek literature, *'Periplus of the Erythrian Sea'*. Monsoon winds guided the sailors and Arabian traders to reach the western coast of India. The Monsoon winds are of great significance to India. Indian sub-continent gets 80 percent of its rainfall from the Southwest Monsoon. So the economy and even the indices of the stock exchange largely depend on the nature of monsoon winds in a year. Moreover Indian life-style, culture and even literature are intimately linked and emotionally aligned with the monsoon winds. Monsoon circulation is observed not only in India, but also over the maritime regions of Southeast Asia, Africa and America.

Though the Monsoon circulation resembles the land and sea breeze, it possesses a few other characteristics. These characteristics distinguish the Monsoon circulation from other periodic winds. Renowned meteorologist C.S. Ramage identified four characteristics of the Monsoon circulation. These are:

1. There shall be at least 120 degree reversal between wind circulation in July and January.
2. This 120 degree reversal shall be at least for 40 days in a year for the months of July and January.
3. The average wind speed in the months of July and January shall be at least 3 metre per second.
4. Every two years in one month of July and January there shall be less than one cyclone or anticyclone.

The Monsoon circulation in India is considered to be best developed amongst the monsoon regions of the world. Triangular shape of the Indian landmass tapering into seas to the south appears to be one of the favourable conditions. In summer the Inter-Tropical Convergence Zone (ITCZ) moves north over the northern part of Indian Ocean. The Bay of Bengal and the Arabian Sea are the northern extension of Indian Ocean. Due to shifted location of low pressure cells (ITCZ) depressions tend to form over the warm seas.

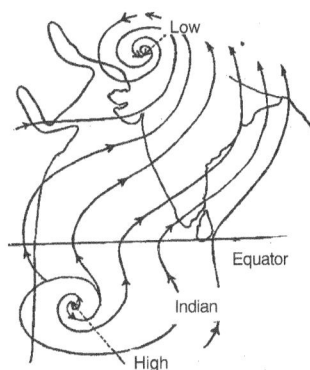

**Figure 3.19:** Advance of Indian Monsoon

Before the advent of Southwestern Monsoon warm and dry winds flow over Indian landmass from the west and northwest. Thereafter synoptic conditions change, which help the penetration of monsoon circulation. The following changes are noted:

1. Due to intense heating in the months of April and May thermal low pressure zone develops over the northwestern India and adjoining areas of Pakistan.
2. To the south over the seas a low pressure area emerges near the equator (5° North and 5° South).
3. In the southern hemisphere to the Madagascar Island a high pressure zone develops. This is known as Mascarenes High.
4. The winter jet stream flowing towards east and north east over the 22–24 degree north latitude becomes weaker. The jet stream moves north beyond the Himalayas and stays over the Tibetan Plateau near 30° north latitude.
5. Consequent to the shifting of westerly jet stream, another jet stream, known as easterly jet, appears over 13° north latitude (above Chennai). This Jet Stream continues to flow till the month of September and draws monsoon circulation towards Indian landmass.
6. Tibetan plateau stands to the north of the Himalayan ranges. In April and May thermal convections in high altitude occur to precipitate associated with thunderstorms. Latent heat is released to the atmosphere due to such occurrence. *Heat Island* is formed over the Tibetan Plateau, say at an altitude exceeding 5000 metre. This altitude is nearly the middle of the troposphere where rarified air and freezing temperature are observed. So Tibetan High develops surrounding the Tibetan Plateau.
7. It is also observed that the wind speed increases over the Arabian sea just before the advent of monsoons.

Monsoon circulation is activated within a very short time. In no case the time and magnitude of monsoon circulation are uniform over India. Monsoon winds first reach the Malabar coastal areas in Kerala in the first week of June from the Arabian Sea. Another circulation from the Bay of Bengal moves to the secondary low region of Northeastern India following the arrival of monsoon in Kerala. Afterwards the monsoon circulation spreads all over India. By July the monsoon circulation is well established over Northern India.

*Genesis of the Monsoon*

With the application of modern scientific techniques of monitoring such as satellite imaging, our knowledge of the dynamics of the monsoon has vastly improved. But the genesis and mechanism of monsoon circulation are yet to be conclusively established.

Nevertheless, it is generally accepted that the monsoon circulation is a result of three physical processes:

(i) The uneven heating of land and sea, which causes a differential pressure, drives the winds from high pressure to low pressure;

(ii)     The rotation of the earth, which forces moving wind to veer towards the right in the northern hemisphere and to the left in southern hemisphere;

(iii)    The transition in the state of water from liquid to vapour, determines the strength and location of the monsoon rains.

Originally, theories of the causes of monsoon in the nineteenth and eighteenth century were based on an understanding of the effects of the non-uniform heating of the land and sea. The earliest concept of Edmund Halley, showed in his 'sea breeze-land breeze' model of 1686 the tendency for the northeasterly trade winds on the northern side of the equator and the southeasterly trades to converge over the equatorial seas, south of the equator.

Differential heating creates pressure differences in the atmosphere, and thus the winds blow from high pressure to low pressure. As the land is heated to a temperature higher than that of surrounding oceans, the ascending warm air causes an in-draft of cooler sea air towards the land interior, bringing the rains associated with the monsoon. Consequently, in the summer, cooler, oceanic air would blow towards warmer land masses, while in winter as the continents cool to temperatures lower than the surrounding oceans, the wind flow would reverse, blowing from the land toward the ocean. This theory was refined by George Hadley, who observed that the wind flow between land and ocean was in an oblique direction (southwest in summer, northwest in winter). Hadley explained that moving air is deflected as a result of rotation of the earth on its axis, veering towards the right in the northern hemisphere and forming the southwesterly monsoon, and to the left in the southern hemisphere creating the monsoon northwesterly. While this theory of the monsoon was further refined during the 1950s and 1960s, the development of space observatory technologies and upper-air measuring devices has lent weight to theories that emphasize the seasonal shifting of thermally induced belts of pressure and winds due to a differential heating of air over the Tibetan High Plateau.

As this differential heating creates warm air, which then rises and spreads out southwards and gradually sinks over the equatorial regions of the Indian Ocean, it generates a reverse circulation of air in the lower troposphere, creating a low-level pole-ward flow known as the easterly jet stream. This concept is known as Heat Engine Concept and was proposed by P. Koteswaram. As this flow gains momentum, the winds pick up the moisture from the warm sea surface during their northward movement across the Indian Ocean, bringing the rains of the south-west monsoon. For this reason, the onset of the southwest monsoon is associated with low depressions, manifested in the form of cyclones with associated thunderstorms, in the Indian Ocean and Bay of Bengal, which move westerly into the Indian sub-continent. By September, as the sun retreats to the south of the equator, the monsoon gradually withdraws, in marked contrast to its abrupt arrival, leading to the reappearance of the subtropical westerly jet stream over the northwestern end of the Himalayas and the disappearance of the easterly jet stream.

Till the advent of satellites, the land–sea contrast theory was the dominant concept. The radiance data from satellites showed the role played by the net radiation at the top of the atmosphere on the energy budget of the monsoon. The satellite imagery highlighted the eastward migration of clouds along the equator and the northward migration from the equator to the Indian latitudes resulting in the inter-annual variation of the Indian monsoon rainfall. It is observed that variations in both the equatorial Pacific Ocean and the equatorial Indian Ocean influence the Indian summer monsoon rainfall.

In recent years the role of ITCZ in monsoon circulation has been highlighted. Monsoon is off-equator ITCZ and its associated circulation. A monsoon is in temporal understanding (say, monthly) a continental-size convective system in the tropics. It is also considered as predominating southwesterly flow at the low levels converging towards the continental-scale precipitation region. At the upper level the wind direction is reverse, northeasterly. Large vertical wind shear exists, not found elsewhere in the tropics. A monsoon contains a sizeable precipitation area and its associated circulation field. One experiment by the Chinese scientists clearly shows the characteristics of monsoon circulation i.e., southwesterly at low-level, converging towards the precipitation area and large vertical wind shear in monsoon continental-scale precipitation region. At the upper level the wind direction reverse, northeasterly. Large vertical wind shear exists, not found elsewhere in the tropics. A monsoon circulation contains a sizeable precipitation area and its associated circulation field. So it can be concluded that earth's rotation itself is sufficient to generate monsoon. It can be found that monsoon is interpreted as circulation associated with ITCZ substantially away from the equator. The existence of ITCZ and therefore monsoon does not rely on land-sea contrast. Land-sea contrast can only provide a favourable longitudinal location of ITCZ.

### Monsoon Rainfall in India

India is a land of *Monsoon.* The circulation of monsoon wind was known in ancient India and had been referred in many texts and literature. Kalidasa in his classical book, *'Megha Dutam'* described vividly the occurrence of monsoon. He estimated the movement of monsoon winds in Central India (Ujjaini) to be around the middle of June (on the first day of *ashada)*. His estimation appears to be nearly accurate even today. But the monsoon circulation is complex and varies. The monsoon rainfall is not also uniform all over the 35 climatic subdivisions of India. Moreover the circulation is not continuous through the season (June–September). There are certain breaks in the circulation. Also there is erratic distribution of monsoon rainfall in different parts of India in different years. Some areas receive incessant rainfall causing extensive floods and some areas experience unprecedented droughts in some years. As Indian economy is greatly dependent on monsoon we need to study the nature of Indian Monsoon more intricately. Even the BSE (Bombay Stock Exchange) index becomes *bull* or *bear* depending on the prospect of good or bad monsoon.

The onset of monsoon occurs over the Indian subcontinent in phases. It never enters India at the same time. The break of monsoon not only causes uncertainty, but also the uncertain behaviour of monsoon rainfall results in climatic hazards on many occasions. Monsoons, particularly South West Monsoon displays uncertain behaviour in occurrence in different

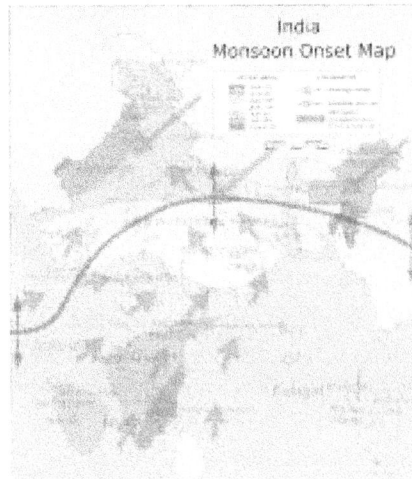

**Figure 3.20:** Onset of Indian Monsoon

years. The uncertainty in monsoon accounts for variable rainfall, which in turn affects the whole economy over the monsoon lands. This is sometimes referred as vagaries of monsoon. In some years scanty rainfall results in the occurrence of drought conditions. In some years rainfall above normal causes unprecedented floods. Droughts are manifested by rainfall less than two standard deviation ($< 2$ s.d.), whereas the floods are manifested by rainfall more than two standard deviation ($> 2$ s.d.). So the behavioural sequence of monsoon rainfall has gained importance in the study for the genesis of monsoon circulation. Many hypotheses have been proposed to explain the development of monsoon and the debate still continues. It will be useful to review various hypotheses to understand the behavioural pattern of Indian monsoon.

## Differential Heating of Land and Water

Differential heating of the land and water surface is considered the traditional hypothesis for the development of monsoon. Differential heating between land and waterbodies of the ocean results in the development of a situation where pressure and potential temperature of the surfaces do not coincide. Available potential energy is then obtainable for conversion of kinetic energy. The conversion of potential into kinetic energy is controlled by the pressure gradient force as developed due to variable temperatures over land and sea illustrated in a synoptic chart. During the period of northward shifting of the positions of the sun from vernal equinox to summer solstice, the deviation from the equilibrium condition starts to rise until a state is reached when available potential energy generated by differential heating is balanced by conversion to kinetic energy of monsoon winds. The condition again changes when after the autumnal equinox the sun moves its position into the southern hemisphere.

That is the beginning of the winter in the northern hemisphere. During this time of the year the oceans are warmer than the surrounding lands and there is decline in ocean-land temperature contrast in the northern hemisphere. This is caused as the thermal capacity of water is greater and the rate of radiative cooling over the land is faster. The land being cooler than the ocean, available potential energy is again caused. This is transformed into kinetic energy, but reverse to the conditions in summer. Anti-cyclonic cold surface air sweeps from Siberia and Central Asian landmass towards the equator. The southward motion of the cold surface air is balanced by northward moving warmer air in the upper troposphere. Being deflected due to *Coriolis* force this upper air actually moves from west to east and is known as *Westerly Jet Stream*. After the winter solstice the sun again begins to move north, after the vernal equinox the land in the northern tropics becomes warmer than the adjoining ocean, the summer or south west monsoon emerges again in the northern hemisphere. The cycle of differential heating thus causes the alternating wind circulations in summer and winter, which are known as monsoons over the tropics.

This concept of monsoon circulation is a simple convective cycle and supposed to occur regularly with certainty. But in practice the behaviour of the monsoon is complex and in many cases very difficult to predict its occurrence. So we can consider a few other hypotheses to understand the causes of monsoon.

*Heat Engine Concept*

P. Koteswaram (1958) attempted to explain the genesis of Indian/Asian monsoon with reference to the unique role played by the Tibetan plateau. The average height of the Tibetan plateau is 5000 m. This height appears to be in the middle troposphere. Being a landmass the Tibetan plateau becomes heated in the northern summer and turns to be *heat island* in the middle troposphere. This heating over the unusually high plateau results in the development of upper air easterlies near 15° N latitude and shifting of the westerly jet stream further north near 30° N latitude. He suggested that the southward gradient aloft becomes stronger due to radiational gain and latent heat of condensation. This results in a southward motion in the upper air from the Tibetan plateau, which subsequently subsides south of the equator over the Indian Ocean. The surface wind moves from this centre of subsidence firstly as southeasterly circulation, which after crossing the equator turns to be the south west monsoon current.

*Diabatic Heat Sources and Sinks for Monsoon Circulation*

A new concept for the monsoon circulation has been suggested in recent years in terms of diabatic heating rates. By computations it has been estimated that to maintain a steady flow of monsoon over India, i) warming at the rate of 3.2° C per day over northeast India and ii) diabatic cooling at the rate 2.4° C per day over northwest India, are required. The following table shows the diabatic heating rates over northeast India and northwest India. It is also indicated that the warming over northeast India is associated with the ascent of moist air, as it releases latent heat on a macroscopic scale. On the other hand there is descent of air over northwest India. The descent of air causes cooling over northwest India due to radiative divergence. *Radiometresonde* observations also indicated that the northwest India appears to be a radiative sink with a cooling rate of 1.6 C per day. This was further substantiated from the data available from MONEX 79.

**Table 3.6:** Diabatic Heating Rates (°C / 12 Hours) over India

| Mean Pressure (hPa) | Northeast India (Warming) | Northwest India (Cooling) |
|---|---|---|
| 900 | 3.7 | 3.7 |
| 800 | 2.4 | 2.7 |
| 700 | 1.4 | 1.4 |
| 600 | 1.2 | 1.2 |
| 500 | 1.2 | 0.5 |
| 400 | 0.8 | 0.8 |
| 300 | 0.6 | 0.3 |
| 200 | 0.4 | 0.2 |
| 100 | 2.4 | 0.1 |
| Mean | 1.6 | 1.2 |

**Table 3.7:** Diabatic Heating Pattern for Summer–1979

| Sl. No. | Phase | Warming zone | Temp. °C/day | Cooling zone | Temp. °C/day |
|---|---|---|---|---|---|
| 1. | Pre-onset (June 1-4) | Malaysia Indonesia Tibet | 0.8 0.8 3.2 | India East Arabian Sea | 0.5 0.5 |
| 2. | Active Monsoon (June 26-30 & August 5-10) | New Guinea (N of equator) Indo-China North-east India | 3.2 3.2 1.6 | Arabian Sea | 1.6 |
| 3. | Weak (break) Monsoon (July 12-15 August 22-27) | North-east India Indonesia | 1.6 1.6 | South Indian Ocean | 1.0 |

Diabatic warming and cooling rates were also computed for the summer and winter monsoons in 1979. On planetary scale an east-west zonal circulation was observed extending from the eastern Pacific to Papua New Guinea during the northern winter. Net warming over Papua New Guinea exceeded 1.5° C per day, where the net cooling over the eastern Pacific was 1° C per day. With the advent of summer a shift towards north-west was observed in the zone of maximum heating. The maximum heating rate of 1.5° C per day was observed over Philippines, whereas radiative cooling zone was located over the arid and semi-arid regions of the Middle East. It is clearly evident that the winter and summer monsoons are associated with diabatic heat sources and sinks, which are aligned in an east-south-east to a west-north-west direction. It is further observed that the source-sink pattern usually shifts to the north-west from winter to summer and is related to sun's declination. We will also note that the *El Nino* phenomena developing over the eastern Pacific in some years during the northern winter affect the monsoon circulation as the heat sink conditions change with sudden warming of the ocean surface. The diabatic heating patterns for summer, 1979 have been shown in Table 3.7.

The diabatic heating patterns in Table 3.7 indicate the south-east to the north-west orientation in diabatic heating and variations in intensity during different phases of the monsoon. The overturning occurs on a vertical plane during the summer monsoon. The ascending limb of the overturning lies along the south-eastern part of Asia, whereas the descending motion is observed over north-west India and the Middle East.

In 1979, the Monsoon Experiment (MONEX 79) was conducted by the World Meteorological Organisation in collaboration with the International Council of Scientific Unions to study the behavioural pattern of the monsoon. Many countries including the USA, USSR, India, China etc. participated in the programme. The programme did not yield desirable result as 1979 was the bad year of monsoon. However, a lot of available information threw new light on the genesis of monsoon. MONEX data were used to derive diabatic warming (Krishnamurti and Ramanathan, 1982). Diabatic heating are classified into three components:

a)  Condensation,
b)  Radiative flux divergence, and
c)  Sensible heating resulting from small scale turbulent transfer of heat from the surface to the atmosphere.

Higher estimate of warming was done by Krishnamurti & Ramanathan (Table 3.8). However their estimate could explain the south-east to north-west pattern of diabatic heat sources and sinks, with variation during different phases of the summer monsoon.

From the study of MONEX 79 we can summarise the main features of diabatic heating:

a)  Large scale heating rates indicate a south-east to north-west orientation with warming over Papua New Guinea and in adjoining region and cooling over north-west India, Pakistan and Middle East;
b)  A steady north-westward shift in the warming and cooling pattern as the summer monsoon spurs from its pre-onset to onset and mature stage;

c) Radiative sinks developing over north-west India and the Middle East are characterised by subsiding motion.

d) The plateau of Tibet stands as a heat source in the middle troposphere for most of the summer monsoon resulting in flow of upper Easterlies.

**Table 3.8:** Diabatic Heating Pattern (after Krishnamurti and Ramanathan, 1982)

| Sl. No. | Phase | Warming zone | Temp. °C/day | Cooling zone | Temp. °C/day |
|---|---|---|---|---|---|
| 1. | Pre-onset (June 1-5, 1979) | South Indian Ocean<br>Tibet<br>New Guinea | 4.0<br>3.0<br>2.0 | North-west India | 1 |
| 2. | Onset (June 12-16, 1979) | Burma<br>South Arabian Sea<br>South Indian Ocean<br>New Guinea | 4.0<br>4.0<br><br>3.0<br><br>3.0 | North-west India | 0.5 |
| 3. | Post-onset (June 18-22, 1979) | North Bay of Bengal & North -east India<br>New Guinea | 5.0<br><br><br>4.0 | North-west India | 0.5 |

*Other Concepts*

In recent years, the roles of Wave Energetics, Upper Tropospheric Vorticity Budget and Inter-tropical Convergence Zone are considered to explain the monsoon circulation. We have already mentioned the relationship of the jet streams—both westerly and easterly, with the development of the monsoon circulation. However, there are variations in the circulation of these jet streams in different years. It has been observed that if there be delay in the shift of the westerly jet stream the easterly jet stream appears to be weaker, which affects the normal rainfall. The year to year deficient rainfall over Central India, Africa and Australia is related to: a) warm Sea Surface Temperatures (SST) over the equatorial Pacific ocean, b) a weaker easterly jet associated with weaker than normal anti-cyclone over Tibet, and c) a general south-eastward movement of principal circulation patterns.

In recent years the mean velocity potential and stream function in the upper troposphere have been computed. The seasonal mean vorticity appears to be the minimum (over Tibet) when the divergence is at maximum. These are related to fluctuations of the monsoon system. With a large scale divergence having strong fluctuations with time a correlation between divergence and vorticity fluctuations may initiate a damping mechanism.

ITCZ is the zone of convergence of trade winds from both hemispheres. The ITCZ is usually located between 5° to 15° N and S of the equator, whereas the equator is normally free from clouds over the oceans. Two specific cloud bands, one to the north of 15° N and another slightly north of the equator have been identified in the summer monsoon region (70° E to 90° E). However, year to year variation is observed in the location of the ITCZ. The lower troposphere in this zone is usually conditionally unstable. So the ascending air possesses sufficient energy to rise further up with the release of latent heat of condensation. Many of the depressions and tropical cyclones develop close to ITCZ. Monsoon depressions, particularly developing over the Bay of Bengal in the monsoon months, move landwards and follow the isobaric trough to precipitate. During the Bay of Bengal Monsoon Experiment in July-August 1999, several active and weak spells of convection were observed over the Bay of Bengal representing intra-seasonal oscillations in the North Bay of Bengal. The Convective Available Potential Energy of the surface air decreased following convection, and recovered in one to two days. Even though the local thermodynamic instability was comparable during different phases of the intra-seasonal oscillation, the monsoon systems formed only during the active phase.

In recent years many studies indicate the influence of ENSO (El Nino-Southern Oscillation) on the monsoon circulation. The fluctuating barometric conditions over the Pacific Ocean, referred by Sir Gilbert Walker as *Southern Oscillation,* play significant role in modulating the monsoon circulation over India and Asia-Pacific monsoon region. El Nino and La Nina are components of the El Nino-Southern Oscillation (ENSO) system, which has been found to influence the inter-annual variability of the monsoon in a major way. It has been observed that the monsoon rainfall is more strongly correlated with the convection over the Central Pacific Ocean (120°-170° W, 5° N-5° S) than over the Eastern Pacific Ocean. Normally, El Nino condition has been found to result in deficit *Indian Summer Monsoon Rainfall* (ISMR) and La Nina in an above-normal *Indian Summer Monsoon Rainfall* (ISMR). The El Nino of 1982 and 1987 were associated with droughts and the La Nina of 1988 was associated with excess rainfall. *(See also Chapter VIII on Climatic Anomalies and Extreme Climate).*

---

## Further Readings

Barry, R.J. and R.G. Chorley (1968) *Atmosphere, Weather and Climate,* Methuen & Co., London.

Critchfield, H.J. (1975) *General Climatology,* Prentice Hall India Ltd., New Delhi.

Corby, G.A. (ed.) (1970) *The Global Circulation of the Atmosphere,* Royal Meteorological Society, London.

Crowe, P.R. (1971) *Concepts in Climatology,* Longmans, London.

Das, P.K. (1970) *The Monsoons,* National Book Trust, New Delhi.

Flohn, H. (1969) *General Climatology,* Elseveir, Amsterdam.

Hare, P.K. (1965) 'Energy Exchanges and the General Circulation', *Geography,* **50**.

I.M.D. (1960) *Monsoons of the World,* India Meteorological Department, New Delhi.

McDonald, J.E. (1952) 'The Coriolis Effect', *Scientific American,* **186**, 5.

Namias, J. (1952) 'The Jet Stream', *Scientific American, Vol. 187*.

Palmen, E. (1951) 'The Role of Atmospheric Disturbances in the General Circulation', *Quarterly Journal of Royal Meteorological Society,* **77**, London.

Riehl, H. (1962) 'General Atmospheric Circulation', *Science,* **135**.

Riehl, H. (1969) 'On the Role of the Tropics in the General Circulation of the Tropics', *Weather,* **24**.

Riehl, H. (1969) *Jet Streams of the Atmosphere,* Colorado State University, Colorado.

Rossby, C.G. (1949) 'On the Nature of the General Circulation of the Atmosphere', in Kuiper, G.P. (ed.), *The Atmosphere of the Earth and Planets,* University of Chicago Press, Chicago.

Saha, P.K. and P.K. Bhattacharya (1994) *Adhunik Jalavayu Vidya,* West Bengal State Book Board, Calcutta.

Sawyer, J.S. (1957) 'Jet Stream Features of the Atmosphere', *Weather,* **12**.

Starr, V.P. (1956) 'The General Circulation of the Atmosphere', *Scientific American,* **195**.

Trewartha, G.T. (1968) *An Introduction to Climate,* McGraw Hill Kogakusha Ltd., Tokyo.

Tucker, G.B. (1962) 'The General Circulation of the Atmosphere', *Weather,* **17**.

Walker, J.M. (1972) 'Monsoon and the Global Circulation', *Meteorological Magazine,* **101**.

# Chapter 4

# Moisture in the Atmosphere: Forms and Characteristics

## Defining the Atmospheric Moisture

*Humidity* is a measure of the water vapour content of the air. The amount of water vapour in the air depends on the controls over evaporation. There are several ways to express the humidity of the air. Each humidity measure is controlled to some degree by air temperature.

*Absolute humidity* is the weight of water vapour per unit volume of air, usually measured in units of grams of water vapour per cubic metre of air. Absolute humidity is not often used to express the moisture content of air because it is subject to changes in both the temperature of the air and atmospheric pressure. For instance, let's say that a 1 cubic metre parcel of air at the surface has 2 grams of water in it. Now lift the parcel of air upwards into the atmosphere. As the air rises upward the decrease in atmospheric pressure on the parcel allows it to expand outward occupying more space. Let's say that the parcel doubles in size as a result of uplift. Before rising, the absolute humidity was 2 gm/m$^3$. As the air doubles in volume the new absolute humidity is 1 gm/ m$^3$. In actuality the parcel still has the same weight of water in it, i.e. 2 grams. But given the way absolute humidity is calculated it appears the amount of water in the air has decreased.

*Specific humidity* is measured as the weight of water vapour in the air per unit weight of air, which includes the weight of water vapour. The unit of measurement is grams of water vapour per kilogram of air. Given that weight is not significantly influenced by temperature or atmospheric pressure, specific humidity is much more useful as a measure of humidity. Another measure very similar to specific humidity is the mixing ratio. *Mixing ratio* is the weight of water vapour per unit weight of dry air. Because the atmosphere is made up of so little moisture by volume, the mixing ratio is virtually the same as the specific humidity.

Humidity is not only measured as a weight, but also by the pressure it exerts. *Vapour pressure* is the partial pressure exerted by water vapour. Vapour pressure, like atmospheric pressure, is measured in millibars and is relatively insensitive to volumetric expansion or temperature. *Saturation vapour pressure* is simply the pressure that water vapour exerts when the air is fully saturated.

*Dew point temperature* is the temperature at which condensation takes place. The dew point temperature depends on the amount of moisture in the air. The more moisture in the air, the higher will be the dew point temperature. It is defined as "dew point" because dew will form on surfaces when the air reaches saturation.

*Relative humidity* is defined as the ratio of the amount of water vapour in the air to its saturation point. Often relative humidity is considered to be the amount of water vapour as the air can hold at a given temperature. Regardless, the saturation level of the air with respect to water vapour also depends on air temperature. We know that as air temperature increases, the ability for the air to keep water in its vapour state is easier. So, as the air temperature rises it can hold more water in the vapour state.

**Moisture in Air**

When we think of air as being saturated with moisture we often say that the air is 'holding all the moisture it can'. This implies that once the air has reached saturation it can not 'accept' anymore water by evaporation. This is wrong. So long as there is water available evaporation will continue even when the air is fully saturated. Let's examine the concept of saturation in more detail.

Say a beaker is filled halfway with water. Let's put a lid on it to constrain the movement of water molecules and eliminate the influence of wind on evaporation. As the water absorbs heat it begins to change phase and enter the air as water vapour. Above the surface, water vapour molecules dart about suspended in the air. However, near the surface water molecules are attaching themselves back the surface, thus changing back into liquid water (condensation) (A). As evaporation occurs the water level in the beaker decreases (B). This occurs because evaporation exceeds condensation of water back onto the surface. After some time, the amount of water entering the air from evaporation is equal to that condensing (C). When this occurs the air is said to be saturated.

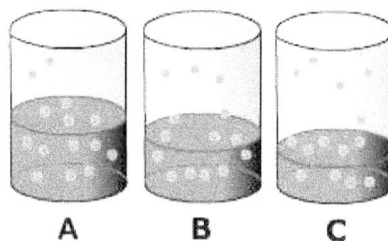

**Figure 4.1:** Evaporation and Condensation in an Enclosed Beaker of Water

The saturation level of the air is directly related to air temperature. As air temperature increases, more water can exist in a gas phase. As temperature decreases, water molecules slow down and there is a greater chance for them to condense on to surfaces. The graph below shows the relationship between air temperature and vapour pressure, a measure of the humidity, at saturation.

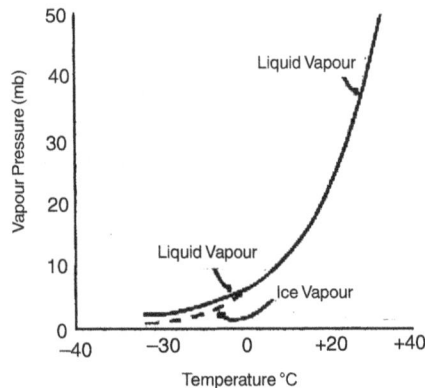

**Figure 4.2:** Relationship between Air Temperature and Vapour Pressure

Below the freezing point (0° C) the curve splits, one for the saturation point above a liquid surface (liquid-vapour) and one for a surface of ice (ice-vapour). Water can remain a liquid below the freezing point. Water that is not frozen below 0° C is called "super-cooled water". For water to freeze, the molecules must become properly aligned to attach to one another. This is less likely to occur especially with small amounts of water, like cloud droplets. Thus in clouds with temperatures below the freezing point both super-cooled liquid water and ice crystals exist together.

It is noted that the saturation vapour pressure at –20° C is lower for ice than for a liquid surface. To convert water from a liquid to a gas requires about 600 calories per gram and conversion of water from a solid to a gas demands about 680 calories. So it is more difficult to "liberate" a molecule of water from ice than water. Therefore, when the air is saturated, there are more molecules (as super cooled) above a water surface (i.e. more vapour pressure) than an ice surface (i.e. less vapour pressure).

**Processes for the Change of State**

In understanding the state of water in the atmosphere the role of different processes shall be outlined. The principal processes are: 1. Melting, 2. Evaporation, and 3. Condensation. All the processes are related to varying temperature conditions in the atmosphere. Ice (solid) melts into water (liquid) at 0° C. Conversely water (liquid) turns to be ice (solid) at 0° C and is known as freezing. Water is changed into vapour state at 100° C and the process is called the evaporation. Also, the atmospheric vapour condenses in the atmosphere under specific temperature and pressure conditions. The process is known as condensation. Of these processes the role of evaporation and condensation is of immense importance for the state of water in the atmosphere. We will discuss the processes of evaporation and condensation.

## *Evaporation*

Free water molecules in air are vapourised and transformed into water vapour. This process is known as *Evaporation.* On earth's surface evaporation mostly occurs over the water bodies of the ocean, and also from other water-bodies like lake, river, ponds etc. Moreover the plant surface adds water vapour to the atmosphere through a process known as *Evapo-transpiration.*

Evaporation is intimately linked to vapour pressure. When the vapour pressure on water surface exceeds the vapour pressure of surrounding air evaporation occurs. Specific heat of water exceeding the specific heat of land surface helps build up moisture in air. As evaporation proceeds, the surrounding air becomes gradually saturated and the process will slow down and might stop if the wet air is not transferred to the atmosphere. When the vapour pressure difference is off-set the evaporation process is stopped. However, the condition changes, with more heating evaporation process starts again. We know that the water vapour holding capacity of air increases with the addition of heat. So it happens. Saturated water vapour condition also changes with the differential heating of land and water. The process of evaporation will proceed until there are as many molecules returning to the liquid as there have been escaping. At this point the vapour is said to be saturated, and the pressure of that vapour (usually expressed in mmHg) is called the saturated vapour pressure.

Evaporation from the oceans is the primary mechanism supporting the surface-to-atmosphere portion of the water cycle. After all, the large surface area of the oceans (over 70 percent of the earth's surface is covered by the oceans) provides the opportunity for large-scale evaporation to occur. On a global scale, the amount of water evaporating is about the same as the amount of water delivered to the earth as precipitation. Evaporation is more prevalent over the oceans than precipitation, while over the land, precipitation routinely exceeds evaporation. Most of the water that evaporates from the oceans falls back into the oceans as precipitation. Only about 10 percent of the water evaporated from the oceans is transported over land and falls as precipitation. Once evaporated, a water molecule spends about 10 days in the air. The process of evaporation is so great that without precipitation runoff, and ground-water discharge from aquifers, oceans would become nearly empty.

## *Factors Influencing the Rate of Evaporation*

The rate of evaporation on earth's surface depends on three important factors:
1. Temperature of the evaporating surface, 2. Relative humidity of air and 3. Wind speed. These factors are further defined as:
   • If the air already has a high concentration of water molecules, then the evaporation will be slower. Air with lower relative humidity helps evaporation than the air with higher relative humidity.
   • Flow rate of air determines the rate of evaporation. This is in part related to the concentration

of water molecules. If fresh air is moving over the substance all the time, then the concentration of the substance in the air is less likely to go up with time, thus encouraging faster evaporation. This is the result of the boundary layer at the evaporation surface decreasing with flow velocity, decreasing the diffusion distance in the stagnant layer.

- If the temperature of the water surface is hotter, then evaporation will be faster.
- Stronger the *Inter-molecular* forces keeping the molecules together in the liquid state the more energy that must be input in order to evaporate them.
- A substance which has a larger surface area will evaporate faster as there are more surface molecules which are able to escape. So vast ocean bodies are favourable places for evaporation.
- The thickness of the water bodies also decides the rate of evaporation. If the thickness of the body being heated is thick at a time of heating, the heat being delivered for evaporation of the water would be reduced. If the thickness is very low the heat might have been delivered more to the evaporation of the water. Thus the shallow waterbodies will be evaporated faster than the deeper waterbodies.

*Distribution of Water Vapour on Earth's Surface*

The amount of water vapour depends largely on temperature conditions, so the potential evaporation declines normally from lower to higher latitudes, with one exception in the wet tropics (10°–30° N or S). The amount of variation of potential evaporation on global scale may be evidenced from the following table.

**Table 4.1:** Distribution of Water Vapour on Earth's Surface

| Latitude | North (Evaporation in cm.) | South (Evaporation in cm.) |
|----------|----------------------------|----------------------------|
| 70–60    | 18                         | –1                         |
| 60–50    | 40                         | 30                         |
| 50–40    | 70                         | 60                         |
| 40–30    | 110                        | 100                        |
| 30–20    | 130                        | 130                        |
| 20–10    | 140                        | 130                        |
| 10–0     | 120                        | 120                        |

Source: Landsberg

The actual evaporation depends not only upon the temperature of the air and oceans, direction of the wind and degree of dryness of air, but also upon the extent of the water surface of the oceans. From the distribution of actual evaporation (Table 4.1) the following observations are made:

I)  Actual evaporation is more over the oceans than over the continents, as the extent of water surface is large over the oceans. But the amount of actual evaporation, however, is recorded much over the land surface in latitudes 10° N to 10° S. It is caused due to large supply of water vapour to the atmosphere by the process of evapotranspiration in the tropical rainforest region in this latitudinal extent.

II) Actual evaporation is greatly controlled by the temperature of air. Hence the lower latitudes having high temperature record more actual evaporation than the higher latitudes with lower temperature.

**Table 4.2:** Actual Evaporation on the Earth's Surface

**Northern Hemisphere**

*(in inches)*

|           | 60°–50° | 50°–40° | 40°–30° | 30°–20° | 20°–10° | 10°–0° |
|-----------|---------|---------|---------|---------|---------|--------|
| Continent | 14.2    | 13      | 15      | 19.7    | 31.1    | 45.3   |
| Ocean     | 15.7    | 27.6    | 37.8    | 45.3    | 47.2    | 39.4   |
| Mean      | 15      | 20.1    | 28      | 35.8    | 42.9    | 40.6   |

**Southern Hemisphere**

|           | 60°–50° | 50°–40° | 40°–30° | 30°–20° | 20°–10° | 10°–0° |
|-----------|---------|---------|---------|---------|---------|--------|
| Continent | 7.9     | 19.7    | 20.1    | 16.1    | 35.4    | 48     |
| Ocean     | 9.1     | 22.8    | 35      | 44.1    | 47.2    | 44.9   |
| Mean      | 8.8     | 22.8    | 35.5    | 39      | 44.5    | 45.7   |

*Source: Wust*

### Evapotranspiration

*Evapotranspiration* (ET) is defined as the sum of evaporation and plant transpiration from the earth's land surface to air. Evaporation accounts for the movement of water to air from sources such as the soil, canopy interception, and water bodies. Transpiration accounts for the movement of water within a plant and the subsequent loss of water as vapour through stomata in its leaves. Evapotranspiration is an important part of the water cycle.

Evapotranspiration is a significant water loss from a plant. Types of vegetation and land use significantly affect evapotranspiration. Because water transpired through leaves comes from the roots, plants with deep reaching roots can more constantly transpire water. Thus herbaceous plants transpire less than woody plants because herbaceous plants usually lack a deep taproot. Coniferous forests tend to have much higher rates of evapotranspiration than deciduous forests. This is caused as their needles give them superior surface area, resulting in more pores for transpiration, and allowing for more droplets of rain to be suspended in and around the needles and branches, where some of the droplets can then be evaporated.

The rate of evapotranspiration at any location on earth's surface is controlled by several factors:

- More the energy available, the greater the rate of evapotranspiration. It takes about 600 calories of heat energy to change 1 gram of liquid water into a gas.
- The rate and quantity of water vapour entering into air both become higher in drier air.
- The wind speed immediately above the surface—the process of evapotranspiration moves water vapour from ground or water surfaces to an adjacent shallow layer that is only a few centimetres thick. When this layer becomes saturated evapotranspiration stops. However, wind can remove this layer replacing it with drier air which increases the potential for evapotranspiration. Winds also affect evapotranspiration by bringing heat energy into an area. Winds having higher speed cause more evapotranspiration.
- Evapotranspiration cannot occur if water is not available.
- Physical attributes of the vegetation such as vegetative cover, plant height, leaf area index and leaf shape and the reflectivity of plant surfaces can affect rates of evapotranspiration.
- Plants regulate transpiration through adjustment of small openings in the leaves called stomata. As stomata close, the resistance of the leaf to loss of water vapour increases, decreasing to the diffusion of water vapour from plant to the atmosphere.
- Soil characteristics can affect evapotranspiration depending on its heat capacity, and soil chemistry and *albedo*.

Minimum evapotranspiration rates generally occur during the coldest months of the year. Maximum rates generally coincide with the summer season. However since evapotranspiration depends on both solar energy and the availability of soil moisture and plant maturity the seasonal maximum evapotranspiration actually may precede or follow the seasonal maximum solar radiation and air temperature by several weeks.

## Condensation

*Condensation* is the change of the physical state of aggregation of matter from gaseous phase into liquid phase. Condensation is the phase change of water vapour into a liquid. Condensation commonly occurs when a vapour is cooled to its dew-point, but the dew-point can also be reached through compression. During the condensation process, water molecules lose the 600 cal/gm of latent heat. This latent heat was stored during the process of evaporation changing water into water vapour. When latent heat is released (called the "*latent heat of fusion*"), it is converted into sensible heat which warms the surrounding air. Due to warming the air increases its buoyancy and energises the development of storms. Condensation takes place in the presence of condensation nuclei and when the air is nearly saturated.

Water vapour is moving around so fast in the air that the molecules tend to bounce off one another without bonding. Even if a few pure water molecules were to collide and bind together, the surface tension created by such a tiny sphere is so great that it is extremely difficult for additional water molecules to become incorporated into the mass. Hence *condensation nuclei* act as a platform for condensation to take place, increasing the size of a droplet and decreasing surface tension. Condensation nuclei are essential for condensation. Water absorbent dust particles and sea salt are good condensation nuclei. Sulfates and nitrates are water absorbent and are responsible for precipitating acid rain.

The air must be at or near its saturation point for condensation to take place. Air can become saturated in two ways, 1) adding water vapour to the air by evaporation thus reaching to saturation point given its present temperature, or 2) cooling the air to its dew point temperature. Cooling the air is the most common cause for condensation process in the atmosphere for the formation of clouds. Air can also be cooled through contact with a cold surface or by uplift. *Contact cooling* occurs when air comes in contact with a cooler surface and conduction takes away heat out of the air. Cooling by contact is called *diabatic cooling*.

## The Role of Condensation Nuclei

In the atmosphere it is, however, very difficult in initiating condensation by the casual grouping together of a number of water molecules of water vapour. It is estimated that an aggregation of 100 molecules of water vapour would form a droplet of radius $10^{-7}$ cm. and the equilibrium vapour pressure at the surface of such a droplet would be 300 percent more of the saturation vapour pressure as normally defined. Condensation in the free atmosphere in the absence of a central core would demand higher degree of super-saturation than under normal conditions.

Hence condensation occurs with much difficulty in a purified clean air. Normally condensation results on a foreign surface like land and plant surface, as is the case for dew and frost; but in the free atmosphere condensation is initiated around a complex particles of dust, smoke, sulphur dioxide, salts (NaCl) or similar microscopic substances which possess the property of *wettability*. These are called the *Hygroscopic Nuclei*. These are sometimes referred as condensation nuclei or hygroscopic aerosols, which are soluble. This is important as the saturation vapour pressure over the solution of hygroscopic nuclei particles is less than over pure water of the same size and temperature. So condensation is initiated when the vapour pressure in the atmosphere exceeds the equilibrium vapour pressure over the saturated solution of the condensation nuclei. The amount of condensation depends on the size and nature of the molecules. The nucleus, if sufficiently hygroscopic, may start to absorb moisture from the atmosphere when the air is well below saturation, even at relative humidity of 74 percent. In the initial stages small drops will grow faster in size than the larger ones. But its growth rate by condensation will decrease with the increase in droplet size. The radial growth rate is reduced with increasing droplet size, because of the increasingly greater surface area to add to with every increment of radius. However, the condensation rate is determined by the speed at which the liberated latent heat is lost from the drop by conduction to the air. Such liberation of latent heat is responsible for the reduction of the vapour gradient. This means that the liberation of latent heat will lessen the degree of super-saturation, as competition for the available moisture exists between the droplets.

However, the growth rate of droplet size by simple process for condensation can not account for precipitation in free air. We know that in most clouds precipitation starts within an hour, but the droplet growth by condensation to reach the rain drop size (>1 mm diameter) would take nearly 24 hours. Moreover the falling drops would be reduced in size due to evaporation in the unsaturated air below the cloud base. So it appears that the cloud droplets by simple condensation process are not necessarily the immediate source of raindrops. There exist some other complex mechanisms for rain making.

### Adiabatic Temperature Change and Stability

In the atmosphere air temperature usually decreases with an increase in elevation through the troposphere. The decrease in temperature with elevation is called the *normal lapse rate of temperature,* also known as *Environmental Lapse Rate (ELR)*. The *normal lapse rate of temperature* is the average lapse rate of temperature of 6.5° C/1000 metres. The normal lapse rate of temperature is the actual vertical change in temperature on any given day and can be greater or less than 6.5° C/1000 metres. We know that the decrease in temperature with height is caused by increasing distance from the source of energy that heats the air, i.e. the earth's surface. Air is warmer near the surface because it is closer to its source of heat. The further away from the surface, the cooler the air will be. Temperature change caused by an exchange of heat between two bodies is called *diabatic temperature change.* Another important way to change the temperature of air is called the *adiabatic temperature change*.

*Adiabatic temperature change* of air occurs without the addition or removal of energy. So there is no exchange of heat with the surrounding environment to cause the cooling or heating of the air. The temperature change is due to work done on a parcel of air by the external environment, or work done by a parcel of air on the air that surrounds it. It is caused due to expansion or compression of air.

Assume a parcel of air is moving vertically through the troposphere. We know that air pressure decreases with increasing height. When a parcel of air rises upwards it expands in volume due to drop of pressure with increasing height. This is called the *Adiabatic Expansion* (without addition of heat). For such expansion no heating is required. In order to expand the parcel must use its innate energy to do so. As the air expands, the molecules spread out and ultimately collide less with one another. Due to expenses of energy for expansion the temperature of the air drops instantly. This is known as *Adiabatic Cooling*.

Normally cooling rate will be higher for the unsaturated air. This cooling rate for the unsaturated air is called the *Dry Adiabatic Lapse Rate (DALR)*. Dry Adiabatic Lapse Rate is 10° C per 1000 metres. It means that there will be a drop of 10° C temperature with a rise of 1000 metres height. Condensation occurs when the air becomes saturated. Due to condensation latent heat is added to the air and the rate of cooling drops. Cooling rate is always lower in the saturated air than in the unsaturated air. This is known as *Saturated Adiabatic Lapse Rate (SALR)*. This rate is not uniform and it varies. Normally Saturated Adiabatic Lapse Rate is 6° C per 1000 metres. Even this rate may drop to 4° C in high temperature condition or rise to 9° C per 1000 metres under very cold condition (–40° C).

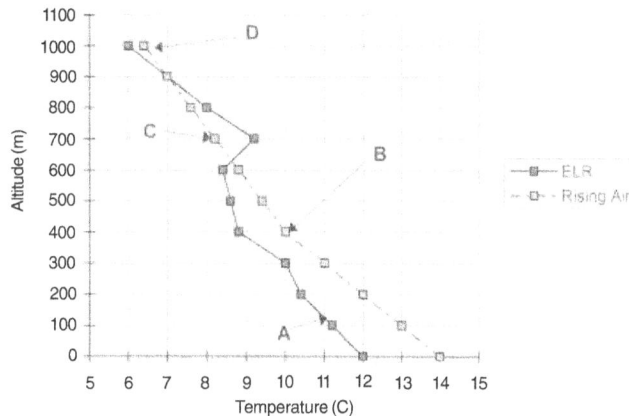

**Figure 4.3:** Rising Air and Adiabatic Cooling

As a parcel of air sinks through the troposphere it is subject to increasing atmospheric pressure. This causes the volume of the air parcel to shrink in size, squeezing the air molecules closer together. In this case, work is done on the parcel. As the volume shrinks, air molecules bounce off one another more often turning with greater speed. The increase in molecular movement causes an increase in the temperature of the parcel. This process is known as *Adiabatic Heating*.

The rate at which air cools or warms depends on the moisture status of the air. The DALR remains constant, i.e. 10° C/1000 metres. The SALR varies with the moisture status in the air, however it is assumed to be a constant value (6° C/1000 metres). The reason for the difference in the two rates is due to the liberation of latent heat as a result of condensation. As saturated air rises and cools, condensation takes place. As water vapour condenses, latent heat is released. This heat is transferred into the other molecules of air inside the parcel causing a reduction in the rate of cooling.

*Stable and Unstable Air*

Adiabatic temperature change is an important factor in determining the stability of the air. We consider air stability as the tendency for air to rise or fall through the atmosphere under its own 'power'. Stable air tends to resist upward movement in the atmosphere. On the other hand, unstable air favours to rise.

Air is unstable when the normal lapse rate is greater than the dry adiabatic lapse rate. Under these conditions, a rising parcel of air is warmer and less dense than the air surrounding it at any given elevation. Figure 4.3 depicts unstable conditions. See the graph for the rising parcel of air. Note that at any elevation above the surface the temperature of air parcel is higher than the air that surrounds it. Even as it reaches the dew point temperature at 2000 metres, the air remains warmer than the surrounding air. As a result it continues to rise and cool at the saturated adiabatic rate. Vertically developed clouds are likely to develop under unstable conditions such as this.

The conditions of stability/instability/conditional instability in the atmosphere depend on the relationship of ELR with DALR/SALR.

*Stability*

Stability occurs when the ELR is *less* than the DALR. Under this condition the air parcel cools faster than the surrounding air. So if there be a rise of 1000 m the rising air might have cooled more than the surrounding air. As it is colder, it is denser and will sink down (if there be no orographic barrier forcing it to rise). This air becomes stable as the dew point might not have been reached and the cumulus clouds usually form which will not produce rain. Stability is observed with anti-cyclonic weather conditions, that restricts convection. As a result dry and sunny weather persists.

*Instability*

When the ELR exceeds the DALR (and also greater than the SALR) the parcel cools down more slowly with height from the ground than the surrounding air. So the air parcel will continue to rise further upward as it becomes warmer (and lighter) than its surrounding air. As the dew point is reached the latent heat is released due to condensation of vapour. The released latent is added to the air parcel warming it. The air parcel then rises even more rapidly as the SALR is even lower than the DALR (SALR always remains lower than the DALR). The rise may continue even up to the limit of tropopause, when only the vertical rise could be restricted by temperature-inversion. Usually under very high unstable condition cumulo-nimbus clouds develop, and the anvil shape at their top occurs due to reaching the limit of tropopause.

*Conditional Instability*

This occurs when the ELR is lower than the DALR but exceeds the SALR. The rising air remains stable in the lower layers of the atmosphere, but if it is forced to rise then it may rise and cool at its dew point. After attaining the SALR it will tend to rise freely being unstable. The weather found with the conditional instability is usually fine and sunny in low lying areas (i.e. those below the condensation level) but rainy and cloudy in higher areas.

### *Diabatic Cooling*

Diabatic cooling processes are associated with the release of heat. These include the processes: 1. *Radiational cooling,* 2. *Cooling by conduction* and 3. *Cooling by mixing.* Due to radiation cooling occurs in the radiating body. Fog and clouds develop due to such cooling of the moist air. For such occurrence *condensation nuclei* play significant role. During night hours, particularly in winter nights, land surface becomes cooler due to terrestrial radiation. So the moisture laden air enveloping the cool land surface becomes cooler to initiate the process of condensation to form dew, fog, frost etc. In the long winter night the process is mostly favoured as the rate of cooling is higher. When two contrasting air masses, warm moist and cool dry, meet condensation process starts due to saturation of warm moist air. In this way fog and clouds form. But it must be specifically noted that diabatic cooling helps only the formation of dew, fog, frost etc. in the atmosphere close to the cool land surface. However by this process major cooling of the atmosphere is not possible.

### Measurement of Water in Air

The amount of water in the air is measured in different ways. The *specific humidity* of air is a measure of the amount of water in the air. Warmer air can hold more water than colder air. When the air reaches its capacity, it is considered *saturated.* This capacity doubles for about every $11°$ C rise in temperature. The *relative humidity* is the measure of the amount of water in the air divided by its holding capacity. The relative humidity reading is expressed as percentage. The relative humidity for saturated air is 100 percent.

### *Measuring the Relative Humidity*

The relative humidity of air can be measured by two different methods. One can measure the relative humidity by the use of a *hygrometer*. This is a pointer attached to a piece of hair. As the humidity increases, the hair stretches out. When the humidity drops, the hair shrinks, causing the needle to point in a different direction. The other method is the use of Wet and Dry Bulb Thermometers along with the use of a chart. The instrument with the wet and dry-bulb thermometers is called a *psychrometer*. The thermometers and chart are all in the Celsius scale. Dry bulb thermometer measures the air temperature. This is the *dry-bulb* reading. The other thermometer has a wet wick on the bottom of the bulb. Water evaporating from the wick into the air takes energy with it, cooling off the thermometer. As the relative humidity increases, less water can evaporate into the air. This makes the temperature readings between the two thermometers closer. In case of more dryness the thermometer readings will be more apart. The difference between the wet and dry-bulb temperatures is called the *wet-bulb depression*. The wet-bulb temperature is always lower or the same as the dry-bulb temperature. If the temperatures are the same, the relative humidity is 100 percent. The wet-bulb depression is used with the dry-bulb temperature and a chart to determine the relative humidity. Table noted below shows this chart.

**Table 4.3:** Chart to Calculate Relative Humidity %

*Relative Humidity (%)*

| Dry-Bulb Temperature (°C) | Difference between Wet-Bulb and Dry-Bulb Temperatures (°C) | | | | | | | | | | | | | | | |
|---|---|---|---|---|---|---|---|---|---|---|---|---|---|---|---|---|
| | 0 | 1 | 2 | 3 | 4 | 5 | 6 | 7 | 8 | 9 | 10 | 11 | 12 | 13 | 14 | 15 |
| −20 | 100 | 28 | | | | | | | | | | | | | | |
| −18 | 100 | 40 | | | | | | | | | | | | | | |
| −16 | 100 | 48 | | | | | | | | | | | | | | |
| −14 | 100 | 55 | 11 | | | | | | | | | | | | | |
| −12 | 100 | 61 | 23 | | | | | | | | | | | | | |
| −10 | 100 | 66 | 33 | | | | | | | | | | | | | |
| −8 | 100 | 71 | 41 | 13 | | | | | | | | | | | | |
| −6 | 100 | 73 | 48 | 20 | | | | | | | | | | | | |
| −4 | 100 | 77 | 54 | 32 | 11 | | | | | | | | | | | |
| −2 | 100 | 79 | 58 | 37 | 20 | 1 | | | | | | | | | | |
| 0 | 100 | 81 | 63 | 45 | 28 | 11 | | | | | | | | | | |
| 2 | 100 | 83 | 67 | 51 | 36 | 20 | 6 | | | | | | | | | |
| 4 | 100 | 85 | 70 | 56 | 42 | 27 | 14 | | | | | | | | | |
| 6 | 100 | 86 | 72 | 59 | 46 | 35 | 22 | 10 | | | | | | | | |
| 8 | 100 | 87 | 74 | 62 | 51 | 39 | 28 | 17 | 6 | | | | | | | |
| 10 | 100 | 88 | 76 | 65 | 54 | 43 | 33 | 24 | 13 | 4 | | | | | | |
| 12 | 100 | 88 | 78 | 67 | 57 | 48 | 38 | 28 | 19 | 10 | 2 | | | | | |
| 14 | 100 | 89 | 79 | 69 | 60 | 50 | 41 | 33 | 25 | 16 | 8 | 1 | | | | |
| 16 | 100 | 90 | 80 | 71 | 62 | 54 | 45 | 37 | 29 | 21 | 14 | 7 | 1 | | | |
| 18 | 100 | 91 | 81 | 72 | 64 | 56 | 48 | 40 | 33 | 26 | 19 | 12 | 6 | | | |
| 20 | 100 | 91 | 82 | 74 | 66 | 58 | 51 | 44 | 36 | 30 | 23 | 17 | 11 | 5 | | |
| 22 | 100 | 92 | 83 | 75 | 68 | 60 | 53 | 46 | 40 | 33 | 27 | 21 | 15 | 10 | 4 | |
| 24 | 100 | 92 | 84 | 76 | 69 | 62 | 55 | 49 | 42 | 36 | 30 | 25 | 20 | 14 | 9 | 4 |
| 26 | 100 | 92 | 85 | 77 | 70 | 64 | 57 | 51 | 45 | 39 | 34 | 28 | 23 | 18 | 13 | 9 |
| 28 | 100 | 93 | 86 | 78 | 71 | 65 | 59 | 53 | 47 | 42 | 36 | 31 | 26 | 21 | 17 | 12 |
| 30 | 100 | 93 | 86 | 79 | 72 | 66 | 61 | 55 | 49 | 44 | 39 | 34 | 29 | 25 | 20 | 16 |

**Table 4.4:** Chart to Calculate the Dew Point

*Dew Point Temperatures (0°C)*

| Dry-Bulb Temperature (°C) | Difference between Wet-Bulb and Dry-Bulb Temperatures (°C) | | | | | | | | | | | | | | | |
|---|---|---|---|---|---|---|---|---|---|---|---|---|---|---|---|---|
| | 0 | 1 | 2 | 3 | 4 | 5 | 6 | 7 | 8 | 9 | 10 | 11 | 12 | 13 | 14 | 15 |
| –20 | –20 | –33 | | | | | | | | | | | | | | |
| –18 | –18 | –18 | | | | | | | | | | | | | | |
| –16 | –16 | –24 | | | | | | | | | | | | | | |
| –14 | –14 | –21 | –36 | | | | | | | | | | | | | |
| –12 | –12 | –18 | –28 | | | | | | | | | | | | | |
| –10 | –10 | –14 | –22 | | | | | | | | | | | | | |
| –8 | –8 | –12 | –18 | –29 | | | | | | | | | | | | |
| –6 | –6 | –10 | –14 | –22 | | | | | | | | | | | | |
| –4 | –4 | –7 | –12 | –17 | –29 | | | | | | | | | | | |
| –2 | –2 | –5 | –8 | –13 | –20 | | | | | | | | | | | |
| 0 | 0 | –3 | –6 | –9 | –15 | –24 | | | | | | | | | | |
| 2 | 2 | –1 | –3 | –6 | –11 | –17 | | | | | | | | | | |
| 4 | 4 | 1 | –1 | –4 | –7 | –11 | –19 | | | | | | | | | |
| 6 | 6 | 4 | 1 | –1 | –4 | –7 | –13 | –21 | | | | | | | | |
| 8 | 8 | 6 | 3 | 1 | –2 | –5 | –9 | –14 | | | | | | | | |
| 10 | 10 | 8 | 6 | 4 | 1 | –2 | –5 | –9 | –14 | –28 | | | | | | |
| 12 | 12 | 10 | 8 | 6 | 4 | 1 | –2 | –5 | –9 | –16 | | | | | | |
| 14 | 14 | 12 | 11 | 9 | 6 | 4 | 1 | –2 | –5 | –10 | –17 | | | | | |
| 16 | 16 | 14 | 13 | 11 | 9 | 7 | 4 | 1 | –1 | –6 | –10 | –17 | | | | |
| 18 | 18 | 16 | 15 | 13 | 11 | 9 | 7 | 4 | 2 | –2 | –5 | –10 | –19 | | | |
| 20 | 20 | 19 | 17 | 15 | 14 | 12 | 10 | 7 | 4 | 2 | –2 | –5 | –10 | –19 | | |
| 22 | 22 | 21 | 19 | 17 | 16 | 14 | 12 | 10 | 8 | 5 | 3 | –1 | –5 | –10 | –19 | |
| 24 | 24 | 23 | 21 | 20 | 18 | 16 | 14 | 12 | 10 | 8 | 6 | 2 | –1 | –5 | –10 | –18 |
| 26 | 26 | 25 | 23 | 22 | 20 | 18 | 17 | 15s | 13 | 11 | 9 | 6 | 3 | 0 | –4 | –9 |
| 28 | 28 | 27 | 25 | 24 | 22 | 21 | 19 | 17 | 16 | 14 | 11 | 9 | 7 | 4 | 1 | –3 |
| 30 | 30 | 29 | 27 | 26 | 24 | 23 | 21 | 19 | 18 | 16 | 14 | 12 | 10 | 8 | 5 | 1 |

**Dew Point**

Water vapour in the free air condenses into liquid water at a temperature which is called the *dew point*. If the dew point stands below 0° C, it is then called the *frost point*. Dew point temperature may vary for a parcel of air depending on amount of cooling in the free atmosphere. The dew-point temperature is measured in a similar way to that of relative humidity. The dry-bulb and wet-bulb temperatures are determined. The wet-bulb depression and dry-bulb temperature are referred with the chart as per Table 4.4 to find the dew point.

If the water vapour comes in direct contact with the cooler surface, it can condense onto it. Fog may result when warm air moves into an area that has a cold surface temperature. It results due to advection of air and is called the *advected fog*. The situation is reversed where cooler air moves over a warmer surface. *Ground fog* is formed by radiational cooling at night. These phenomena are commonly found in humid valleys and near rivers and lakes.

*Forms of Moisture*

*Dews*

During night the land surface re-radiates the heat gained during the day hours. Due to radiation back to the atmosphere temperature of the land surface markedly drops. The process of condensation starts at a certain temperature (*Dew Point*) depending on the nature of air close to the land surface. Water vapour at dew point condenses on ground surface, grasses, leaves of the trees, mosses as tiny water droplets. These are called *dews*.

Dews are formed under certain conditions. Dews are formed usually in the mornings of winter months. Clear sky favours high rate of cooling due to high incidence of re-radiation from the land surface. So dews are observed in a cool morning having clear sky in the night. Dew formation is usually caused in an enclosed area with prevailing gentle breeze. If the wind speed is high it may favour evaporation of the atmospheric moisture. In all cases the amount of moisture in air shall be high. Dews rarely form over the deserts.

*Fogs*

Fogs closely resemble the clouds. They have nearly the same form and structure. Water droplets constitute both the fog and cloud. But the clouds are found high above the ground surface, where fogs remain close to the ground surface. Fogs are formed mostly due to either contact cooling or mixing of warm and cold air. Fogs, floating near the ground surface, affect the visibility condition over land and water surface affecting all forms of transport, navigation and aviation.

Fogs may occur for different reasons. Fogs are classified into: 1. Ground fog due to radiational cooling, 2. Contact cooling and 3. Frontal cooling by mixing of warm and cold air masses.

Radiational fog occurs due to cooling of the land surface in the night hours through radiation of the earth's surface. Moist air enveloping the land becomes cooler due to contact cooling. The condensation process is initiated and tiny water droplets form. As this type of fog lies near the ground surface it is also called the *ground fog*.

A few particular factors favour the formation of ground fog. As the ground surface becomes cooler during long winter night inversion of temperature is observed close to the ground surface that restricts the convection. Fog formation is usually favoured in the night with cloud-free sky. Cloud-free sky helps the terrestrial radiation back to the space and so cooling occurs at a faster rate. Light winds having speed not exceeding 10–16 km. per hour help in the formation of fog. Fog is formed better in the mountain terrain, particularly in the valleys, than over the plains. Air pollution over the cities and industrial locations causes fog. The pollutants form a canopy which restricts the convection and helps radiational cooling. Sometimes this type of fog is called a smog (smoke + fog), commonly initiated in the evening and persists through the night till early morning.

In the upper air radiational cooling may also sometimes occur initiating the formation of fog. When cold air subsides and comes into contact of the warm layer below such condition may develop. However, such occurrence is not very common.

Fog may develop due to advection. When cold air mass moves over the relatively warm land and water surface fog develops locally over the land and water surface. Similarly as the cold ocean current from the polar region meets the warm ocean current from the tropics, fog develops along the front of two contrasting ocean currents. Atmospheric condition also turns to be unstable due to such mixing. Along the Labrador coast such fog is commonly observed throughout the year, particularly in the winter months. In the mid-latitude region in winter polar cold air mass meets the traveling warm maritime air mass. Due to such meeting around a centre of low front develops between two contrasting air masses. Fogs are developed along the front along with other characteristics of instability condition in the atmosphere. This type of fog is known as *Frontal fog*.

*Frosts*

Frost is also formed due to contact cooling of the lower atmosphere close to the land surface. When temperature drops below the *freezing point* (0° C) atmospheric moisture drops as *frost* on the ground, shaded slope and leaves of the tree and shrubs. In this case due to freezing temperature water vapour turns to be frost instead of dew by the process of sublimation. Frost is usually formed in the early morning when minimum temperature is recorded. In high latitude and highland regions frost is common in winter morning. In highland region where temperature tends to be lower than 0° C in the morning frost may occur throughout the year. In mountainous region when cold gravity wind rolls down the slope during night hours fog forms in the valley. Frost occurs when the water droplets in the fog are cooled below the freezing point. Frost melts steadily with dissipation of fog after the sunrise. Water droplets in the fog may be deposited on the edge of a sharp body and form frosts. These are known as *Rime*. Frosts on many occasions damage the orchards in the valleys and mountain slope.

### *Formation of Clouds*

We know that the cloud is the source of rainfall on earth's surface. The formation of clouds is related to the ascent of moisture bearing air parcel. As the air moves upward the pressure becomes lower with ascending heights. So there is loss of weight of the atmosphere upon it and the parcel of air expands in volume. The rising air is thus subject to expansion by which some energy is expended through work. Hence the temperature of the air parcel drops. Consequently with the fall of temperature at dew point the process of condensation is initiated. As a result the latent heat of condensation is released in the atmosphere, which again acts as trigger in uplifting the air parcel to a higher level. This process of expansion is called the *adiabatic expansion*, and the cooling as *adiabatic cooling*. The adiabatic expansion and cooling in the atmosphere are considered important weather making processes. The instability or stability of the atmosphere is closely related with the adiabatic expansion and cooling. The rate of cooling in the atmosphere largely depends on the amount of water vapour present in air and the speed of ascent. The nearly saturated air with a steep ascent will markedly be transformed into cumulonimbus cloud and results in the development of very unstable weather condition. All weather disturbances are usually related to developments of such atmospheric conditions.

A cloud is formed when air is cooled to its dew-point temperature. The air cools as it rises above from the earth's surface. Water droplets form in the cloud after reaching the freezing level, i.e. at a temperature 0° C or below. If the cloud forms below 0° C, the cloud is made from ice and snow crystals and super-cooled water. Super-cooled water remains as all the water droplets are sometimes not converted to ice by cooling below 0° C. When the air is warmer than the air surrounding it and is less dense becomes *buoyant*. That is why clouds appear to float in the sky. The clouds may continue to develop vertically. Eventually, a cumulonimbus cloud is formed. These are clouds associated with thunderstorm, heavy rain, hail, strong winds, and tornadoes and known as *thunder clouds*. These clouds are developed in an unstable air mass with air moving up due to density differences.

A cloud is also formed in a stable air mass, but it rises for other reasons. These are layered clouds that form from air that is forced upward by the land (mountains) or by radiational cooling as the air mixes with a cooler layer of air. Some clouds are formed having a flat base and waving out on top. The bottom of the cloud is the place where the air temperature is the same as the dew-point temperature. This point is referred as the *condensation level*. The height of the cloud base can be calculated with a simple formula or a chart. To use the formula, take the difference between the temperature and dew point at the surface and divide it by 0.8° C (the amount that the dew-point temperature gets closer to the air temperature in 100 m). The result is multiplied by 100, which indicates the *lifting condensation level* or the height that a cloud can reach. Eventually the cloud and air temperatures become equal. The cloud is not buoyant at this point and begins to spread out. This creates the classic anvil-shaped tops that are seen at cloud tops.

The clouds in general indicate the physical processes prevailing in the atmosphere. As such the study of clouds has become relevant synoptic analysis of weather at any place. The understanding and recognition of cloud types help in the prediction of weather. From any *Cloud Atlas,* different types of clouds can be identified. On the basis of their distinctive characteristics the clouds are grouped into 4 principal *families* and 10 *classes*.

Cloud formations are categorised into three types. *Cirrus* clouds are very high clouds and are formed of ice crystals. They are thin, feathery clouds observed on a fair weather day. *Stratus* clouds are the layered, sheet-like clouds. They are found at lower altitudes. *Cumulus* clouds are the puffy, cotton-like clouds formed by upward rising of air. Other clouds are made from combinations and variations of these clouds. The name of a cloud may also contain a prefix or suffix that indicates about the cloud. *Alto* means high and *nimbus* indicates rain.

### Classification of Clouds

Clouds are usually classified either on the basis of elevation or by the form or character.
- Classified on the basis of elevation:
    - High Clouds     = above 7,000 m (23,000 ft)
    - Middle Clouds  = 2,000 m–7,000 m (6,500 ft–23,000 ft )
    - Low Clouds     = 0–2,000 m (0–6,500 ft)
- Classified by the form, or character:
    - Stratiform > layered
    - Cumuloform > vertically developed
    - Nimbo.... or .....nimbus > precipitating

*Family A: High Clouds (Mean Base Level 6000 m.)*

1. Cirrus (Ci)—This type of clouds appears like feather and is formed of ice crystals. These are thin and of fibrous structure. When they are detached and arranged irregularly, indicate fair weather. But when arranged in bands systematically and connected with cirrostratus or altostratus, may foretell bad weather.
2. Cirrostratus (Cs)—This type of clouds looks like a thin whitish sheet covering the whole sky. The sky thus assumes a milky appearance. The clouds produce a halo around the sun or moon, which is the harbinger of a storm.
3. Cirrocumulus (Cc)—These appear as white flakes or small globular masses and normally do not cast shadow. These are usually arranged in groups, lines or ripples characterising the sky as 'Mackarel' sky.

*Family B: Middle Clouds (Mean Upper Level 6000 m. and Mean Base Level 2000 m.)*

4.  Altostratus (As)—These clouds appear as a uniform sheet cloud of grey and bluish colour and normally have fibrous structure. They in many cases resemble the cirrostratus and gradually merge with them. Widespread and continuous precipitation occurs as a consequence with the formation of these clouds.
5.  Altocumulus (Ac)—These clouds have flattened globular form and arranged in lines or waves. Unlike the cirrocumulus they cast shadow and have larger globules.

*Family C: Low Clouds (Mean Upper Level 2000 m. and Base Level Close to the Ground)*

6.  Stratocumulus (Sc)—These are large globular masses or sprawls of soft grey clouds having brighter interstices. These are usually arranged in a regular pattern.
7.  Stratus (St)—This type of clouds with uniform layers almost resembles the fog, but being detached from the ground.
8.  Nimbostratus (Ns)—These low clouds are dense and shapeless. In most cases precipitation occurs from the ragged layers of low clouds.

*Family D: Clouds with Vertical Development (No Specific Height, but may be as High as > 6000 m. and as low as 480 m.)*

9.  Cumulus (Cu)—These are the thick clouds with vertical development. These have dome structure with cauliflower appearance but the base being horizontal. Normally these clouds indicate fair weather.
10. Cumulonimbus (Cb)—These are sometimes referred as thunder clouds. They have large masses of vertical development, where the summits may look like mountains, towers or anvils. They characterise the atmospheric instability resulting in thunderstorms, storm rainfall, squalls, and even hails.

*A Few Types of Common Cloud*

(a) Cirrus

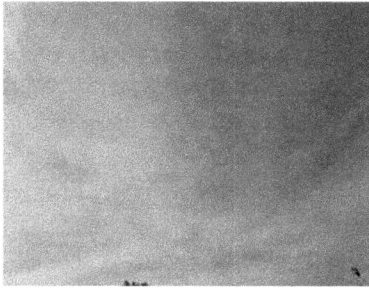
(b) Cirrostratus

(c) Altocumulus

(d) Altostratus

(e) Stratus

(f) Nimbostratus

(g) Cumulus

**Figure 4.4:** Different Forms of Clouds

## Formation of Raindrops and Rain-making Processes

The clouds floating overhead contain water vapour and cloud droplets, which are small drops of condensed water. These droplets are usually too small to fall as precipitation, but they are large enough to form visible clouds. Water is continually evaporating and condensing in the sky. If one looks closely at a cloud he can see some parts disappearing (evaporating) while other parts are growing (condensation). Most of the condensed water in clouds does not fall as precipitation because their fall speed is not large enough to overcome updrafts which support the clouds. For precipitation to happen, first tiny water droplets must condense on even tinier dust, salt, or smoke particles, which act as a nucleus. Water droplets may grow as a result of additional condensation of water vapour when the particles collide. If enough collisions occur to produce a droplet with a fall velocity which exceeds the cloud updraft speed, then it will fall out of the cloud as precipitation. This is not a simple task since millions of cloud droplets are required to produce a single raindrop.

### Mechanisms of Precipitation

It has been observed that simple condensation process and simple growth of cloud droplets are apparently inadequate mechanism to initiate precipitation. Different hypotheses have been postulated to explain the mechanism of precipitation. Of different hypotheses the traditional ones are collision and coalescence and sweeping theories. A new concept of rain making has been proposed by Bergeren and Fiendisen.

### Collision, Coalescence and Sweeping Theories

It was earlier thought that atmospheric turbulence by making cloud particles collide would result in significant proportion to coalesce. If there be no coalescence the cloud is considered to be colloidally stable. The coalescence of droplets is so the result of colloidal instability in the cloud. The colloidal stability/instability of the cloud depends on five factors, namely–

1. The electric charge of the droplets,
2. The size of the droplets,
3. The temperature of the droplets,
4. The motion of the droplets, and
5. The presence or non-presence of ice crystals in the cloud.

### Electric Charge of the Droplets

Droplets of the cloud carry electric charges. On the poles of the droplet positive and negative charges exist. When the opposite poles (+ or –) of two adjoining droplets come closer, they are attracted to each other. In this way two droplets can coalesce. But later it has been found that the distance between the two is so far and the electrical force is so weak appreciable attraction may not occur. Moreover it is difficult to understand how the pattern could evolve the arrangement of positive and negative charges face to each other within a very short time in the atmosphere.

*Size of the Droplets*

Saturated vapour pressure changes according to the length of curvature of the cloud droplets. If the size of the droplet is large the saturated vapour pressure will be lower. On the other hand in case of smaller size droplets the saturated vapour pressure will be high. So the saturated vapour pressure varies for multiple droplets in the cloud. It could be found that the large droplets become saturated as it is higher than the saturated vapour pressure, and *vice-versa* the smaller droplets remain to be unsaturated. As a result smaller droplets evaporate and condense over the larger droplets. This hypothesis is also not very tenable as the variation of droplet size is meagre in the cloud. Most of the cloud droplets have the diameter 10 to 15 micron, only a few have larger diameter of 40 micron.

*Temperature of the Droplets*

Saturated vapour pressure varies with the temperature condition in the droplets. In cold cloud the variation of temperature is negligible, but increases in the warm clouds. Due to turbulence and convection warm droplets of the low clouds come into contact with cold droplets of high cloud. Cold droplets are saturated and warm droplets remain unsaturated. As a result warm droplets evaporate and condense over the cold droplets. In this process the size of the water becomes larger to precipitate. Normally this process is effective for the warm tropical clouds having temperature above 10° C.

*Motion of the Droplets*

In the atmosphere due to turbulence and convection the motion of the droplets increases. There may be collision of droplets due to increased motion in the atmosphere. As a result of collision droplets may attract each other and remain buoyant in the atmosphere. When the droplets attain the size of a raindrop they begin to fall due to gravitational force. During fall colder droplets also attract warmer droplets below to condense over the larger droplets. Langmuir explained this attraction in a different way. According to him the terminal velocity of the droplets is related to the diameter of the droplets. So the larger droplets result in coalescence of the smaller droplets. This is caused just like sweeping and is known as *sweeping theory* of rainmaking by Langmuir. Initially the process may be slow, but within 50 minutes the droplet size attains 200 micron diameter.

*Presence of Ice Crystal in the Droplets*

The above stated process may not adequately explain the complex process of rainmaking. A new concept for the presence of ice crystal with the super-cooled water droplets has been suggested. The colloidal instability increases with the presence of ice crystal in the cloud droplets. Bergeron-Fiendisen explained this concept in rainmaking process.

Two questions arise in explaining the mechanism of precipitation:
a)  How to explain the coalescence of droplets and
b)  How to explain the sudden release of precipitation that is usually observed.

It is found that collisions in many cases may lead to fragmentation of the droplets instead of coalescence. There is often no precipitation from highly turbulent clouds.

### *Bergeron-Findeisen Theory*

The role of ice crystals present in clouds along with water droplets has been highlighted in recent years to explain the mechanism of precipitation. Bergeron-Findeisen considered that the colloidal instability would result in a cloud consisting of a mixture of ice particles and super-cooled water droplets. The foundation of Bergeron-Findeisen theory of rain making lies in the following facts.

The relative humidity of air is greater with respect to an ice surface than with respect to water surface. As the temperature drops below 0° C the atmospheric vapour pressure falls very sharply than over the water surface. This induces Saturation Vapour Pressure (SVP) over water becoming greater than that over ice, particularly at temperature, –5° to –25° C. When the difference of SVP exceeds 0.2 mb., ice crystals and super-cooled water droplets exist together in a cloud, the super cooled water droplets tend to evaporate and deposit over the ice crystals (by the process of sublimation). Small water droplets can be super-cooled in pure air to –40° C. However spontaneous freezing takes place normally when temperature drops below –22° C. The role of the freezing nuclei is very important in the formation of ice crystals. The freezing nuclei are less numerous in a cloud than the condensation nuclei, only 10/litre at –30° C. Fine soil particles, aerosols, volcanic dust particles etc. favour the formation of freezing nuclei. The freezing nuclei grow by deposition. Being detached from the parent one a new freezing nucleus may be formed separately, whereas the freezing of the super cooled water may lead to the formation of ice splinters. The ice crystals aggregate upon collision, particularly at temperature 0° C to –5° C.

The following table illustrates the differences in saturation vapour pressure of water relative to ice.

**Table 4.5:** Saturation Vapour Pressure of Water Relative to Ice

| Temperature | Relative Humidity with respect to water | Relative Humidity with respect to ice |
|---|---|---|
| 0° C | 100% | 100% |
| –05° C | 100% | 105% |
| –10° C | 100% | 110% |
| –15° C | 100% | 115% |
| –20° C | 100% | 121% |

As the ice crystals grow larger they can not be suspended further in the free air and fall downwards. At the base of the cloud where the temperature exceeds 0° C the ice crystals form large water droplets and ultimately drop as rain on earth's surface.

### *Artificial Rain-making*

The basic concept of Bergeron and Findiesen has been applied in the artificial rain making. In a cloud silver dioxide or carbon dioxide is injected either by battery charging from the ground or by an aircraft or rocket in the air. These particles when injected into a cloud act as freezing nuclei and account for rainfall. However, the mechanism of artificial rainfall is expensive and in some cases the experiment may not yield desirable result.

Production of rain by artificial means now generally disregarded, though it is probable that rainmaking hastens or increases rainfall from clouds suitable for natural rainfall. Interest in rainmaking has been spurred by factors including drought and the need for irrigation water. Until recent times it was thought that rain might be induced by explosions, updrafts from fires, or by giving the atmosphere a negative charge. Research during the 1930s showed that rain forms in warm clouds when larger drops of condensed water grow at the expense of smaller ones until they are big enough to fall; also that in cold clouds super-cooled water below –15° C freezes into ice crystals that act as nuclei for snow. On this basis the American physical chemist Irving Langmuir and his associates carried on Project Cirrus from 1940 to 1952 to find ways to produce rain. Three methods resulted, including spraying water into warm clouds; dropping dry ice into cold clouds, where the dry ice freezes some water into ice crystals that act as natural nuclei for snow; and wafting silver iodide crystals or other similar crystals into a cold cloud from the ground or from an airplane over the cloud, with the crystals hastening the freezing of super-cooled water between –2.8° C and –15° C. Over-seeding can dissipate a cloud. These techniques are only moderately successful; they cannot be relied upon in case of drought.

### *Uplift Mechanisms*

Uplift mechanisms may be identified as:
1. Convective Uplift,
2. Convergent Uplift,
3. Orographic Uplift, and
4. Frontal Uplift.

#### *Convective Uplift*

*Adiabatic cooling* occurs when air is uplifted from the surface causing the air to lose heat through the work of expansion. A parcel of air is uplifted when it initially gains heat from the surface causing *convective uplift*. When the air is warmed by the surface it will expand and become less dense relative to air that surrounds it. Being less dense than the air that surrounds it, the air becomes buoyant and begins to rise. Because atmospheric pressure decreases with height, the parcel of air expands and cools. If the air cools to its dew point temperature saturation occurs and condensation begins. The elevation above the surface where condensation begins is called the *condensation level*.

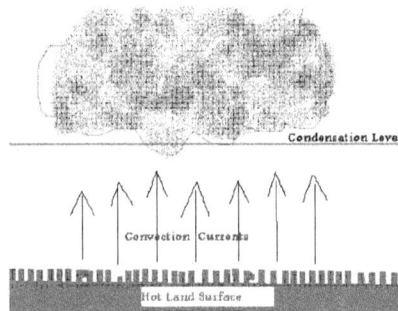

**Figure 4.5:** Convective Uplift—Heated Land Surface Causes Convection Currents to Move Upward

*Convergent Uplift*

*Convergent uplift* occurs when air enters a centre of low pressure. As air converges into the centre of a cyclone it is forced to rise off the surface. As the air rises it expands, cools, and water vapour condenses. Convergent and convective uplift are the two most important uplift mechanisms for c copious rainfall in the wet tropics as well.

*Orographic Uplift*

*Orographic uplift* is the forced ascent of air when it collides with a mountain. As air strikes the windward side, it is uplifted and cooled. Windward slopes of mountains tend to be the rainy sides while the leeward side is dry. Dry climates like steppes and desert are often found in the "rain shadow" of tall mountain systems that are oriented perpendicular to the flow of air.

*Frontal Uplift*

*Frontal uplift* occurs when greatly contrasting air masses meet along a weather front. For instance, when warm air collides with cool air along a warm front, the warm air is forced to rise up and over the cool air. As the air gently rises over the cool air, horizontally developed stratus-type clouds form. If cold air collides with warm air along a cold front, the more dense cold air can force the warmer air ahead to rise rapidly creating vertically developed cumulus-type clouds.

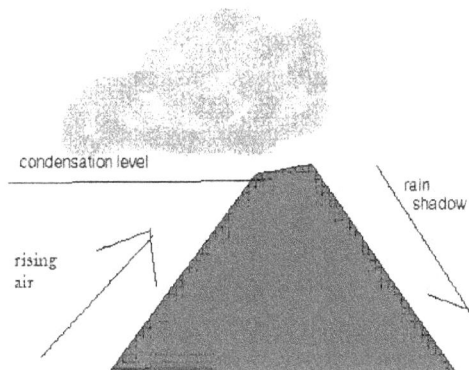

**Figure 4.6:** Orographic Uplift—Air Rising over the Windward Slope to Reach the Condensation Level

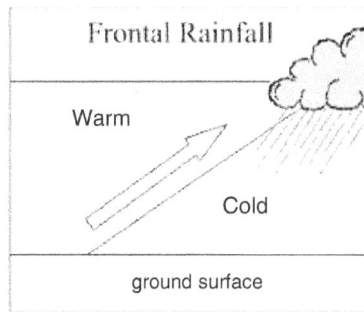

**Figure 4.7:** Frontal Uplift—Warm Air Rising over the Cold Air

### Types of Rainfall

Air rising upwards is activated by different ways and condenses in the free atmosphere. Rainfall may be caused by different processes acting as the lifting agent. We can identify three important processes lifting air. These are: 1. *Convective Uplift*, 2. *Orographic Uplift* and 3. *Frontal Uplift.* So rainfall are classified into categories: 1. *Convectional Rainfall,* 2. *Orographical Rainfall* and 3. *Cyclonic* or *Frontal Rainfall.* Another type also occurs, known as *Convergent Rainfall* due to convergent uplift.

### Convectional Rainfall

When the land surface becomes heated air turns to be warmer and rises upwards. As the air rises vertically adiabatic expansion occurs resulting in cooling of the rising air parcel. When the dew point is reached latent heat is released. This released latent heat warms the rising air more and provides energy to move further up. Ultimately it attains the freezing point and precipitation starts. Convection currents helps the air rising and to become buoyant.

Convection current has the components of *updraft* and *down draft cells.* As long as the energy is provided by heated land surface updraft continues. After reaching the freezing level downdraft starts and comes down with associated rainfall, squall, thunder and even hail storm. If the convection process is strong cumulonimbus (Cb) cloud forms. With the formation of cumulonimbus clouds weather becomes unstable. Convectional rainfall is usually associated with the formation of thunder clouds and characterised by lightning and thunder.

As stated earlier, the convection process is related to heating of the land surface. So we find the occurrence of convectional rainfall in the afternoon of summer months in the tropics and almost daily in the equatorial areas. Convectional rainfall may also occur in the hot summer afternoon in the middle latitudes. The convectional rainfall is of short duration and very much localised. The convectional rainfall may not always occur even when strongly heated. If it be so there would have been daily rainfall in summer months in the tropics. Convectional process in many cases is favoured by some other factors which act as a trigger. These include: 1. Relief barrier, 2. Friction over land, 3. Formation of eddies, and 4. Lift due to convergence.

*Orographic Rainfall*

Mountain slope acts as a barrier to the rising air. Such barrier acts as a trigger to help the rising air more upwards. So this process is known as *forced ascent.* Due to ascent the rising air reaches the dew point and even move upwards for saturation and precipitation.

Moist air strikes the windward slope of the mountain. So the air rides up slope on windward side and precipitates more on this slope. When the air slides down the opposite slope little moisture is available for precipitation. Moreover air subsides here instead of rising. So precipitation rarely occurs on the leeward slope of the mountain. Classical example may be cited from the rainfall pattern over the Western Ghat or Sahyadri Mountain along the west coast of India. The coastal areas extending from Malabar to Konkan get adequate rainfall from the moisture laden winds during the southwest monsoon. Whereas the eastern slope of the Western Ghat receives meagre rainfall to create the semi-arid conditions in Maharashtra, Karnataka and Tamilnadu. This is designated as the *rain-shadow area.*

The location and alignment of mountain ranges play significant role in orographic rainfall. It is also found that rainfall is high over the mountain terrain than over the plains. The role of the mountain can be specified as follows:

1. Location of the mountain helps the turbulence in the atmosphere generating mechanical and convective currents.
2. Location of the mountain restricts the movement of cyclone and favours the cyclone to be slower and sometimes stagnant to precipitate over the mountain slope.
3. Due to hindrance by the mountain slope convergence may occur.
4. Such hindrance may act as a trigger for further lift of the rising air to reach the condensation level.

The orographic rainfall may be of shorter duration causing heavy precipitation. In high altitude mountain slope intense unstable conditions may develop for the formation of thunderstorms in the afternoon. Such thunderstorms are difficult to predict and cause immense hazards to the mountaineers. Orographic rainfall may also be less intense and persists longer.

*Cyclonic or Frontal Rainfall*

When two contrasting air masses, warm moist and cold dry, converge at the centre of low fronts are formed. The air masses are separated by the fronts, known as *cold front* and *warm front.* Of the two fronts cold front is pervasive and ultimately swaps the warm front.

Due to convergence air is lifted. Rising warm moist air slides over the cold dry air. Cold air forms wedge to push the rising warm air more upward. Being lifted instability conditions prevail. Weather conditions differ along the two fronts. Rainfall occurs over the both fronts— cold and warm. However the nature of the rainfall differs. Rainfall appears to be heavy and of short duration along the cold front. On the other hand rainfall is light, actually drizzle continues for a longer period along the warm front.

In case of tropical cyclone due to absence of two contrasting air masses no such fronts develop. Here the cyclonic rainfall is caused due to convection, on some occasions favoured by orographic barrier.

## Hydrologic Cycle

The *Hydrologic Cycle* is a conceptual model that states the storage and movement of water between the *biosphere, atmosphere, lithosphere*, and the *hydrosphere*. Water on this planet can be stored in any one of the following reservoirs: *atmosphere, oceans, lakes, rivers, soils, glaciers, snowfields,* and *groundwater.*

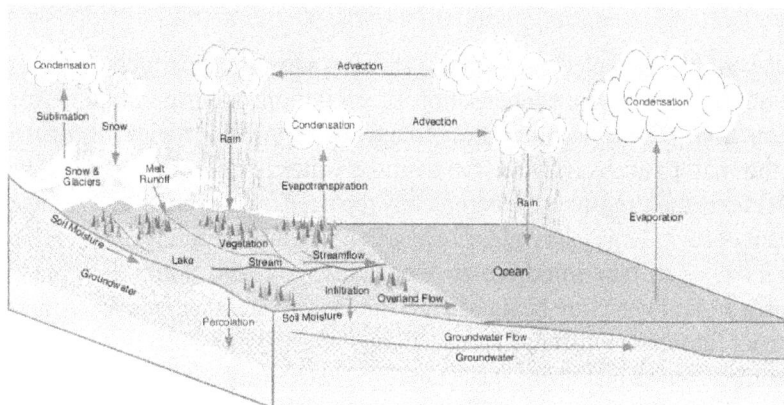

**Figure 4.8:** Hydrologic Cycle

Water moves from one reservoir to another by way of processes like *evaporation, condensation, precipitation, deposition, runoff, infiltration, sublimation, transpiration, melting*, and *groundwater flow*. The oceans contribute the supply of most of the water in the atmosphere by evaporation. Of this evaporated water, only 91 percent return to the ocean basins by way of precipitation. The remaining 9 percent is transported to areas over landmass where climatic factors induce the formation of precipitation. The resulting imbalance between rates of evaporation and precipitation over land and ocean is adjusted by runoff and groundwater flow to the oceans.

Nearly all planetary water is contributed by the oceans (Table 4.6). Approximately 97 percent of all the water on the earth is in the oceans. The other 3 percent is found as freshwater in glaciers and icecaps, groundwater, lakes, soil, the atmosphere, and within biosphere life. So the available freshwater is very scarce on earth's surface.

**Table 4.6:** Inventory of Water at the Earth's surface

| Reservoir | Volume (cubic km × 1,000,000) | Percent of Total |
|---|---|---|
| Oceans | 1370 | 97.25 |
| Ice Caps and Glaciers | 29 | 2.05 |
| Groundwater | 9.5 | 0.68 |
| Lakes | 0.125 | 0.01 |
| Soil Moisture | 0.065 | 0.005 |
| Atmosphere | 0.013 | 0.001 |
| Streams and Rivers | 0.0017 | 0.0001 |
| Biosphere | 0.0006 | 0.00004 |

Water is continually cycled between its various reservoirs. This cycling occurs through the processes of evaporation, condensation, precipitation, deposition, runoff, infiltration, sublimation, transpiration, melting, and groundwater flow. The typical residence times of water in the major reservoirs vary. On average water is renewed in rivers once every 16 days. Water in the atmosphere is completely replaced once every 8 days. Slower rates of replacement occur in large lakes, glaciers, ocean bodies and groundwater. Replacement in these reservoirs can take hundreds to thousands of years. Some of these resources (particularly groundwater) are being used by humans at rates that far exceed their renewal times. In near future our existence will be at stake due to wrong use.

One estimate shows that the oceans annually contribute $4.49 \times 10^{17}$ kg. of water to the atmosphere through the process of evaporation, but get back only $4.12 \times 10^{17}$ kg as precipitation. Evaporation from the water bodies of land surface and evapotranspiration from plants add annually $0.71 \times 10^{17}$ kg. water to the atmosphere, but get about $1.05 \times 10^{17}$ kg. water as precipitation. This is in excess of evaporation, of which about $0.37 \times 10^{17}$ kg. water flow back to the oceans as runoff by different rivers and streams. The rest of the water is retained by the land surface, which is essential for the sustenance of life on the earth's surface. The operation of hydrologic cycle is crucial to the availability of fresh water to the biosphere especially to the land-based ecosystem.

## Residence Times

The residence time of a reservoir within the hydrologic cycle is the average time a water molecule will be spent in that reservoir. It is a measure of the average age of the water in that reservoir, though some water will be spent much less time than average, and some much more.

Groundwater can remain over 10,000 years beneath earth's surface before being spent. Particularly old groundwater is called *fossil water*. Water stored in the soil remains there very briefly, because it is spread thinly on the earth's surface and is readily lost by evaporation, transpiration, stream flow, or groundwater recharge. After evaporating, water remains in the atmosphere for about 8 days before condensing and falling to the earth as precipitation.

---

## Further Readings

Barry, R.G. and R.J. Chorley (1968) *Atmosphere, Weather and Climate,* Methuen & Co., London.

Braham, R.R. (1959) 'How does a raindrop grow?' *Science,* **129**.

Critchfield, H.J. (1975) *General Climatology,* Prentice Hall India Ltd., New Delhi.

Durbin, W.G. (1961) 'An introduction to cloud physics', *Weather,* **16**.

Ludlam, F.E. (1956) 'The structure of rain clouds', *Weather,* **11**.

Mason, B.J. (1962) *Clouds, Rain and Rainmaking,* Cambridge University Press, Cambridge.

Mason, B.J. (1971) *The Physics of Clouds,* Oxford Press, New York.

Petterssen, S. (1969) *Introduction to Meteorology,* McGraw Hill Book Co., New York.

Simpson, C.G. (1941) 'On the formation of clouds and rain', *Quarterly Journal of Royal Meteorological Society,* **67**, London.

Trewartha, G.W. (1968) *An Introduction to Climate,* McGraw Hill Kogakushu Ltd., Tokyo.

World Meteorological Organization (1956) *International Cloud Atlas,* Geneva.

# Chapter 5

# AIR MASSES

## Defining the Air Mass

Air spreads horizontally hundreds of kilometres and maintains uniform physical characteristics, particularly temperature, moisture and lapse rate. Such vast expanse of air over a place, whether land and seas is called the air mass. Study of the air mass is important in understanding the synoptic climatology. Synoptic climatology means the study of climates in relation to atmospheric circulations. It explains the relationships between the circulations, determines the weather types and outlines the regional climatic differences. We know that for general circulation exchanges of energy occurs between the low latitudes and high latitudes. Air masses play significant role in such transfer of energy from low to high latitude and *vice-versa*.

## Classification of Air Masses

Air masses are classified according to their approximate place of origin with specific weather characteristics. An air mass is classified as Arctic, Polar, Tropical, or Equatorial. It is also classified as either Maritime or Continental. Maritime air mass is a moist air mass, whereas continental air mass is relatively dry. Air masses of oceanic origin are marked with a lower-case 'm' (maritime), while air masses of continental origin are marked with a lower-case 'c' (continental). Air masses are also denoted as either Arctic (upper-case 'A', or 'AA' for Antarctic air masses), polar (upper-case 'P'), tropical (upper-case 'T'), or equatorial (upper-case 'E'). These two sets of attributes are used in combinations depending on the air mass being specified. For instance, an air mass originating over the desert of African Sahara in summer may be designated 'cT'. An air mass originating over northern Siberia in winter may be indicated as 'cA'.

An upper case 'S' is occasionally used to denote something called a *superior* air mass. This was regarded as an adiabatically drying and warming air mass descending from aloft. In South Asia, an upper case 'M' (for monsoon) has been occasionally used to denote an air mass within the summer monsoon regime in that region.

The stability of an air mass (thermal condition) may be indicated using a third letter, either 'k' (air mass colder than the surface below it) or 'w' (air mass warmer than the surface

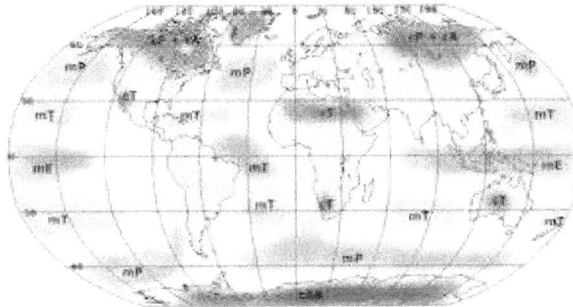

**Figure 5.1:** Global Air Masses

below it). An example of this might be a polar air mass blowing over the Gulf Stream, denoted as 'cPk'. Another convention is to indicate the modification or transformation of one type to another combining two. For instance, an Arctic air mass blowing out over the Gulf of Alaska may be shown as 'cA-mPk'. Yet another convention indicates the layering of air masses in certain situations. For instance, the overrunning of a polar air mass by an air mass from the Gulf of Mexico over the Central United States might be shown with the notation 'mT/cP' (sometimes using a horizontal line as in fraction notation).

### *Modification of Air Mass*

Air masses are characterised by their dynamic movements. From the source region an air mass moves with the specific weather characteristics, whether warm or cold and moist and dry. During its travel the air mass may be modified losing its temperature and moisture characteristics. The basic characteristics will be modified the farther it moves from its source region. To indicate the type a third letter is used to suggest the modification. If the traveling air mass is warmer than over the surface it is moving, then 'w' will indicate the warmer condition. Similarly colder air mass will be indicated by the letter 'k'.

The modification of air mass is significant initiating weather conditions over the region. Even the stable air mass will turn to be unstable and weather may change with the arrival of air mass over the region from distant source region. Both cPk and mTw air masses are important in understanding the regional climatology. Say, cP air mass is moving over the great lakes in winter. Water of the lake is relatively warmer than the traveling air mass. So there will be evaporation of the water and moisture will be added to the air mass. The temperature and moisture status of the air mass will be modified and also weather condition will become unstable. Similarly modification may also occur in mT air mass, originating from the subtropical seas and traveling inland over the cold continent in winter.

So we find that the traveling air masses change their characteristics through modification of temperature and moisture conditions at a place, far away from their source regions. As a result barotropic air mass changes to a modified baroclinic air mass. Such change may be of thermodynamic nature. Modification may also be evidenced in a different terrain. If the underlying relief is hilly, friction at lower layer may cause turbulence. As a result the traveling air mass may be modified. Modification of the air mass is also related to its age. After a long period of movement the traveling air mass loses its original characteristics and accommodates the characteristics of the surrounding air mass.

**Table 5.1:** Classification of Air Masses in Terms of Place of Origin

| Name of the air mass | Place of origin | Properties | Symbol |
|---|---|---|---|
| Polar continental | Subpolar continental areas | Low temperatures (increasing with southward movement), low humidity (nearly uniform) | cP |
| Polar maritime | Subpolar and arctic oceans | Low temperatures (increasing with movements), higher humidity | mP |
| Tropical continental | Subtropical high pressure landmass | High temperatures, low moisture content | cT |
| Tropical maritime | Southern borders of subtropical oceans, high pressure areas and specific humidity | Moderately high temperatures, high relative | mT |
| Equatorial | Equatorial and tropical seas | High temperatures and high humidity | E |

### Barotropic and Baroclinic Air Mass

In a region where the isobars and isotherms lie parallel the air mass over that region is called the *barotropic air mass.* Barotropic air masses are stable and extend over wide areas. They possess uniform characteristics. Two barotropic air masses, one cold and the other warm, may exist separated by a plane extending more than 100 kilometre. To the contrary in a region where the isobars and isotherms cross each other baroclinic conditions prevail. The air mass lying over this region is called the *baroclinic air mass.* Baroclinic conditions denote unstable conditions where the mixing of two barotropic air masses takes place. The mixing of cold and warm air at the plane of separation of two barotropic air masses may occur. This plane of separation is called the *baroclinic front.* When the barotropic air masses move from the place of their origin they become modified and turn to be moderately baroclinic air mass.

Barotropic systems are characterised by: 1. Cold-core barotropic low (occlusion), 2. Warm-core barotropic low (tropical cyclone), 3. Warm-core barotropic high (cutoff high),

4. Cold-core barotropic high (polar source), 5. Cold-core barotropic low (cutoff low) 6. Warm-core barotropic low (heat low) and 7. Warm-core barotropic high (subtropical ridge).

Baroclinic systems have different characteristics:

1.   A weak low pressure area in a baroclinic region is affected by an upper-level disturbance. This causes convergence of the low-pressure area, which draws the warm and cold air masses together and makes it more baroclinic. This in turn strengthens the upper-level features, which in turn produces stronger convergence at the surface.

2.   Coastal regions are more favourable for baroclinic development during the winter when cold land temperatures contrast sharply with warm oceanic temperatures.

3.   The rain above the warm sector will cause extensive cooling by evaporation in the dry warm sector air. This will reduce the thermal contrasts (the baroclinicity) available to the low pressure area, retarding its development.

4.   The rapid deepening will cause stronger convergence of temperature gradients, strengthening the upper-level features and possibly causing further self-development.

5.   Baroclinic low development is most likely in areas where thermal gradients (thickness packing) are closest together.

6.   The term for a baroclinic low that undergoes very rapid deepening is called a *bomb*.

7.   The final stage in the life cycle of a baroclinic low is known as an occlusion (which by this time is largely barotropic).

8.   A baroclinic high is different from a barotropic high. It is initially associated with temperature contrasts, which strengthen upper level features which in turn strengthen the high.

9.   A baroclinic high is most likely to dissipate quickly in mountainous areas, where the increased friction reduces air velocity, and in turn reduces the Coriolis force. This allows air to flow more directly away from the high pressure area.

10.   The dissipation of a baroclinic high can be evaluated by seeing how rapidly the thermal contours (thickness lines) are spreading apart with time in the high.

### *Traveling Air Masses*

Air masses acquire characteristics from a given region which they may occupy for any significant length of time. For example, Polar air masses develop during the northern winter as intense night time radiation and loss of daylight strongly cools the air to great depths. On the other hand maritime air masses generally form over oceans and seas where widespread evaporation takes place. When air masses move into regions with different weather conditions, they are modified. Even without movement, air masses may differ gradually over distance, with one type gradually becoming another. The boundary between two air masses may also be sharply marked. This boundary of separation between two contrasting air masses is called a *front*. Fronts are often characterised by unstable and inclement weather. Fronts are usually associated with areas of low atmospheric pressure called frontal systems.

Fronts are defined in terms two groups of warring soldiers and the term was coined after the World War I. A *weather front* is a boundary separating two air masses of different densities, and causes typical weather phenomena, particularly in the mid-latitude region. V.J. Bjerkness, H. Solberg and T. Bergeron studied in detail on the formation of fronts and analysed their effect on weather conditions in Western Europe. The air masses separated by a front usually differ in temperature and humidity. Cold fronts may signify narrow bands of thunderstorms and severe weather, and occasionally may be preceded by squall lines or dry lines. To the contrary warm fronts are initially characterised by stratiform precipitation and fog. The weather usually becomes fair quickly after a front's passage. Some fronts produce no precipitation and little cloudiness, although there is invariably a wind shift.

Cold fronts and occluded fronts generally travel from west to east, while warm fronts move pole-ward. Because of the greater density of air in their wake, cold fronts and cold occlusions move faster than warm fronts and warm occlusions. In case a front becomes stationary, and the density contrast across the frontal boundary vanishes, the front can degenerate into a line which separates regions of differing wind velocity, known as a *shearline*. It occurs most commonly over the wide ocean.

## Mid-latitude Cyclones

In the mid-latitude region close to the sub-polar low region two contrasting air masses, namely *cold polar* (cP) and maritime tropical (mT) air masses converge. These air masses build along their margin fronts, which ultimately develop into a mature cyclone.

The cyclone model was illustrated by J.Bjerkness in 1937 and is known as Polar Front theory. In this model at initial stage the cold air mass of polar origin and warm air mass of tropical oceans form a line of separation between the two along a low pressure centre. The temperature contrasts result in the development of wind shear and also wind shift. As a consequence there will be rising light warm air, pushed back as wedge from the rear by cold heavy air. Hence two fronts are formed—*cold front* and *warm front*—having distinctive characteristics. The cold front is more active and faster than the warm front in movement. The warm front is ultimately overwhelmed by the cold front, when there will be formation of *occluded front*. The occluded front will indicate the decay of the cyclone.

*Characteristics*

1.  Two fronts are formed—*warm front* and *cold front*.
2.  The gradient along the front is steeper over the cold front (slope–1 : 50 to 1 : 100) than over the warm front (slope–1 : 100 to 1 : 400).
3.  The mid-latitude cyclones move from west to east. The formation of warm front to the east is the first indication of approaching cyclone. At first warm condition prevails which is replaced very fast by cold wave condition with marked drop of temperature.

4. Along the cold front the rainfall is heavy, but of short duration. Whereas along the warm front the rainfall is in the form of drizzle, but continues for longer time.

5. Unlike tropical cyclone barometric pressure difference is not very marked in the mid-latitude cyclone. So no 'eye' is formed at the centre. But temperature contrast is very marked in this cyclone.

6. Veering and backing wind shifts are observed in different quadrants of the cyclonic zone. At the initial phase warm air blows from south-east and cold air from north-east. The wind direction is later changed, warm air from south and cold air from north. Lastly the cold air moves from north-west and warm air from south-west.

7. The mid-latitude cyclones mostly develop during the winter months and are stronger. Whereas the cyclones developing in summer months are fewer and weaker.

### *Development of Mid-latitude Cyclones*

Mid-latitude cyclone has a life history. Two stable contrasting air masses move and come closer with the formation of line of separation, known as front. Initially there is development of undulating waves between the two contrasting air masses. Warm air mass forms wedge into the cold air mass. Warm air mass being lighter and less dense rides over the cold air mass. Cold air mass pushes the warm air mass to move further up. A front is formed along the cold air mass and has a shape like nose penetrating the warm air mass. This is known as *cold front.* Due to such penetration of cold air mass the contact cooling occurs within the warm moist air mass. Warm air mass becomes progressively unstable and results in the formation of warm front to precipitate. Cold front appears to be very aggressive, penetrates deeply into warm air mass and finally swaps the warm air mass. Warm front disappears and the process of occlusion starts. After the occlusion contrast between the two air masses, cold and warm, recedes and finally taken over by the cold air mass.

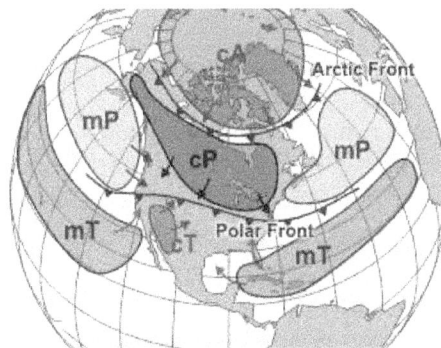

**Figure 5.2:** Different Air Masses and Fronts

### Cold Front

A *cold front* is defined as the advancing edge of a cooler and drier mass of air. The air with greater density wedges under the less dense warmer air, lifting it, which results forming a narrow line of showers and thunderstorms when enough moisture is present. This upward motion causes lowered pressure along the cold front. A cold front's location lies at the leading edge of temperature fall. Isotherm analysis indicates the leading edge of the isotherm gradient, and it normally exists within a sharp surface trough. Cold front moves faster, twice as fast and produces sharper changes in weather than the warm fronts. Since cold air is denser than warm air it rapidly replaces the warm air wiping out the boundary. Cold front is usually characterised with a low pressure cell, and sometimes, a warm front on other edge.

Cold front is formed when a cold air mass moves into an area of warmer air. Warmer air interacts with the cold air mass, and precipitation occurs. On weather maps, the surface position of the cold front is marked with the symbol of a blue line of triangle-shaped pips pointing in the direction of travel, and it is placed at the leading edge of the cooler air mass.

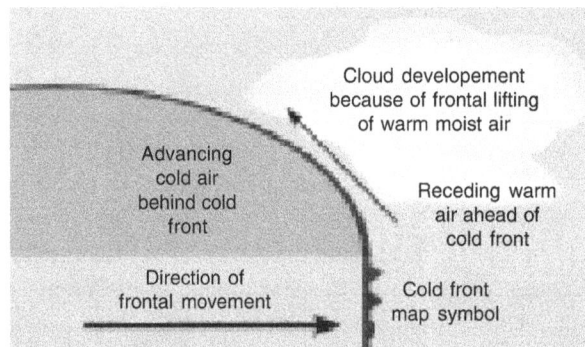

**Figure 5.3:** Example of a Cold Front

### Weather Characteristics along the Cold Front

A cold front commonly forms a narrow band of precipitation that follows along the leading edge of the cold front. These bands of precipitation are often very strong in nature, and cause heavy precipitation accompanied by severe thunderstorms and or tornadoes. In spring months these cold fronts become very strong, and cause strong winds when the pressure gradient is steeper than normal. In summer months, cold fronts can cause severe thunderstorms and hailstorms. In winter months cold fronts can bring severe cold spells, and heavy snowstorms. In autumn months cold fronts rarely cause severe thunderstorms, but cause heavy and widespread rainstorms. Very often, cold fronts are associated with very severe cold weather. The passage of the cold fronts is usually associated with a quick, yet strong gust of wind, indicating that the cold front is passing. The effects of a cold front can last only a few hours to several weeks, depending on the emergence of the next weather front.

Cold fronts are very often associated with a warm front, squall line, or other weather front. Usually cold fronts have an adjacent warm front that is ahead of the cold front. This forms an area where warm air is penetrating and interacting with the cold front. This area is known as a *warm sector*. In the warm sector, very often severe thunderstorms, tornadoes, and hailstorms occur due to contrast between the warm air and cold air.

### Warm Front

Warm fronts are the leading edge of a homogeneous warm air mass, located on the equator-ward edge of the gradient in isotherms, and lie within broader troughs of low pressure than cold fronts. A warm front travels more slowly than the cold front.

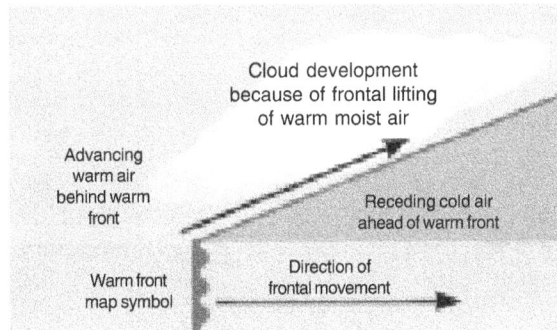

**Figure 5.4:** Example of a Warm Front

*Weather Characteristics along the Warm Front*

As the cold air is dense and hardly to be removed the traveling warm air forms heavy cloud and causes precipitation. This also induces temperature differences across warm fronts to be broader in scale. Clouds ahead of the warm front are mostly stratiform, and rainfall gradually increases with the advancing front. Near the earth's surface stratus, after that strato-cumulus, in the middle alto-stratus and finally cirro-stratus and cirrus clouds are formed. Fog can also occur preceding a warm frontal passage. Clearing and warming is usually fast after frontal passage. If the warm air mass is unstable, thunderstorms may be embedded among the stratiform clouds ahead of the front, and after frontal passage thundershowers may continue.

### Occluded Front

An *occluded front* is formed during the process of cyclogenesis when a cold front overtakes a warm front. When this occurs, the warm air is separated (occluded) from the centre of low of the mid-latitude cyclone. There are two types of occlusion, the warm and the cold. In a cold occlusion, the air mass overtaking the warm front is cooler than the cool air ahead of the warm front, and turning up under both air masses. In a warm occlusion, the air mass overtaking the warm front is not as cool as the cold air ahead of the warm front, and rides over the colder air mass while lifting the warm air.

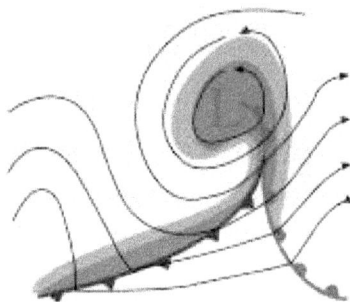

**Figure 5.5:** Example of an Occluded Front

*Weather Characteristics along the Occluded Front*

A wide variety of weather can be found along the occluded front. Thunderstorms may occur, but usually their passage is associated with drying of the air mass. Occluded fronts usually form around mature low pressure areas.

**Stationary Front and Shear Line**

A *stationary front* is a boundary between two different air masses, neither of which is strong enough to replace the other. So they remain almost in the same area for extended periods of time, and waves sometimes propagate along the frontal boundary. A wide variety of weather can be found along a stationary front, but usually clouds and continuous precipitation are observed. Stationary fronts will either dissipate after several days or devolve into shear lines, but can change into a cold or warm front if conditions aloft change.

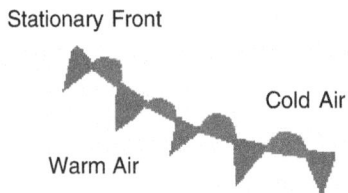

**Figure 5.6:** Stationary Front and Shear Line

A stationary front becomes a *shearline* when the density contrast across the frontal boundary vanishes. Usually it results due to equalisation of temperature. A narrow zone of wind shift persists for a time. This is very often observed over the open ocean as the temperature of the ocean surface is usually the same on both sides of the frontal boundary. The air masses on either side of it are modified corresponding to its own temperature.

## Development of Clouds in Mid-latitude Cyclones

Cold air is heavy. When warm air moves over the cold air clouds are formed due to condensation and rainfall occurs. Along the warm front close to the surface stratus clouds are formed. Above it the strato-cumulus, alto-stratus are formed and on the top lie the cirro-stratus and cirrus. The cold front on the other hand forms wedge like a nose and penetrates the warm front. Finally the wedge will collapse and warm air locked in the cold air may form cumulonimbus or thunder clouds. Heavy rain for short duration may occur along the cold front for this reason.

When the warm front and cold front are developed over a region, cirrus and cirro-stratus clouds are observed first to the right of the warm front. Then the sky will be overcast, but rainfall may not occur immediately. Alto-stratus cloud will appear in the sky to initiate rainfall. Steadily dense dark nimbostratus cloud will emerge and heavy rainfall will start. It will indicate the dissipation of the warm front. Afterwards stratus and strato-cumulus clouds reappear and rainfall ceases. No clouds could be seen in the sky and fog covers the area. It indicates the takeover of the warm front by the cold front. There will be no rainfall and only scattered low clouds may be observed. Suddenly thunder clouds (cumulonimbus) will emerge and thunder rain with lightning will occur. After a few spells of heavy rain cold front will dissipate and the sky will be clear. It indicates the occlusion of fronts.

## Dry Line

The dry line is the boundary between air masses with significant moisture differences. The increase of westerly winds aloft on the north side of surface highs areas of low pressure area generates downwind of north–south oriented mountain chains, leading to the formation of a lee trough. Near the surface during day hours, warm moist air is denser than dry air of higher temperature, and thus the warm moist air forms wedges under the drier air like a cold front. At higher altitudes, the warm moist air is less dense than the dry air and the boundary slope reverses. In the vicinity of reversal aloft severe weather may be experienced. It is favoured when a triple point is formed with a cold front. The lee trough is the weaker form of the dry line which displays weaker differences in moisture status. When moisture pools along the boundary during the warm season, it can become the centre of diurnal thunderstorms. The dry line may occur in regions between the arid lands and warm seas. The dry line normally moves eastward during the day and westward at night.

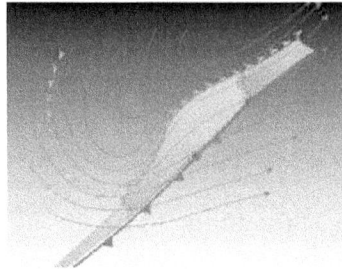

**Figure 5.7:** Initial Frontal Wave (or Low Pressure Area) Formation.

## Cyclogenesis

*Cyclogenesis* is the development or strengthening of cyclonic circulation in the atmosphere (a low pressure area). Cyclogenesis is a general term for several different processes, all of which result in the development of some sort of cyclone. It can occur at various scales, from the micro-scale to the synoptic scale. Extra-tropical cyclones form as waves along the weather fronts before occlusion later in their life cycle as cold core cyclones. Tropical cyclones form due to latent heat driven by significant thunderstorm activity, and are warm core. Meso-cyclones form as warm core cyclones over land, and can lead to tornado formation. Waterspouts can also form from meso-cyclones, but more often develop from the conditions of high instability and low vertical wind shear. Cyclogenesis is the opposite of cyclolysis, and has an equivalent anticyclonic (high pressure) system.

## Further Readings

Barett, E.C. (1964) 'Satelltite meteorology and the geographer', *Geography,* **49**.

Belasco, J.E. (1952) 'Characteristics of airmasses over the British Isles, *Geographical Memoirs,* Meteorological Office, London.

Gentilli, J. (1949) 'Airmasses of the Southern Hemisphere', *Weather,* **4**.

Hare, F.K. (1960) 'The Westerlies', *Geographical Review,* **50**.

Lamb, H.H. (1951) 'Essay on frontogenesis and frontolysis', *Meteorology Magazine,* **12**.

Oliver, J.E. and John J. Hidore (2002) *Climatology,* Pearson Education Inc., Delhi.

Petterssen, S. (1968) *Introduction to Meteorology,* McGraw Hill Book Co., New York.

Petterssen, S. (1956) *Weather Analysis and Forecasting,* McGraw Hill Book Co., New York.

Pothecary, I.J.W. (1956) 'Recent research on fronts', *Weather,* **22**.

Trewartha, G.W. (1968) *An Introduction to Climate,* McGraw Hill Kogakushu Ltd., Tokyo.

## Chapter 6

## WEATHER DISTURBANCES

**Phenomena**

Temperature and moisture conditions in the atmosphere occasionally give rise to specific weather conditions associated with disturbances. These phenomena may be very much localised or may be extensive over wide areas. A typical weather system may develop. These are: 1. Thunderstorms, 2. Tropical Cyclone, 3. Extra Tropical Cyclone or Mid-Latitude Cyclone and 4.Western disturbances.

The atmospheric disturbances may occur for several reasons. According to their genesis they are classed into different types of disturbances. Different types of disturbances include:

1. Thunderstorms and tornadoes, which mostly develop over the land surface as turbulent convectional storms.
2. Tropical storms, cyclones, hurricanes or typhoons, which usually originate over the tropical seas and invade the coastal areas causing wide spread damage over the land surface.
3. Mid-latitude or extra-tropical cyclones which result due to convergence of two contrasting air masses over the mid-latitude land surface around a centre of low. Normally the mid-latitude cyclones are less severe than the tropical cyclones.

**Thunderstorms**

A thunderstorm is characterised by lightning and thunder and results due to intense convectional instability. Fundamentally the thunderstorm is a thermo-dynamic machine in which the potential energy of latent heat of condensation and fusion in moist conditionally and convectively unstable air is rapidly converted into kinetic energy of violent rising air associated with torrential rain, hail, gusty surface squall winds, lightning, thunder etc.

*Characteristics*

The thunderstorms are identified by chimneys of updrafts and downdrafts of air over a heated land surface. When the land surface becomes heated very fast, as in the summer months and noon hours of the day, air tends to rise vertically. Where the air is considerably moist or nearly saturated, the rising air will reach the dew points very shortly and at a lower height. After condensation the released latent heat will also act as a trigger for further rise of the air. The rise will continue until it reaches the freezing layer and afterwards the saturated air will move downwards with associated torrential rain, gusty squall winds, hails, lightning and thunder. Actually the rising air is related to the steepening of lapse rate under moist and convectible conditions. So the thunderstorms are caused by the processes which steepen the lapse rate. The intensity of thunderstorm depends on the supply of latent energy and the rate at which this energy is expended. A thunderstorm is so considered an intense instability outbreak.

### Factors for Development

The favourable conditions for the development of thunderstorms are:

I.    The air must be appreciably warm, moist and unstable. The relative humidity of air must exceed 75 percent. The supply of moisture should be large and regular to provide energy in maintaining the storm. The lapse rate must be steepened and should be as high as 14°F. The degree of instability determines the intensity of the storm—larger the instability stronger the storm, lower the instability weaker the storm.

II.   The thickness of the cloud should be high lying within the cloud base and icing level. This thickness should normally exceed 3000 m. The thunderstorms are associated with the development of a particular cloud namely, the cumulonimbus cloud. This anvil or cauliflower shaped cloud is referred as thunder cloud. "The initial cumulus cloud is built up by the ascent of successive 'bubbles' of warm air rising either directly from the heated ground, especially from slopes of favourable aspect, or, more usually, from the top of a layer in convective equilibrium, which has been developed and deepened by surface turbulence" (Crowe, 1997). It is of paramount importance for the unstable air to reach the icing level, as ice crystals are necessary for sudden release of abundant precipitation. Such release of abundant precipitation produces the electrical phenomena in the cloud and violent convective currents accompanying the thunderstorms. It can be said—higher the freezing level above the cloud base, i.e. thicker the cloud, the more intense be the convective activity.

III.  An agent is favoured to trigger the activity by rendering the air unstable at some level.

IV.   A vertical ascent @ 750 metre/minute is desirable. Higher the temperature, higher the specific humidity, higher will be the potential for the thunderstorm.

### Life Cycle of a Thunderstorm

The thunderstorms develop in stages. In the initial stage the cumulus cloud is formed. With the heating of the ground surface the updrafts strengthen and the successive bubbles of air rise as convective cells like chimneys. In the mature stage the mixture of supercooled droplets and ice crystals produce raindrops in the central portion of the cloud and rain or hail pours down below the base of the cloud. Thus an active downdraft is set off within and below the cloud. The drop of downdraft often reaches a speed of 4–8 metre per second towards the cloud base and two important developments occur. Firstly in the lower layer of the cloud, in particular horizontal contrasts of temperature may be 4–5° C and the updraft is accelerated. The cloud top will further rise reaching another 3000 m. height. Secondly at ground level the downdrafts spread outward forming a micro cold front. The toe of such downdraft advances as violent gusts often against surface wind and renews supplies of warm air aloft in the updraft chimneys like a shovel. The process thus becomes self sustaining and is likely to continue depending upon the supply of unstable air. The dissipating stage is reached when the updraft ceases and there is no more supply of unstable warm air to sustain the thunderstorm. For sometimes the downdraft may continue, but weakens and ultimately ceases.

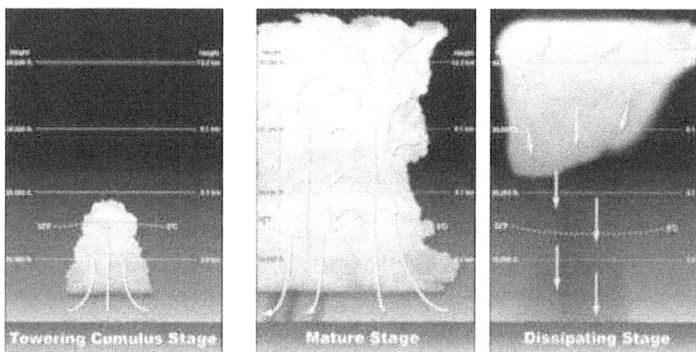

**Figure 6.1:** Airflow Diagrams Showing Three Stages of a Thunderstorm Life Cycle

All thunderstorms, regardless of type, pass through three stages: the *cumulus stage*, the *mature stage*, and the *dissipation stage*. Depending on the conditions present in the atmosphere, these three stages can take about 20 minutes to several hours to occur.

*Cumulus Stage*

The first stage of a thunderstorm is the cumulus stage, or developing stage. In this stage, masses of moisture are lifted upwards into the atmosphere. The trigger action for this lift appears to be intense heating of the ground surface creating a low, where two winds converge forcing air upwards, or where winds blow over terrain of high altitude. The water vapour rapidly cools into water molecules, and the *cumulus* clouds appear. As the water vapour condenses into liquid, latent heat is released warming the air, causing it to become less dense than the surrounding dry air. The air tends to rise in an *updraft* through the process of convection.

*Mature Stage*

In the mature stage of a thunderstorm, the warm air continues to rise until it reaches existing air which is warmer, and the air can rise no further. This level is the tropopause, which acts as 'cap'. The air can not rise further upwards and is forced to spread out, giving the storm cloud a characteristic anvil shape. So *cumulonimbus* cloud is formed. The water droplets coalesce into heavy droplets and freeze to become ice crystals. The ice crystals during fall melt to become rain. If the updraft is strong enough, the droplets are held aloft long enough to be so large that they do not melt completely and fall as hail. While updrafts are still present, the falling rain creates *downdrafts* as well. The simultaneous presence of both an updraft and downdraft marks the mature stage of thunderstorm. During the mature stage considerable internal turbulence may occur, which sometimes form squalls of strong winds, severe lightning, and even tornadoes.

**Figure 6.2:** Anvil Shaped Thundercloud in the Mature Stage

Under a condition of little wind shear, the storm will rapidly enter the dissipating stage. But if there is sufficient change in wind speed and/or direction the downdraft will be separated from the updraft, and the storm may turn to be a super-cell, and the mature stage can sustain itself for several hours.

In certain cases however, even with little wind shear, but with enough atmospheric support and instability the storm may even maintain its mature stage a bit longer.

*Dissipating Stage*

In the dissipating stage, the thunderstorm is dominated by the downdrafts only. If there be no super cellular development, this stage occurs rather quickly, some 20-30 minutes within the life of a thunderstorm. The downdraft will push down out of the thunderstorm, hit the ground and spread out. The cool air carried to the ground by the downdraft cuts off the energy inflow to the thunderstorm, the updraft ceases and the thunderstorm will dissipate finally.

**Classification of Thunderstorms**

There are four main types of thunderstorms: single cell, multi-cell, squall line (also called multi-cell line) and super-cell.

*Single Cell*

This refers to a single thunderstorm with one main updraft. Within a cluster of thunderstorms, the term "cell" refers to each separate principal updraft. Thunderstorm cells can and do form in isolation to other cells. Such storms are rarely severe and caused due to local atmospheric instability. These are the typical summer thunderstorms found in many temperate locations. They occur in the cool unstable air which often follows the passage of cool air from the sea during winter months.

*Multi-cell Cluster*

**Figure 6.3:** A Single Cell Thunderstorm

Multi cell storms form as clusters of storms but may then evolve into an organised line or lines of storms. They often arise from convective updrafts in or near mountain ranges and linear weather boundaries, usually strong cold fronts or troughs of low pressure.

*Multi-cell Lines*

Multi-cell line storms are commonly known as *squall lines.* These occur when multi cellular storms form in a line rather than clusters. They may be hundreds of kilometres long, move swiftly, and be preceded by a gust front. Heavy rain, hail, lightning, very strong winds and even isolated tornadoes can occur over a large area in a squall line. An unusually powerful type of squall line, called a *derecho,* occurs when an intense squall line travels for several hundred kilometres, causing widespread damage over thousands of square kilometres.

Occasionally, squall lines also form near the outer rain band of tropical cyclones. The squall line is propelled by its own outflow, which reinforces continuous development of updrafts along the leading edge.

*Super-cell*

Super-cell storms are large, severe quasi-steady-state storms with characteristic wind speed and direction varying with height ("wind shear"), separate downdrafts and updrafts (i.e., precipitation is not falling through the updraft) and a strong, rotating updraft (a meso-cyclone). These storms normally have such powerful updrafts that the top of the cloud (or anvil) can break through the tropopause and reach the lower levels of the stratosphere and can be 24 km wide. Destructive tornadoes, extremely large hailstones (4 inch or 10 cm diameter), straight-line winds in excess of 130 km/h, and flash floods may be caused. In fact, most tornadoes occur from this kind of thunderstorm.

*Locations of Thunderstorms*

Thunderstorms occur throughout the world, even in the polar region, with the greatest frequency in tropical rainforest climate, where they may occur almost daily. Thunderstorms are also associated with the monsoon seasons around the globe, and they are found in the rain-bands of all tropical cyclones. In temperate regions, they are most frequent in spring and summer, although they can occur along or ahead of cold fronts at any time of year. They may also occur within a cooler air mass following the passage of a cold front over a relatively warmer body of water. Thunderstorms are almost rare in polar region due to cold surface temperatures.

*Lightning*

*Lightning* is an electrical discharge always associated with a thunderstorm. It can be seen in the form of a bright streak (or bolt) from the sky. Lightning occurs when an electrical charge is built up within a cloud. When a large enough charge is built up, a large discharge will occur and can be seen as lightning. The temperature of a lightning bolt can be hotter than the surface of the sun. Although the lightning is extremely hot, the short duration makes it not necessarily fatal.

A                                                                B

**Figure 6.4:** Cloud to Ground Lightning (A) and Fork Lightning (B)

There are several types of lightning:

- In-cloud lightning is the most common. It is lightning within a cloud and is sometimes called intra-cloud or sheet lightning.
- Cloud to ground lightning happens when a bolt of lightning from a cloud strikes the ground. This is known as thunder and poses the greatest threat to life and property.
- Ground to cloud lightning occurs when a lightning bolt is induced from the ground to the cloud.

- Cloud to cloud lightning is rarely seen and is formed when a bolt of lightning arches from one cloud to another.
- Ball lightning is extremely rare and has no known scientific explanation. It is seen in the form of a 20 to 200 centimetre ball.
- Cloud to air lightning results when lightning from a cloud hits air of a different charge.

### *Tornado*

Tornadoes are unpredictable and deadly disaster. The violent wind speed can twist every thing with a cyclonic spin to a degree beyond imagination and almost flouts description. Hence in America the tornado is termed the *twister.* The term tornado has been derived from the Spanish word *tronada,* which means thunderstorm.

Actually the tornado is associated with the development of an intense thunderstorm. As the updraft moves up in a large swells of rising warm air it rotates. If the rotation grows sufficiently intense, it can turn into a tornado or funnel cloud. Associated with the intense development of thunderstorm sometimes the base of the cumulonimbus cloud touches the ground surface as a trunk and tremendous suction results. The funnel cloud spins and moves along the ground for some distance with unusual wind velocity, even sometimes reaching up to 400 km. per hour. Air rushing to fill in the void of low pressure creates additional fierce, potentially destructive winds. However, the extent of this funnel cloud or tornado is very limited—only 200–500 m. in width and 15–25 km. in length.

Most tornadoes form within a specific intense weather system known as *super-cell.* Super-cell thunderstorm develops when the warm updraft drives through an overlying stable layer and still rises upward into a zone of cool, dry air. In this way the resulting instabilities produce powerful vortex motion, the driving engine of the tornado.

The forecasting of the tornado is even today an impossible task and can not minimise the risk of damage with suitable protection measures. Sometimes an updraft gives rise to tornado and sometimes it fails. The time to form a tornado from a thunderstorm is not more than few ten minutes. So it provides little time for operational observation, identification and warning. Many studies have been made, including the VORTEX during 1994–95, but no key to issue warning for the approaching tornado has been identified. However, *The Optical Transient Detector*, an experimental NASA detector recently placed into earth's orbit, indicates that longer advance warning may be issued. H.J. Christian used the satellite to tally the rate of lightning flashes in large storms. It was found that the flash rate often reached the peak shortly before the tornadoes appear. The Optical Transient Detector also observed far more flashes than were observed from the ground, which signifies that the storms that develop into tornados, create cloud to cloud strike.

Tornados, though rare on a global scale, are found to occur in border areas of Orissa and West Bengal and in Bangladesh in hot summer season (March–May). A severe tornado hit

Bangladesh on May 13, 1999 killing at least 500 and injuring more than 50,000 people. In May, 1998 tornado affected the Balsore district of Orissa and Midnapore district of West Bengal causing widespread damage including loss of human life. Tornados are also frequent in the central part of U.S.A., particularly from Kansas to Indiana states. On average nearly 150 tornadoes develop in a year. These storms usually develop during the months of March, April and May between 14–00 hours to 22–00 hours. The following table will indicate the occurrences of tornados in U.S.A. in the months of March, April and May in 1996.

**Table 6.1:** Deadly Tornado Statistics, U.S.A. (March–May, 1996)

| Date | Time | Location | Deaths |
|------|------|----------|--------|
| March 6 | 04-01 | W Selma, Alabama | 4 |
| March 6 | 05-15 | Montgomery, Alabama | 2 |
| April 14 | 18-36 | Allison, Arkansas | 5 |
| April 14 | 18-42 | Sylamore, Arkansas | 2 |
| April 19 | 19-00 | NR Ogden, Illinois | 1 |
| April 20 | 07-30 | Carrollton, Massachussets | 1 |
| April 21 | 22-20 | VanBuren, Arkansas | 2 |
| April 21 | 23-15 | St. Paul, Arkansas | 2 |
| May 12 | 20-12 | Okeechobee, Florida | 1 |

*Source:* National Oceanic and Atmospheric Administration, U.S.A.

**Tropical Cyclones**

A *tropical cyclone* is a storm characterised by a centre of low pressure and numerous thunderstorms that generate strong winds and heavy rainfall causing widespread floods. A tropical cyclone feeds on the heat released when moist air rises and the water vapour condenses to release the latent heat. Released latent heat makes the air more buoyant to provide more energy for violent storm. They are fueled by a different heat mechanism than other cyclonic windstorms and called the *warm core* storm systems.

The term cyclone was given to a rotating storm in the tropics by Piddington. The name cyclone has been derived from a Greek word 'cyclo', which means the coil of a snake. These disturbances normally originate over warm tropical seas around a centre of low. This centre of low is usually formed in the months of May to September north of the equator ranging 5–10° N. Here the trade winds from both hemispheres converge at the equatorial low in summer months. This zone is called the *Inter-Tropical Convergence Zone (ITCZ)*. The position of ITCZ shifts and is related to the declination of the sun.

When the ITCZ becomes active the low pressure area develops. This low pressure area may turn to a depression, deep depression and ultimately to a severe cyclone. Based on their intensity the tropical disturbances are classified as:

1. Tropical disturbance—identified by one or no closed isobar, wind circulation poorly developed and velocity of winds weak.
2. Tropical depression—identified by one or more closed isobars, wind force equal to or less than Beaufort scale 6 (up to about 25 to 30 miles/hour).
3. Tropical storms—identified by closed isobars and a wind force in the range of Beaufort scale 6 to 12 (under 75 miles/hour).
4. Hurricane or Typhoon—identified by intense low pressure, very steep pressure gradient and wind force greater than Beaufort scale 12 (>75 miles/hour). The severe cyclones which develop over the Caribbean Sea are called the Hurricanes and those developing over the Pacific Ocean are called the Typhoons. Their characteristics are identical.

In most situations, Sea Surface Temperatures (SST) of at least 26.5° C (79.7° F) are essentially required to a depth of at least 50 metres (160 ft); waters at this temperature cause the overlying atmosphere to be unstable enough to sustain convection and thunderstorms. Another factor is the steep lapse rate, which favours rapid cooling with increasing height and allows the release of latent heat of condensation that drives a tropical cyclone. High humidity is needed, especially in the lower-to-mid troposphere; when there is a great deal of moisture in the atmosphere, conditions are more favourable for disturbances to develop. Low amounts of wind shear are needed, as high shear is disruptive to the storm's circulation. Lastly, a formative tropical cyclone needs a pre-existing system of disturbed weather, although without a circulation no cyclonic development will take place.

### *Conditions for Development*

Several factors appear to be generally necessary for the development of tropical cyclones. However, the tropical cyclones may occasionally form without meeting all of the following conditions.

I. It is observed that the cyclones are normally not formed over the equator as the Coriolis parameter (f) is zero over there. Hence the Coriolis force is essentially required for a spin of the wind circulation. Tropical cyclones generally need to form more than 5 degrees of latitude away from the equator, allowing the Coriolis force to deflect winds blowing towards the low pressure centre and creating a circulation.

II. The Sea Surface Temperature (SST) must be appreciably high, at least 26.5° C up to a depth of 50 metre. If the sea surface temperature drops below this critical temperature no cyclone could develop. Waters at this temperature cause the overlying atmosphere to be unstable enough to sustain convection and thunderstorms. Such favourable temperature conditions can be found over the tropical seas— particularly in Bay of Bengal, Arabian Sea, Pacific Ocean close to Philippines,

Caribbean sea in North Atlantic Ocean and Pacific Ocean around New Zealand in southern hemisphere.

III.   Another factor is the steep lapse rate, which favours rapid cooling with increasing height and allows the release of latent heat of condensation that drives a tropical cyclone. High humidity is needed, especially in the lower-to-mid troposphere as high humidity status favours the development of disturbances.

IV.   The Barometric Pressure (BP) must be very low—at the centre of the depression the BP may be ranging between 850–900 mb. The intensity of the storm depends on the barometric pressure at the centre of low—lower the barometric pressure higher the intensity of storm.

V.   Where the barometric pressure drops critically no cloud is formed in the centre. This is called the *eye* of the storm. Just outside the 'eye' of the storm wind velocity rises steeply, clouds form and rainfall occurs. The diameter of the cloud free zone or 'eye' of the cyclone extends from 20 to 30 km.

VI.   Intense tropical cyclone displays the formation of cumulonimbus cloud, where the thickness of the cloud exceeds 12000 m. Radar studies also show that the convection cells are usually organised in bands which spiral inward towards the centre.

VII.   Low amounts of wind shear are needed, as high shear is disruptive to the storm's circulation.

VIII.   Lastly, a formative tropical cyclone needs a pre-existing system of disturbed weather, although without a circulation no cyclonic development will take place.

### Structure of a Tropical Cyclone

All tropical cyclones occur in areas of low atmospheric pressure (850–900 mb.) near the earth's surface. The pressures recorded at the centres of tropical cyclones appear to be the lowest on earth's surface at sea level. Tropical cyclones are characterised and driven by the release of large amounts of latent heat of condensation, which occurs when moist air is carried upwards and its water vapour condenses. This heat is distributed vertically around the centre of the storm. Thus, at any given altitude (except close to the surface, where water temperature dictates air temperature) the condition inside the cyclone is warmer than its outer surroundings.

### Banding

Rain-bands are usually formed in the tropical cyclones. Theses are the bands of showers and thunderstorms that spiral cyclonically toward the storm centre. High wind gusts and heavy downpours often results in individual rain-bands, with relatively calm weather between bands. Tornados often form in the rain-bands of land-falling tropical cyclones. Intense annular tropical cyclones, however, are distinctive for their lack of rain-bands. Instead they form a

**Table 6.2:** Classifications of Tropical Cyclones (all winds are of 10 minutes averages)

| Beaufort Scale | 10 min. sustained Winds | N Indian Ocean (IMD) | SW Indian Ocean (IMD) | Australia (BOM) | SW Pacific (FMS) | NW Pacific (JMA) | NW Pacific (JTWC) | N.E. Pacific & N. Atlantic (NHC and CPHC) |
|---|---|---|---|---|---|---|---|---|
| 0-6 | <28 | Tropical Depression | Tropical Disturbance | Tropical Low | Tropical Depression | Tropical Depression | Tropical Depression | Tropical Depression |
| 7 | 28–29 30–33 | Deep Depression | Depression | | | | | |
| 8-9 | 34–47 | Cyclonic storm | Moderate Tropical storm | Tropical Cyclone | | Tropical storm | Tropical storm | Tropical storm |
| 10 | 48–55 | Severe Cyclonic storm | Severe Tropical storm | Tropical Cyclone | | Severe Tropical storm | | Hurricane |
| 11 | 56–63 64–72 73–85 86–89 90–99 | Severe Cyclonic storm Very severe Cyclonic storm | Tropical Cyclone Intense Tropical Cyclone | Severe Tropical Cyclone | Tropical Cyclone | Typhoon | Typhoon | Hurricane Major Hurricane |
| 12 | 100–106 107–114 115–119 >120 | Very severe Cyclonic storm Super Cyclonic storm | Very Intense Tropical Cyclone | Severe Tropical Cyclone | | Typhoon | Super Typhoon | Major Hurricane |

thick circular area of disturbed weather conditions encircling the centre of low. While all surface low pressure areas require divergence aloft to continue deepening, the divergence over tropical cyclones is directed away from the centre. The upper levels indicate winds directed away from the centre of the storm with an anticyclonic rotation, due to the effect of Coriolis force. So anticyclone remains to be a component along with the cyclonic circulation in tropical cyclone. Winds at the surface blow strongly cyclonic, weaken with height, and eventually reversed. Tropical cyclones possess this unique characteristic to offset a relative lack of vertical wind shear maintaining the warm core at the centre of the storm.

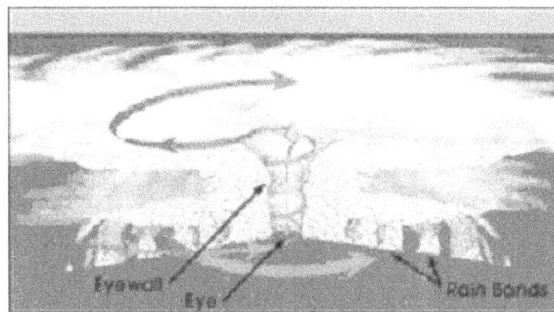

**Figure 6.5:** Eye of a Tropical Cyclone

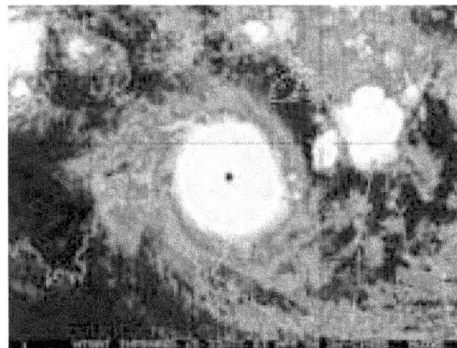

**Figure 6.6:** Infrared Image of Cyclone Monica near Peak Intensity, Showing Clockwise Rotation due to the Coriolis Effect

*Eye and Inner Core*

A strong tropical cyclone has the anchorage in an area of sinking air at the centre of circulation. If this area is strong enough, it can develop into an eye. Weather conditions in the eye remain normally calm and free of clouds, although the sea conditions may be extremely rough. The eye is normally circular in shape, and may range in size from 3 kilometres to 370 kilometres in diameter. An inward curving of the eye-wall's top, resembling a football stadium is observed in the intense and mature tropical cyclones. This phenomenon is thus sometimes known as the *stadium effect*. There are other features that either surround the eye, or cover it. In strong thunderstorm activity the Central Dense Overcast (CDO) zone is found near the centre of a tropical cyclone. In weaker tropical cyclones, the CDO may cover the centre completely. The eye-wall is a circle of strong thunderstorms that surrounds the eye; where the greatest wind speeds are observed, where clouds reach the highest, and precipitation is the heaviest. The heaviest wind damage occurs where the eye-wall passes over land.

Tropical cyclones are distinguished from other meteorological phenomena by deep convection which acts as a driving force. Convection being the strongest in a tropical climate, it remains the initial cause of the tropical cyclone. To continue to drive its heat engine, a tropical cyclone must remain over warm water, which provides the needed atmospheric moisture to maintain the positive feedback loop running. When a tropical cyclone passes over land, it is cut off from its heat source and its strength diminishes rapidly.

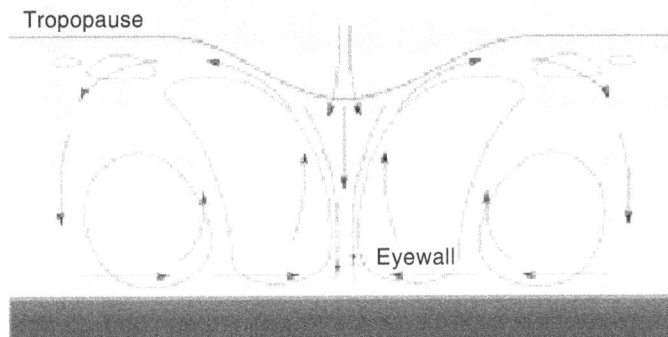

**Figure 6.7:** Tropical Cyclone Formation over Warm Ocean Waters (As the Energy Released by the Condensation of Moisture in Rising Air Causes a Positive Feedback Loop over Warm Ocean Waters)

### Energy

Scientists at the US National Centre for Atmospheric Research estimate that 'a tropical cyclone releases heat energy at the rate of 50 to 200 exajoules ($10^{18}$ J) per day, equivalent to about 1 peta watt ($10^{15}$ watt). This rate of energy release is equivalent to 70 times the world energy consumption of humans and 200 times the world-wide electrical generating capacity, or to exploding a 10-megaton nuclear bomb every 20 minutes.

### Effect of Tropical Cyclones

While tropical cyclones can produce extremely powerful winds and torrential rain, they are also able to produce high waves and damaging storm surges. They develop over large bodies of warm water, and lose their strength if they move over land. This is the reason that coastal regions can receive significant damage from a tropical cyclone, while inland regions are relatively safe from receiving strong winds.

Heavy downpours result in significant flooding inland, and storm surges cause extensive coastal flooding up to 40 kilometres from the coastline. The storm surge or the rise in sea level temporarily due to the cyclone, is typically the worst effect from land-falling tropical cyclones. It is recorded that 90 percent of tropical cyclone deaths result due to storm surge. Over the past two centuries, tropical cyclones have been responsible for the deaths of about 1.9 million persons worldwide. Large areas of standing water caused by flooding lead to infection, as well as contributing to mosquito-borne illnesses and other diseases.

However, tropical cyclones can also relieve drought conditions. Although cyclones take an enormous toll in lives and personal property, they may be important factors in the precipitation regimes of places they hit, bringing much-needed precipitation to otherwise dry regions. They also carry heat and energy away from the tropics and transport it towards temperate lands, playing an important role for the global atmospheric circulation. Tropical cyclones also help maintain the global heat balance by moving warm, moist tropical air to the mid-latitudes and polar regions.

## Fujiwhara Effect

On rare occasion two cyclones approach one another. The centres of the storms start orbiting cyclonically about a point between the two systems. The two vortices will be attracted to each other, and eventually spiral into the central point and merge. If the two vortices are of unequal size, the larger vortex will tend to dominate, and the smaller vortex will orbit around it. Dr. Sakuhei Fujiwhara observed this phenomenon, so it is called the *Fujiwhara effect*.

## Source Regions

Most tropical cyclones are associated with the *Inter-Tropical Front* (ITF), the *Inter-Tropical Convergence Zone* (ITCZ), or the *Monsoon Trough*. These are commonly formed over the Indian Ocean. These form a worldwide band of thunderstorm activity. The other principal source of atmospheric instability is found in *Tropical Waves*, which generate about 85 percent of intense tropical cyclones in the Atlantic Ocean, and become the source of most of the tropical cyclones in the Eastern Pacific Ocean.

## Periods

Tropical cyclones are mostly found in late summer, when the contrast of temperatures aloft and sea surface is the maximum. However, each particular basin has its own seasonal patterns. On a worldwide scale, May is the least active month, while September is the most active. In Bay of Bengal and Arabian Sea the most severe cyclonic storms are usually observed in September, October and early November. In North Atlantic Ocean, hurricane season is marked from June 1 to November 30, sharply rising from late August through September. The Northeast Pacific Ocean has a broader period of occurrence, but nearly similar to the Atlantic. The Northwest Pacific experiences tropical cyclones throughout the year having a minimum in February and March and a peak in early September. In the North Indian basin, storms are most common from April to December, with peaks in May and November.

In southern hemisphere the tropical cyclone season starts in late October and ends in May. Southern hemisphere activity is more intense in mid-February to early March.

**Table 6.3:** Tropical Cyclones—Season Lengths and Seasonal Averages

| Basin | Season start | Season end | Tropical Storms (> 34 knots) | Tropical Cyclones (> 63 Knots) | Category 3+ Tropical Cyclones (> 95 Knots) |
|---|---|---|---|---|---|
| Northwest Pacific | April | January | 26.7 | 16.9 | 8.5 |
| South Indian | October | May | 20.6 | 10.3 | 4.3 |
| Northeast Pacific | May | November | 16.3 | 9.0 | 4.1 |
| North Atlantic | June | November | 10.6 | 5.9 | 2.0 |
| Australia Southwest Pacific | October | May | 10.6 | 4.8 | 1.9 |
| North Indian | April | December | 5.4 | 2.2 | 0.4 |

### *Cyclones over the Bay of Bengal and Arabian Sea*

Each year cyclones and depressions form over the Bay of Bengal and Arabian Sea and affect various parts of India, Bangladesh and Myanmar. Many of these storms turn violent and cause great havoc, damage properties and life. The following table will show the frequency of cyclonic storms over the Bay of Bengal and Arabian Sea.

It appears that the number of storms over the Bay of Bengal is much greater than over the Arabian Sea. Three distinctive periods in a year can be identified for the occurrences of these storms and depressions in the Bay of Bengal and Arabian Sea. These periods are:

(i)    The pre-monsoon season having months of April and May.
(ii)   The monsoon season having months of June, July, August and September.
(iii)  The post-monsoon season having months of October, November and December.

**Table 6.4:** Tropical Cyclones over the Bay of Bengal and Arabian Sea

| Month (Period: 1891–1960) | Bay of Bengal | Arabian Sea |
|---|---|---|
| January | 4  (1) | 2  (0) |
| February | 1  (1) | 0  (0) |
| March | 4  (2) | 0  (0) |
| April | 18  (7) | 5  (4) |
| May | 28  (18) | 13  (11) |
| June | 34  (4) | 13  (8) |
| July | 38  (7) | 3  (0) |
| August | 25  (1) | 1  (0) |
| September | 27  (8) | 4  (1) |
| October | 53  (19) | 17  (7) |
| November | 56  (23) | 21  (16) |
| December | 26  (9) | 3  (1) |
| Total | 314  (100) | 82  (48) |

The figures within brackets indicate the number of storms which reached severe intensity.

The period of January to March records very few storms—both over the Bay of Bengal and the Arabian Sea.

In the monsoon season the disturbances usually form in the North Bay of Bengal and generally move in a north-westerly direction. The storms developing during this period are less severe, but contribute large rainfall over the coastal districts of Orissa, West Bengal and also inland areas of Bihar plateau and eastern Madhya Pradesh. On the other hand, the disturbances in the pre-monsoon and post-monsoon seasons develop in different regions of the Bay of Bengal and move in various directions. The tracks of many of these disturbances undergo re-curvature. In general, these storms are violent and cause widespread damages over the coastal areas of Andhra Pradesh, Northern Tamilnadu, Orissa, West Bengal and sometimes coastal districts of Bangladesh and Arakan coast of Myanmar.

These tropical cyclones cause damages in three ways:
1. The storm surge with an abnormal rise of sea level resulting in inundation of the coastal areas.
2. High velocity stormy winds that damage buildings and other structures.
3. Heavy rain that gives rise to flooding and water logging of vast area. This results in destruction of standing crops in the fields.

## Hurricanes in the Caribbean Sea and North Atlantic

The hurricanes over the Caribbean Sea and North Atlantic Ocean commonly develop in summer months, particularly during August-September. According to the National Hurricane Centre on average 9 storms develop over the Caribbean Sea and North Atlantic Ocean of which 6 turn to be the hurricanes. These hurricanes every year cause damage to properties and life in the Central American Island states and south eastern littoral areas of United States of America. The hurricane named 'Mistral' devastated large areas of Central America and South eastern coastal U.S.A. in 1998. It has been termed one of the worst environmental hazards in the twentieth century. Similarly the Katrina caused wide spread damages in 2005.

### *Disastrous Tropical Cyclones*

Tropical cyclones are the deadliest natural disasters on our planet. Humans have no control yet, except to forecast and predict the nature of disaster. Though it is possible today to locate the growth and movement of tropical cyclone from the study of satellite imageries, in many developing countries technology is too poor to issue the warning in advance and evacuate the people in time. So we find that the most damages are caused in these poor countries with irreparable loss of human life and properties. But in the developed country like U.S.A. the loss of human life could be minimised with prior warning and appropriate action for mitigation.

The *Bhola Cyclone* (1970) struck the coastal region of Bangladesh on November 13, 1970 killing more than 300,000 people and potentially as many as 1 million people. The Bhola cyclone regenerated after refueling its energy over the Bay of Bengal and then re-curved to hit the Bangladesh coast with the power of a super-cyclone. Its powerful storm surge was responsible for the high death toll. We have mentioned earlier that the Bay of Bengal as a part of North Indian Cyclone Basin has historically been the deadliest basin, with several cyclones since 1900 killing more than 100,000 people, all in Bangladesh and Eastern Coast of India. Incidentally the term cyclone was coined by Mr. Piddington after the tropical cyclone of Bay of Bengal. In 1998 the *Super-cyclone of Orissa* hit the coastal districts of Orissa with a super-speed of 250 km. per hour and caused unprecedented damages. The *Cyclone Nargis* stayed over the North Bay of Bengal for three days (May 1–3, 2008). Initially the storm was expected to hit the Bangaladesh coast. It turned right and struck lower Irrawady Basin on May 4, 2008 with devastating wind force and storm surge. Nearly 100,000 people lost their life. It was one of the deadliest tropical cyclone over Myanmar.

Elsewhere, the *Typhoon Nina* killed nearly 100,000 people in China. Heavy rainfall from the Typhoon Nina caused 62 dams including the Banqiao Dam to collapse and resulted in great disaster. The *Great Hurricane* of 1780 was the deadliest Atlantic hurricane on record, killing about 22,000 people in the Lesser Antilles. Tropical Storm *Thelma* in November 1991 killed thousands in the Philippines causing mudslides due to incessant rain. The *Hurricane Katrina* in U.S.A. in August, 2005 was the costliest tropical cyclone in the world, causing

$ 81.2 billion in property damage with overall estimated damage exceeding $100 billion and killing 1836 persons. It was caused due to inundation and failure of sea dykes. Though the loss of life in Katrina appeared to be low compared to many other devastating cyclone events in the developing countries, it was a shocking event from the hurricane in the American history. The Galveston Hurricane of 1900 was the severest natural disaster in U.S.A., killing an estimated 6,000 to 12,000 people in Galveston, Texas. The Hurricane *Iniki* in 1992 was the most powerful storm to strike Hawaii in recorded history outside the mainland of U.S.A. Other destructive Eastern Pacific hurricanes include *Pauline* and *Kenna*, both causing severe damages after striking Mexico as major hurricanes. Over the Indian Ocean off the coast of Africa the *Cyclone Gafilo* struck northeastern Madagascar as a powerful cyclone in March 2004.

In the Northwestern Pacific Ocean the strongest storm on record was *Typhoon Tip*. In 1979, it reached a minimum pressure of 870 mb with a maximum sustained wind speeds of 165 knots or 190 miles per hour (310 km/h). The *Typhoon Keith* in the Pacific and *Hurricanes Camille* and *Allen* in the North Atlantic also recorded the same. Camille was the strongest tropical cyclone on record at landfall. Similarly, a surface-level gust caused by *Typhoon Paka* on Guam was recorded at 205 knots or 235 miles per hour (378 km/h). Perhaps, it would be the strongest non-tornado wind ever recorded on the earth's surface.

The *Typhoon Tip* was the most extensive cyclone on record, with tropical storm-force winds having a diameter of 2,170 kilometres. The *Cyclone Tracy* on the other hand had the least diameter, roughly 100 kilometres wide before striking Darwin, Australia in 1974.

The longest-lasting tropical cyclone was the *Hurricane John*, lasting 31 days in 1994. Reliable data for Southern Hemisphere cyclones is unavailable.

### Case Study of Cyclone over the Bay of Bengal

*Severe Cyclonic Storm: 17th to 20th May, 1998*
*Place of location: Southern parts of Central Bay & adjoining South Bay*
1.   First observed as a trough of low pressure area over SE Bay on 16th May.
2.   Became well marked over Southern parts of Central Bay and adjoining parts of South Bay on 17th May.
3.   Concentrated into a depression in the same evening near 15.5° N/88.5° E and as a deep depression at 0900 UTC of 18th and intensified into a cyclonic storm in the evening of 18th near 19.5° N/90.5° E.
4.   Further intensified into a severe cyclonic storm near 20.5° N/90.5° E in the morning of 20th.
5.   Weakened into a depression, crossed Bangladesh in the evening of 20th and further weakened into a low pressure area moving in a northerly direction over Assam and Meghalaya.

### Case Study of Typhoon Paka

*Typhoon Paka: Originating over Western Pacific Ocean*

Typhoon Paka was the strongest tropical cyclone ever recorded.
1.  Swept across the Island of Guam on 16.12.97, landed around 7 p.m. local time.
2.  The peak gust wind speeds at 235 mph—the highest wind speed ever recorded over land.
3.  The eye of the storm moved on a west-north westerly track.
4.  Passed through the Rota channel, and over the north end of the island where the Anderson Air Base is located.
5.  The storm was traveling @ 19 mph, while approaching the island of Guam.
6.  After landfall the rate of movement slowed to @ 11 mph.
7.  Battered the island for about 8 hrs. Afterwards moved west toward the Philippines.
8.  Paka sustained wind speeds of 100 mph to 150 mph and traveled slowly across the island.
9.  The eye was located along the Rota channel near the Anderson Air Force Base.
10.  Sea levels were reported to rise up to 35 feet above normal.

*Structures*

1.  The primary industries in Guam are tourism, and support services to US Air and Naval Bases.
2.  Commercial buildings are mostly hotels and supporting facilities including wholesale/retail trade or services.
3.  The bulk of the building stock is of reinforced concrete.
4.  Complete structural collapses were limited to light metal frame, warehouse-type structures or wood-frame construction usually with metal roofs.

*Tourist Hotels*

1.  Guam has many large hotels including a number of high-rises.
2.  Essentially, all of them are located in Tumon Bay just east of Agana, or in Agana near the Dungcas Beach.
3.  None of the hotels suffered any significant structural damage, regardless the types of structure.
4.  Damage was limited to window breakage with subsequent water and wind damage to contents, and to support equipment, such as air-conditioning units and back-up power generators.
5.  The impact was serious as such damages effectively shut down the hotels.

*Lifelines*

1. The primary infrastructure damage was to the island's power distributing system.
2. Guam Power Authority estimated that it could take up about three months for full restoration of service.
3. Many coastal roads were damaged, having been washed out or strewn with debris from storm surge.
4. Water and sewer systems were generally operational, except for the loss of power.
5. Portable generators restored pumping capabilities within a few days to all except the most remote areas.
6. Telephones remained operational as the lines are underground.

*Sea Ports*

1. Guam's commercial and military port facilities were severely affected by Typhoon Paka, which included the Port Authority of Guam wharves and U.S. Navy facilities.
2. Damage at Port Authority of Guam was extensive, but within six days the port was operational, *albeit* in emergency mode.
3. The storm surge swept over the seawall, washing out road pavement and depositing rocks, boulders, sand and other debris along the road to the port and Cabras Island.
4. Wharf construction to repair damage from a major earthquake in 1993 was underway when the Paka struck.

*Specific Damages to Sea Port*

1. Damage was sustained to piles and bulkheads along the F-5 wharf. Of three gantry cranes, only Gantry 1 was able to service container ships two days after the typhoon. Gantries sustained damage to windows and salt water damage to electrical systems.
2. Several smaller ships broke their moorings in the harbour and were washed ashore. These included both commercial and navy vessels.
3. The Catamaran, Micronesian Dream, and four fishing vessels were aground in the commercial port.
4. The typhoon caused two of three navy yard tugs, including the Ketchikan and support barges, to separate from moorings and run aground. The floating dry dock AFDM-5 was damaged.

*Damages to Civil Airport*

1. The A.B. Won Pat Guam International Airport was closed for several days due to damages by the typhoon.

2. Runway landing lights, one of three radar systems, the non-directional navigation aids were damaged.
3. Most airport buildings, including the main terminal and hangers, sustained only light damage in the form of isolated window breakage.
4. Airport operations were partly restored in a day, with daylight, Visual Flight Rule (VFR) landings.
5. Emergency electrical power for the airport was provided by generators.

*Damages at Military Airport*

Damages at Anderson Air Force included:

1. The complete destruction of the troop support and Latte Stone refrigeration units.
2. Nearly all bay doors on facilities ripped open or damaged.
3. Hundreds of trees uprooted and snapped.
4. Road signs destroyed.
5. Ceilings ripped open and doors blown off.
6. Cars were flipped over and damaged by flying debris.
7. Aircraft hangars at the base were damaged, including loss of the main doors.
8. Additional damage to stored equipment occurred.

*Extent of Damages by Paka*

1. Approximately 1,500 structures (mostly residential buildings) were completely destroyed.
2. Another 10,000 structures were damaged.
3. Nearly 5,000 people became homeless.
4. Damage cost is estimated at about $ 400 million.
5. Several large petroleum storage tanks were buckled, and one completely destroyed.
6. Standing water remained at three feet and wave action in the loading yards caused shorting and damage to electrical equipment.

## Extra-Tropical Cyclones

In the mid-latitude region close to the sub-polar low region two contrasting air masses, namely *cold polar* (cP) and *maritime tropical* (mT) air-masses converge. These air-masses build along their margin fronts, which ultimately develop into a mature cyclone.

The cyclone model was illustrated by J. Bjerkness in 1937 and is known as Polar Front theory. In this model at initial stage the cold air-mass of polar origin and warm air-mass of tropical oceans form a line of separation between the two along a low pressure centre. The temperature contrasts result in the development of wind shear and also wind shift. As a consequence there will be rising light warm air, pushed back as wedge from the rear by cold heavy air. Hence two fronts are formed—cold front and warm front—having distinctive characteristics. The cold front is more active and faster than the warm front in movement. The warm front is ultimately overwhelmed by the cold front, when there will be formation of occluded front. The occluded front will indicate the decay of the cyclone.

### *Characteristics*

1. Two fronts are formed—warm front and cold front.
2. The gradient along the cold front is steeper over the cold front (slope—1 : 50 to 1 : 100) than over the warm front (slope—1 : 100 to 1 : 400).
3. The mid-latitude cyclones move from west to east. The formation of warm front to the east is the first indication of approaching cyclone. At first warm condition prevails which is replaced very fast by cold wave condition with marked drop of temperature.
4. Along the cold front the rainfall is heavy, but of short duration. Whereas the rainfall is in the form drizzle along the warm front, but continues for longer time.
5. Unlike tropical cyclone barometric pressure difference is not very marked in the mid-latitude cyclone. So no 'eye' is formed at the centre. But temperature contrast is very marked in this cyclone.
6. Veering and backing wind shifts are observed in different quadrants of the cyclonic zone. At the initial phase warm air blows from south-east and cold air from north-east. The wind direction later changes, warm air moves from south and cold air from north. Lastly the cold air moves from north-west and warm air from south-west.
7. The mid-latitude cyclones mostly develop during the winter months and are stronger. Whereas the cyclones developing in summer months are fewer and weaker.

### *Cyclone Model*

We have discussed in the previous chapter on the formation of fronts. The Norwegian Cyclone Model is an idealised formation model of cold-core cyclonic storms developed by Norwegian meteorologists (Bjerkness, Solberg and others) during the First World War. The principal concept of this model states that cyclones progress through a predictable evolution as they move up a frontal boundary, with the most mature cyclone near the northeast end of the front and the least mature near the tail end of the front.

### *Stages in the Development of Extra-tropical Cyclone*

The extra-tropical cyclone has a life-history. A pre-existing frontal boundary is pre-condition for the development of a mid-latitude cyclone. The cyclonic flow begins around a disturbed section of the stationary front due to an upper level disturbance, such as a short wave or an upper-level trough, near a favourable quadrant of the upper level jet.

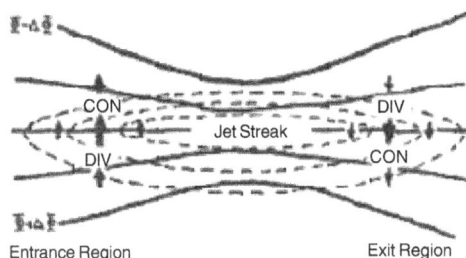

**Figure 6.8:** An Upper Level Jet Streak. DIV Areas are Regions of Divergence Aloft, which will Lead to Surface Convergence and help Cyclogenesis.

## Developing Stage

Initially a low pressure area is formed near the boundary between warm and cold air masses. The cyclone grows in size and intensity as it draws on the energy that is available from the temperature contrast between the two air masses. Initially there is a line of separation between two contrasting air masses, which progressively grows into wave-like form with growing amplitude. As some of the warm air rises up over the cold air mass, and some of the cold air sinks and flows under the warm air mass, the cyclone deepens (the air pressure becomes lower) and the winds around the system increase. The leading edges of these air masses are called fronts. The surface low could be formed for various reasons. Meso-scale convective systems help intensification of surface lows which are initially warm core. The disturbance can grow into a wave-like formation along the front and the centre of low will be located at the crest. Around the centre of low, cyclonic flow will begin. This rotational flow will thrust the polar air equator-ward west of the low along its trailing cold front, and warmer air with push pole-ward low along the warm front. Usually the cold front will move at a faster pace than the warm front and hold it due to the slow decay of higher density air-mass located out ahead of the cyclone and the higher density air-mass converging behind the cyclone. In many cases it results in narrowing the warm sector. At this point an occluded front is formed. The warm air mass is pushed upwards into a trough of warm air aloft.

## Mature Stage

The waves become more intensified and at the centre of low where two fronts meet, warm front forming a wedge over the cold front and cold front pushing under the warm front. Maturity is reached when strengthening of the storm is completed and the cyclonic flow is intensified to its maximum. Thereafter, the strength of the storm diminishes as the cyclone mingles with the upper level trough or upper level low and is transformed increasingly into a cold core.

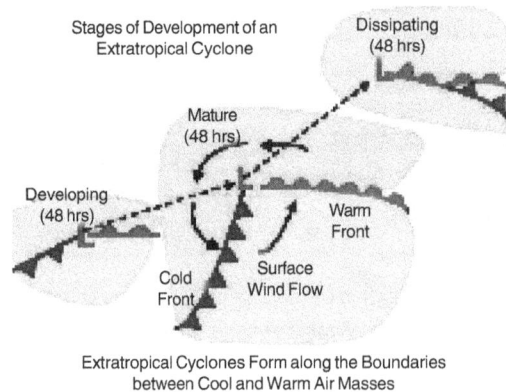

**Figure 6.9:** Stages in the Development of Extra Tropical Cyclone

*Dissipating Stage*

As occlusion starts and with the warm air mass lifted upwards over a cold air mass, the atmosphere becomes increasingly stable and the centre of gravity of the system lowers. As the occlusion process extends further down the warm front and away from the centre of low, more and more of the available potential energy of the system is exhausted. The line of separation between two air masses vanishes and cold air mass swaps the warm air mass finally. The sink of potential energy generates a kinetic energy source which injects a final burst of energy into the storm's motions. After this process develops, the growth of the cyclone or cyclogenesis comes to an end. At this stage the low begins to spin down (fill) as more air is converging into the bottom of the cyclone than is being removed out from the top since upper-level divergence has decreased.

Occasionally, cyclogenesis will re-occur with occluded cyclones. When this happens a new centre of low will form on the triple-point (the point where the cold front, warm front, and occluded front meet). During triple-point cyclogenesis, the occluded parent low will fill as the secondary low deepens into the main weather-maker.

**Difference of the Extra-tropical Cyclones with the Tropical Cyclones**

The Extra-tropical cyclones in many ways differ from the Tropical cyclones. The extra-tropical cyclones originate over the land in the mid-latitude region, whereas the tropical cyclones over the seas in the tropical region. We can distinguish them according to their condition of formations and characteristics.

1. Temperatures within the air masses of the extra-tropical cyclone vary. Temperature structure in the tropical cyclone remains almost the same.
2. The wind speed varies. The wind speed may exceed 200 km. per hour in most cases for the tropical cyclone. Wind speed of the extra-tropical cyclone remains moderate, rarely exceeding 60 km. per hour.
3. For both the extra-tropical and tropical cyclones a centre of low is essentially formed. But the barometric pressure in case of tropical cyclone falls extremely low at the centre of low, in many cases reaching 850 mb. B.P. increases with the distance from the centre of low.
4. Moisture content also varies in two types of cyclones. Moisture content shall be very high with steeper lapse rate for the tropical cyclones.
5. Cloud conditions vary in two cyclones. In tropical cyclones where the *eye* is formed cloud-free zone exists. But around the eye cumulonimbus or thunder clouds are formed. Such clouds are rarely formed in the extra-tropical cyclones.
6. Rainfall pattern varies. Rainfall is very heavy and intense in the tropical cyclones causing floods and inundations in the coastal areas. Whereas in the extra-tropical cyclones rainfall pattern varies between two fronts—torrential shower of short spell along the cold front and continuous light rain and drizzle along the warm front.
7. Tropical cyclones are stronger and cause disaster in many cases. Extra-tropical cyclones rarely cause damages with exception of extreme snowfall along the passage of very cold wave at the rear.
8. Extra-tropical cyclones occur mostly in the winter months. Tropical cyclones are mostly found during the summer months.
9. The occurrences of the extra-tropical cyclones exceed the tropical cyclones.
10. Extra-tropical cyclones usually move from west to east. Occurrence of an extra-tropical cyclone in the west becomes the harbinger of unstable weather and formation of cyclone to the east next. Extra-tropical cyclones can travel a long distance regenerating under favourable conditions further east. Even the extra-tropical cyclones of the Mediterranean region are related to *western disturbances* over India. Tropical cyclones usually move east to west or north-west and decay during their passage inland.

## Western Disturbances

*Western Disturbances* (WD) are considered to be the eastward-moving extra-tropical upper air trough in the subtropical westerlies, often extending down to the lower atmospheric level. The western disturbances are typical weather phenomena, observed in north and northwestern of India in winter and late winter months. Weather events in the northwestern and northern parts of India are influenced by the low-pressure systems originating in the Mediterranean Sea or the adjacent Atlantic Ocean, which travel through Iran and western Pakistan to make their way to India.

These disturbances develop over the Atlantic Ocean and adjoining Mediterranean Sea close to Spain and move further east through Asia Minor. These extra-tropical systems travel in the form of waves in the middle and upper tropospheric westerlies with an average speed of 10 degree longitudes per day. Sometimes, these are observed as closed cyclonic circulations at the sea-level. Analysis of synoptic charts shows that a WD originates usually over the Mediterranean Sea/Black Sea area as an extra-tropical frontal system, but its frontal characteristics are lost while moving eastward towards India across Afghanistan/ Pakistan. In the late winter months (January-February) and pre-monsoon period (March–April) these circulations prevail over large tracts in Punjab, west Uttar Pradesh, Jammu and Kashmir.

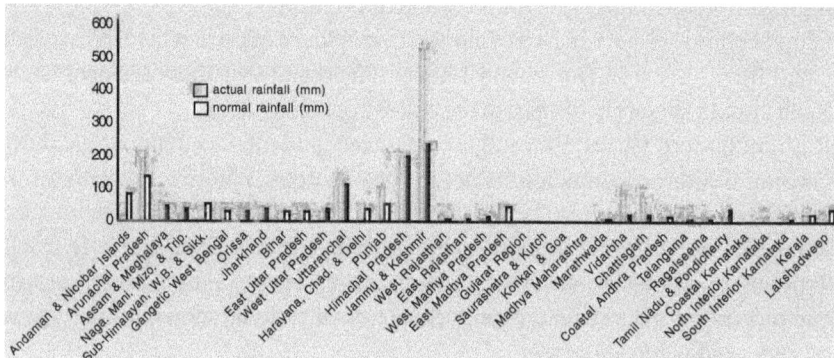

**Figure 6.10:** Rainfall in Winter Season Indicating the Impact of Western Disturbances

These depressions become weaker over the land surface, but when enter north and north-west India turn active. These depressions contribute winter rainfall in Jammu and Kashmir, Punjab, Haryana, Uttar Pradesh and even in north Bihar, West Bengal, Assam and Arunachal Pradesh. In western disturbances we observe the prevalence of occluded front. The cold and warm fronts are weaker in western disturbances. Drizzles or light showers are associated with the warm front and afterwards the weather becomes clear for a short period. Thereafter heavy showers along with thunder rain follow along the cold front. Afterwards the temperature drops markedly and the region comes under the influence of cold wave from rear. Associated with the invading western disturbances ground fogs develop for a longer duration in the morning hours and cause very poor visibility. However the winter rainfall and snowfall (in the hills) contributed by winter disturbances are beneficial for wheat farming in north and north-west India.

Occurrences of Western Disturbances in India have inverse relationship with the formation of depressions in Bay of Bengal in the preceding season. If the frequencies of formation and movement of depression over the Bay of Bengal in a particular year increase, then the frequencies of Western Disturbances in the same year decrease. The secondaries of extra-tropical depressions move northeastward from the eastern Mediterranean and are confined in the latitudinal belt 25° N to 35° N. The frequency of Western Disturbances abruptly decreases from winter to the pre-monsoon season. Over the Indian region, their frequency peaks during winter. About two to three are known to traverse northwest India during a winter month. But their number varies from year to year.

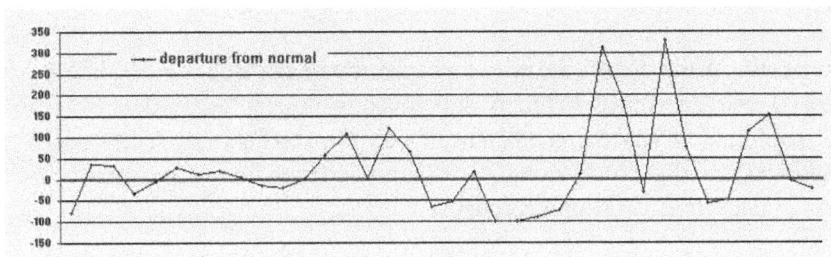

**Figure 6.11:** Departure from Normal Rainfall in Winter Season caused due to Western Disturbances

Even in the hot weather period of April and May, Western Disturbances move across northern India. They quite often interact with monsoonal flows causing widespread rainfall over northwest India during monsoon. During summer, they help moderate the weather in the plains. Heat waves generally occur when Western Disturbances are fewer in number. It is more difficult to predict the occurrences of Western Disturbances than the Monsoon. They travel eastwards in higher latitudes of 30° N to 60° N and barge into the western Himalayas. The Himalayan terrains help extract the moisture and make the belt within 34° N to 36° N receiving maximum rainfall.

### Case Study of a Western Disturbance over Northern India

*Period: 5–8 February, 2002*

*Synoptic Situation*

It entered the Indian sub-continent, a little more to the south than most cases, near latitude 25° N–30° N. This system could be located as a surface low over Southern Iran on 4th February, 2002 with low level cyclonic circulation over North Pakistan and neighbourhood extending up to a height of 4.5 km. An induced low pressure area could be observed on 5th February over central parts of Pakistan and adjoining west Rajasthan with associated low level circulation extending up to 2.1 km. It was seen moving eastward over Jammu & Kashmir and neighbourhood on 6th February whereas the induced low remained stationary. On 7th February both main as well as induced system became more intense with strong surface winds. There was moisture incursion from the Arabian Sea due to southerly/ southwesterly winds and also from Bay of Bengal in association with an anticyclonic circulation over the area. By 8th February both main as well as the induced systems became weak and were observed as low level cyclonic circulations extending up to 2.1 and 0.9 km respectively. There was significant pressure change and fall of barometric pressure with the passage of Western disturbances in the area.

*Mid-tropospheric Westerlies*

With the passage of the system, the mid-tropospheric westerly trough at 500 hPa was located close to 55°E and 25°N on 4th February, subsequently moving eastwards to 62° E and 65° E on February 5 and 6. With the strengthening of the system on February 7, the trough extended south and was found at 20° N along 70° E. It moved then northeastwards and was seen close to 72° E north of 30° N on February 8.

*Precipitation*

The precipitation belt shifted from the Western Himalayan region to central parts and Gangetic plains of India. On February 6, fairly widespread light precipitation occurred over the Western Himalayas and adjoining plains. On February 7 and 8, widespread moderate snowfall with isolated heavy falls was recorded over Western Himalayas. Moderate rainfall occurred over central parts and Gangetic plains.

---

## Further Readings

Barry, R.G. (1968) *Atmosphere, Weather and Climate,* Methuen & Co., London.

Bates, F.C. (1962) 'Tornadoes in the central United States', *Trans. Kansas Academy of Science,* **65**.

Byers, H.R. (1954) *General Meteorology,* McGraw Hill Book Co., New York.

Crowe, P.R. (1971) *Concepts in Climatology,* Longmans, London.

Gallowway, J.L. (1960) 'The three front model, the developing depressions and the occluding process', *Weather,* **15**.

Hare, F.K. (1953) *The Restless Atmosphere,* Hutchinson, London.

Malone, T.F., Ed. (1951) *Compendium of Meteorology,* American Meteorological Society, Boston.

Petterssen, S. (1956) *Weather Analysis and Forecasting (Vol. I & II),* McGraw Hill Book Co., New York.

Petterssen, S. (1969) *An Introduction to Meteorology,* McGraw Hill Book Co., New York.

Reihl, H. (1965) *Introduction to the Atmosphere,* McGraw Hill Book Co., New York.

Saha, P.K. and P.K. Bhattacharyya (1994) *Adhunik Jalavayu Vidya,* West Bengal State Book Board, Government of West Bengal, Calcutta.

Tannehill, I.R. (1950) *Hurricanes,* Princeton University Press, Princeton, New Jersey.

Tepper, M. (1958) 'Tornadoes', *Scientific American,* **198**, **5**.

Trewartha, G.T. (1968) *An Introduction to Climate,* McGraw Hill Book Kogakushu Ltd., Tokyo.

Trewartha, G.T. (1961) *The Earth's Problems Climates,* The University of Wisconsin Press, Madison, Wisconsin.

Vederman, J. (1954) 'The life cycle of jet streams and extra-tropical cyclones', *Bulletin of American Meteorological Society,* **35**.

## Chapter 7

## CLIMATIC ANOMALIES

**Understanding the Climatic Anomalies**

The climatic behaviour in many parts of the globe on many occasions in different periods never remains persistent. A climate *anomaly* is an event in which the magnitude of the deviation from normal conditions is unusually large, occurring infrequently in the historical records. The unusual conditions are regional in extent, involving multi-month periods of extremely high or low temperatures and/or wet or dry conditions. An unprecedented climatic anomaly occurred in the tropics and in the northern hemisphere in 1940–1942. During a strong and prolonged El Niño, extremely cold winters were observed in Europe, accompanied by very warm temperatures in Alaska and a cold North Pacific. The anomalies were strong (comprising the two coldest European winters of the 20th century) and extraordinarily persistent. It has been mentioned in the study of monsoon that certain weather phenomena developing outside the monsoon region even influence the monsoon circulation. Of these El Nino is considered very important. Sir Gilbert Walker in 1925 referred to the changing pressure systems over the Pacific Ocean. In those days no body could estimate the influence of such variations on global climate. Climatic studies in later periods established the facts that there are many other variations deviating from the normal conditions in climatic patterns. In recent years such variation in pressure systems has been also identified over the Atlantic Ocean. We will discuss a few such anomalies which play significant role in moderating global climate.

**El Nino/ENSO and La Nina**

*El Nino* develops when trade winds become weaker over the central and western Pacific Ocean. It results in dramatic rise of surface temperature of the ocean and depression of the thermocline in the eastern Pacific Ocean. Such conditions develop at the end of calendar year (November-December) along the coasts of Ecuador and northern Peru. In Spanish El Nino means 'child' or specifically 'child christ', as the local people calls the phenomenon. El Nino is actually identified with the warm episodes over the tropical eastern Pacific every two to seven years, when much than normal warm condition prevails. It is often associated with high rainfall leading to flooding in Peru and occurrence of drought in Indonesia and

Australia. The emergence of El Nino conditions also cut off the supply of nutrient rich thermocline water. As a result there is reduction in the growth of algae and destruction of the food chain. With such developments resulting from El Nino the commercial fisheries in the eastern Pacific suffer badly.

El Nino is intimately related to global atmospheric oscillation. This was referred as *Southern Oscillation* by Sir Gilbert Walker many years back, as early as 1925. As El Nino condition emerges lower than normal pressure is observed over the eastern tropical Pacific and higher than normal pressure occurs over Indonesia and northern Australia. Such pattern of pressure is associated with weaker than normal near surface equatorial easterly (Trades) winds. This situation reveals the warm phase of the Southern Oscillation, which is commonly described as the El Nino/Southern Oscillation episode.

### Warm ENSO or El Nino Conditions

El Nino is caused due to rise in Sea Surface Temperature (SST) over the Eastern Pacific Ocean. Equatorial warm ocean currents travel towards the Eastern Pacific region replacing the cool ocean currents (Humboldt Current) from the Antarctic region. The cause of such replacement of cool ocean currents by the warm ocean currents from the tropical ocean is still debated, but it is understood that the weakening of trade winds may be responsible for this. The weakening of trade winds is related to shifting pressure belts over the Pacific region.

The occurrence of an El Niño indicates:
1. Rise in air pressure over the Indian Ocean, Indonesia, and Australia,
2. Fall in air pressure over Tahiti and the rest of the central and eastern Pacific Ocean,
3. Trade winds in the south Pacific weaken or move east,
4. Warm air convection is evidenced near Peru, precipitating rain in the northern Peruvian deserts,
5. Warm water moves from the west Pacific and the Indian Ocean to the east Pacific. The rain bearing winds shift to the east causing extensive drought in the western Pacific and rainfall in the normally dry eastern Pacific.
6. Thermocline goes deeper with the loss of nutrient rich ocean water causing disastrous effect on marine life and ecosystem. So El Nino appears to be a curse to the Peruvian people as they lose their livelihood to a great extent.

When the warm ENSO episode domains the normal pressure and atmospheric circulation are changed. The water of the equatorial central and eastern Pacific being warmer than normal gives rise to increased cloudiness and rainfall over the region, particularly during the northern winter and spring seasons. On the other hand the rainfall is reduced over Indonesia, northern Australia and western Pacific regions. The reduction of rainfall sometimes results in drought conditions prevailing over northern Australia, Indonesia and even affects the monsoon circulation over India and south-east Asian landmass having lesser rainfall. South eastern Africa and northern Brazil experience drier than normal conditions during the northern winter season. The monsoon circulation in India and south-east Asia gets disturbed as the atmospheric circulations, such as the Jet Streams in the subtropics and in the temperate latitudes during the northern winter are affected by the ENSO conditions. The Jet Streams over the eastern Pacific Ocean becomes stronger than normal during the warm episodes, i.e. the emergence of El Nino conditions. Moreover under such conditions extra-tropical and frontal systems take different course, quite diverse than normal, ultimately resulting in persistent temperature and precipitation anamoly in many regions. For this reason the importance of ENSO has become important for variability of climatic conditions in many parts of the globe. From the study of different warm episodes in the past over the central and eastern Pacific giving rise to El Nino conditions it appears that precipitation and temperature anomaly patterns are highly consistent from one episode to another.

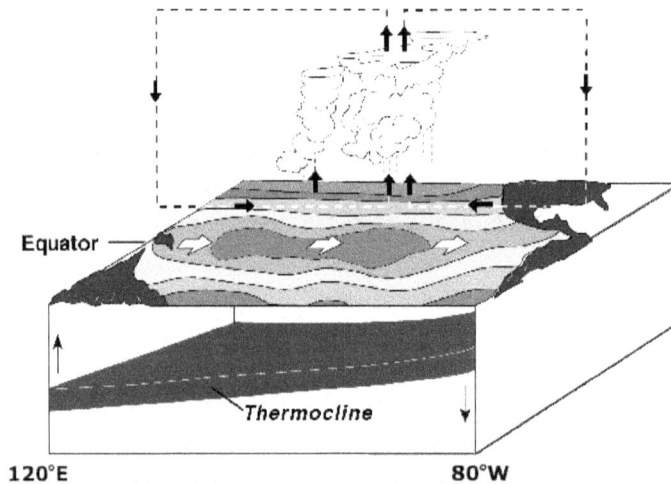

**Figure 7.1:** El Nino Conditions over the Pacific Ocean

*Source:* NOAA

### *Cool ENSO or La Nina Conditions*

The La Nina conditions are just the reverse of the El Nino conditions. La Nina in Spanish means young girl. It represents cold surface temperature of the water in the eastern Pacific Ocean. It occurs when the trade winds are the strongest over the tropical Pacific Ocean, resulting in the flow of cold and nutrient rich water to the surface by upwelling currents. The temperature may be as low as 7° F below normal. With the reduction of surface temperature over the tropical Pacific, particularly along the coast of Peru and Equador a favourable condition emerges with high productivity and good fishing off the west coast of South America.

The pressure belts over the Pacific Ocean shift and affect the global climate just opposite to El Nino situation. La Nina is sometimes referred as *ENSO cold phase* or *counter El Nino* or *El Viejo*. Strong La Nina conditions in 1998/1999 appear to replace El Nino conditions of 1997/1998. The La Nina episode ensures good rainfall in the western Pacific coastal region, in Indonesia and northern Australia. The monsoons over the south and south-east Asia and elsewhere appear to be normal with effective good rainfall.

Actually both the La Nina and El Nino are extreme phases of the ENSO cycle. The ENSO cycle is considered to be a complex interaction between water surface of the ocean and atmosphere in the tropical Pacific. The changing conditions may be related to the heat engine situation developing due to variation in radiational heat budget or flux over the tropical Pacific. The system was first referred as Southern Oscillation by Sir Gilbert Walker. It is just as the pendulum swings at either end of their oscillation back and forth, El Nino and La Nina are the extremes of the ENSO cycle.

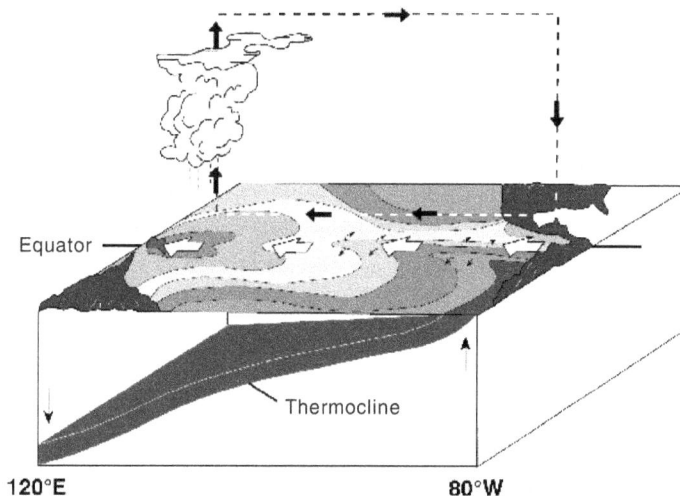

**Figure 7.2:** La Nina Conditions over the Pacific Ocean

*Source:* NOAA

### *Episodes of ENSO*

The El Nino/Southern Oscillation conditions developed during the year 1997/1998. A major warming of the ocean water occurred in the central and eastern Pacific since March, 1997. The El Nino developed rapidly in April–May and was strongest in the month of June, 1997. The episode in 1997/1998 continued up to February–April, 1998 and thereafter cooled down with the emergence of fast developing La Nina conditions during May–July, 1998. The El Nino episode in 1997/1998 is comparable to 1982/1983 episode in its extent and magnitude.

The El Nino/Southern Oscillation (ENSO) episode of 1982/1983 is considered to be the strongest in the twentieth century.

An attempt has been made to monitor the ENSO computing the Multivariate ENSO Index (MEI) on the six principal observed variables over the tropical Pacific. These six variables are: sea level pressure (P), zonal (U) and meridional (V) components of the surface wind, sea surface temperature (S), surface air temperature (A) and total cloudiness fraction of the sky (C). The MEI is calculated separately for each of twelve creeping bi-monthly seasons (Dec/Jan, Jan/Feb, Feb/Mar, Mar/Apr, Apr/May, May/June, June/July, July/Aug, Aug/Sept, Sept/Oct, Oct/Nov and Nov/Dec). After filtering the individual fields on a spatial pattern into clusters the MEI is computed as the first un-rotated Principal Component (PC) of all six observed fields combined. This is done by normalising the total variance of each field first, thereafter extracting the first PC on the covariance matrix of the combined fields. The Multivariate ENSO Index (MEI) has been shown indicating the Standardised Departure from 1950 to 2007 in Fig. 7.3. The MEI for the seven strongest historic El Nino episodes has been represented in Fig. 7.4. Let us compare the 1997/1998 El Nino episode with six other strong El Nino episodes since 1950. The episodes in 1957/1958, 1965/1966 and 1972/1973 experienced early warming in the far eastern Pacific and reached their standardised peak by the end of the first year. The recent episodes like 1982/1983, 1986/1987 and 1991/1992 took longer time to mature. The eastern Pacific did not experience early warming, but attained their peak in the spring and early summer of the second year and even later. The 1997/1998 El Nino episode had more in common with first three episodes (1957/1958, 1965/1966 and 1972/1973) during its development in 1997. Whereas the peak values of nearly +2.9 was surpassed only in early 1983.

The 1997/1998 El Nino episode continued exceeding 2 sigma (from May/June, 1997 through April/May, 1998) and surpassed the 1982/1983 benchmark of nine bi-monthly seasons. Thereafter the MEI of El Nino episode of 1997/1998 descended to La Nina condition.

*Multivariate ENSO Index*

The correlations between the local anomalies and the MEI exist. Each field is noted with a capital letter and the explained variance. The sea level pressure (P), sea surface temperature (S), air temperature (A), zonal component of the surface wind (U), meridional component of the surface wind (V) and cloudiness (C) are computed to calculate the Multivariate ENSO Index.

The sea level pressure (P) loadings indicate the character of Southern Oscillation—low pressure anomalies in the west and high pressure anomalies in the eastern Pacific representing negative MEI values or the La Nina conditions. The meridional wind field (V) features negative loadings north of the equator across the Pacific Ocean and indicate the northward shift of the ITCZ, which commonly occurs during the La Nina events. The condition is reverse with weak positive loadings over north-eastern Australia. The sea (S) and air (A)

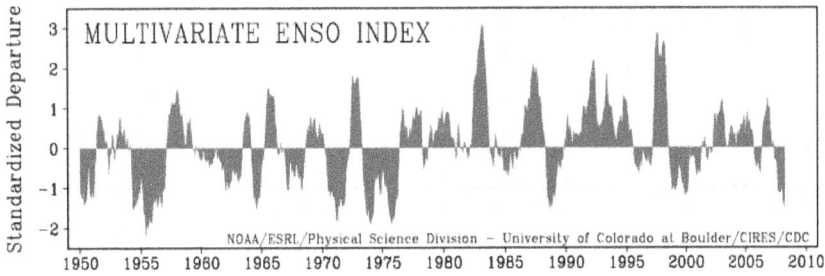

**Figure 7.3:** Multivariate ENSO Index, 1950–2007

*Source:* NOAA

**Multivariate ENSO Index for the 7 strongest El Niño events since 1950 *vs.* 2002–04.**

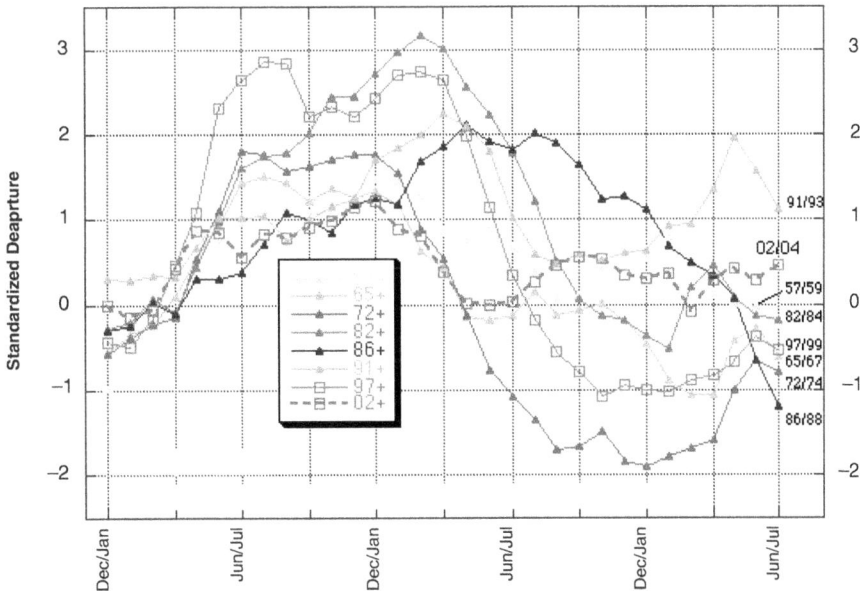

Update: 5 August 2004

NOAA-CRES Climate Diagnostics Center (CDC) University of Colorado at Boulder

**Figure 7.4:** MEI of El Nino Events

*Source:* NOAA

surface temperature also represent the particular ENSO conditions, having a wedge of positive loadings extending from the central and southern American coast or cold anomalies during the mature La Nina episode. During the La Nina event the total cloudiness (C) appears to be restricted in the central and southern America, whereas increased cloudiness is observed from Philippines to Hawaii.

Multivariate ENSO Index for the seven strongest La Nina events since 1949 and of 1998/99 has been compared in Figure 7.5. While the earlier La Nina events took off an early start, the 1998/99 La Nina event emerged so strong and fast that it reached the state of historic La Ninas. Its peak value of –1 sigma lasted since October/November, 1998 to January/February, 1999. During the last two decades only the 1998/99 La Nina has experienced lower MEI values.

**Multivariate ENSO Index for the 7 strongest historic El Niño events since 1949 *vs.* 1998–2000.**

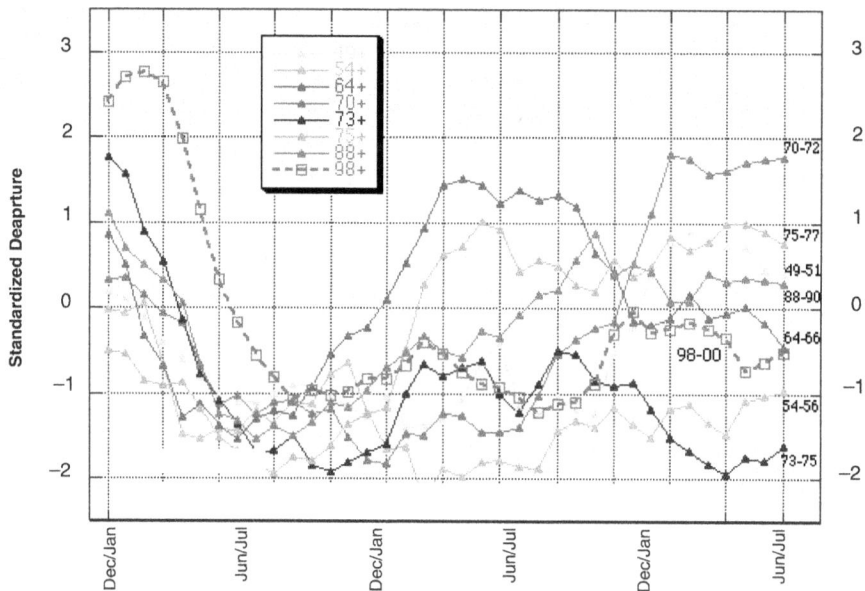

Final Update: 6 February 2001

NOAA-CRES Climate Diagnostics Center (CDC) University of Colorado at Boulder

**Figure 7.5:** MEI of La Nina Events

*Source:* NOAA

## *Records of El Nino (Warm ENSO) and La Nina (Cold ENSO) Episodes*

A list of warm (El Nino) and cold (La Nina) episodes has been prepared by the National Oceanic and Atmospheric Administration (NOAA) to furnish a season-by-season breakdown of conditions over the tropical Pacific. The region has been outlined from 150°W to International Date Line over the Pacific. In the classification, weak periods are designated as –C or –W, moderately strong periods as C or W, and strong periods as +C or +W.

### *El Nino/La Nina as the Climatic Anomalies*

The atmosphere and the adjacent oceans respond to warming and cooling of the equatorial eastern Pacific Ocean region (El Niño and La Niña events) in various ways. The atmosphere,

**Table 7.1:** Records of El Nino and La Nina Events

|  | JFM | AMJ | JAS | OND |
|---|---|---|---|---|
| 1950 | C | C | C | C |
| 1951 | C |  |  | –W |
| 1952 |  |  |  |  |
| 1953 |  | –W | –W |  |
| 1954 |  |  | –C | C |
| 1955 | C | –C | –C | +C |
| 1956 | C | C | C | –C |
| 1957 |  | –W | –W | W |
| 1958 | +W | W | –W | –W |
| 1959 | –W |  |  |  |
| 1960 |  |  |  |  |
| 1961 |  |  |  |  |
| 1962 |  |  |  |  |
| 1963 |  |  | –W | W |
| 1964 |  |  | –C | C |
| 1965 | –C |  | W | +W |
| 1966 | W | –W | –W |  |
| 1967 |  |  |  |  |
| 1968 |  |  |  | –W |
| 1969 | W | –W | –W | –W |
| 1970 | –W |  |  | C |
| 1971 | C | –C | –C | –C |
| 1972 |  | –W | W | +W |
| 1973 | W |  | _C | +C |
| 1974 | +C | C | –C | –C |
| 1975 | –C | –C | C | +C |
| 1976 | C |  |  | –W |
| 1977 |  |  |  | –W |
| 1978 | –W |  |  |  |
| 1979 |  |  |  |  |

| | JFM | AMJ | JAS | OND |
|------|-----|-----|-----|-----|
| 1980 | –W | | | |
| 1981 | | | | |
| 1982 | | –W | W | +W |
| 1983 | +W | W | | –C |
| 1984 | –C | –C | | –C |
| 1985 | –C | –C | | |
| 1986 | | | –W | W |
| 1987 | W | W | +W | W |
| 1988 | –W | | –C | +C |
| 1989 | +C | –C | | |
| 1990 | | | –W | –W |
| 1991 | –W | –W | W | W |
| 1992 | +W | +W | W | W |
| 1993 | –W | W | W | –W |
| 1994 | | | W | W |
| 1995 | W | . | | –C |
| 1996 | –C | | | |

for example, may respond to strong El Niño and La Niña events by shifting the east-west air circulation cells which was referred by Walker and his followers for decades as the Southern Oscillation (SO). The intimate relationship between El Niño/La Niña events and the closely matched atmospheric circulation response (SO) have made many scientists to often refer to the two systems simply as El Niño/Southern Oscillation (ENSO). In such cases, El Niño and La Niña phenomena are simply referred to as the warm and cold ENSO phases respectively. The warming/cooling status of the eastern and central equatorial Pacific Ocean during El Niño/La Niña events is also suggested to trigger world-wide anomalies in Sea Surface Temperatures (SST) and the circulation of the ocean currents. Even its impact on atmospheric and oceanic circulation over the Atlantic Ocean has been recognised. Sir Gilbert Walker also suggested linkage of North Atlantic Oscillation with the ENSO. So the events of ENSO are now universally considered to be climatic anomalies on a global scale.

It has been observed that during periods of El-Niño and La Niña phenomena, world-wide extremes in weather and climate such as droughts, floods, cold/hot spells, tropical cyclones, etc are common, even in distant regions, far off from the Pacific Ocean basin. Such weather and climate extremes are often associated with far reaching socio-economic impacts including loss of life and properties; mass migration of society and animals; lack of water, energy, food and other basic needs of human kind.

## ENSO Cycle

El Nino/La Nina events had been known to the local South American mariners and fishermen for many centuries, back. So these are not new events. Its impact on life and people of western coast of South America was also recognised dating back many centuries. El Nino usually occurs in the month of December, summer in the southern hemisphere. December being the month of birth of Jesus Christ, the people of Peru and Écuador believe that adverse climatic conditions with incessant rainfall with less catch of marine lifes in the warm ocean deem to be the curse of angry young Christ.

**Table 7.2:** Known El-Niño Events Based on Historical Records

| 1877/78 | 1880 | 1885 | 1888/89 | 1891 | 1896/97 | 1899/1900 | 1902/03 |
|---------|------|------|---------|------|---------|-----------|---------|
| 1905/06 | 1911/12 | 1914/15 | 1918/19 | 1925/26 | 1930/31 1941 | 1939/40/ | 1957/58 |
| 1963 | 1965/66 | 1969 | 1972/73 | 1976/77 | 1982/83 | 1986/87 | 1991/92 |
| 1994/95 | 1997/98 | | | | | | |

Source: Coghlan, 2002

El Nino and La Nina events occur alternately. No specific year of rotation of their occurrence could be established. However, El-Niño/La Nina events recur with a period between 2–7 years, and some of the El-Niño events are usually followed by La Niña. Climatologists, now-a-days, have the state of art to forecast the occurrence of El Nino/La Nina events on interpretation of climatic data including satellite imageries and application of mathematical models. Expected occurrence of El-Niño and La Niña could be suggested with lead times of months to over one year in advance. Attempts are now being made in a number of regions of the tropics to develop seasonal climate prediction and regional downscaling of weather and climate expectations using the projected El-Niño and La Niña signals and associated sea surface temperature anomalies. In this respect El-Niño and La Niña events never remain local phenomena but receive a lot of attention at global and regional levels over the last few years.

### Global Consequences of El-Niño and La Niña Phenomena

ENSO events are now known to have severe global climatic implications, particularly in the tropics. During a strong El Niño/La Niña event, it has been observed that there are some displacements of the warm and cold air masses together with the patterns of air convergence/divergence. Such displacements result in anomalous and enhanced convective activities over the central eastern Pacific, central western equatorial Indian Ocean, along the coast of eastern Africa and off/near the Atlantic equatorial coast of Africa, north-western South America, northern part of the Greater Horn of Africa (GHA) sub-region among many other parts of the tropics during a strong El Niño event. Due to such intense convective activities heavy abundant rainfall occurs over these regions. In contrast, over Indonesia, Australia, India, southeast Africa, and some other parts of the tropics, drought conditions are usually experienced. The observed impacts have large degree of variability both temporally and spatially. These depend on the phase of El Niño/La Niña events. El Niño/La Niña phases may be identified as the onset, peak and withdrawal. The onset/withdrawal phases of an El Niño/La Niña corresponds to the months of onset/cessation of the specific El Niño/La Niña event, while peak phase refers to the months when the strength of the El Niño/La Niña event is maximum.

Climate anomalies do occur all over the world each year. Many of them may not be the direct consequence of El Niño and La Niña. However, in some years and in some specific regions and seasons a few consistent extreme climate anomalies tend to occur during some specific El Niño and La Niña phases. In general, weather and climate extremes are caused by regional/ large-scale anomalies in the atmospheric wind circulations and the global ocean currents. El Niño and La Niña are so two of the many systems that can induce such anomalies.

**North Atlantic Oscillation**

The *North Atlantic Oscillation* (NAO) is a climatic phenomenon in the North Atlantic Ocean. The fluctuations in the difference of sea-level pressure are observed between the Icelandic Low and the Azores high. The seasonal shifting of the Icelandic Low and the Azores high, results in modifying the strength and direction of westerly winds and storm tracks across the North Atlantic. It is considered to be a part of the Arctic oscillation. The NAO was identified first by Sir Gilbert Walker in the 1920s similar to his concept of southern oscillation phenomenon in the Pacific Ocean. The NAO appears to be one of the most important drivers of climate fluctuations in the North Atlantic and surrounding humid climates.

The NAO is considered to be a redistribution of atmospheric mass between the Arctic and the subtropical Atlantic that swings from one phase to another resulting in large changes in surface air temperature, winds, storminess and precipitation over the Atlantic as well as the adjacent continents. The NAO also influences the ocean through changes in heat content, gyre circulations, mixed layer depth, salinity, high latitude deep water formation and sea ice cover.

The mT air-mass from the sub-tropical Atlantic Ocean blows as the Westerlies across the Atlantic. It brings moist air to European landmass. In years when the westerlies are strong, summers are cool, winters are mild and rain is common. When the Westerlies are suppressed, the temperature becomes more extreme in summer and winter characterised by heatwaves, deep freezes and reduced rainfall.

A permanent low-pressure system, known as the Icelandic Low, lies over Iceland and a permanent high-pressure system, known as the Azores High exists over the Azores in the Atlantic Ocean bordering the European landmass. The positions of these global pressure systems largely control the direction and strength of the Westerly winds in Europe. The relative strengths and positions of these systems differ from year to year and this variation is known as the NAO. When there exists a large difference in pressure at the Azores High and Icelandic Low it is called a high index year (NAO+). It favours increased Westerlies and, consequently occurrences of cool summers and mild and wet winters in Central Europe and its Atlantic fronts. On the other hand, the low index (NAO–) causes the Westerlies weak in the Central Europe and adjoining areas. As a result the region experiences cold winters and storm tracks move southerly toward the Mediterranean Sea. In a year of low index increased storm activity and rainfall prevail over southern Europe and North Africa. Variation of NAO index usually occurs in the winter months. Particularly during the months of November to April, the NAO is responsible for much of the variability of weather in the North Atlantic region, affecting wind speed and wind direction changes, changes in temperature and moisture distribution and the intensity, number and track of storms. The impact of NAO appears to be less over the Atlantic coast bordering Eastern U.S.A. and Canada than it is over Europe. However, in winter in case of high index (NAO+) the Icelandic low draws a stronger southwesterly circulation over the eastern half of the North American continent which restricts the entry of extreme cold Arctic air from the north. Incidentally if it is combined with El Nino effect in some years significantly warmer winters prevail much in areas of the United States and southern Canada.

### *Characteristics of Positive North Atlantic Oscillation (NAO⁺)*

The Positive NAO index phase shows a stronger than usual subtropical high pressure centre and a deeper than normal Icelandic low.

The Increased pressure difference results in more and stronger winter storms crossing the Atlantic Ocean on a more northerly track.

Warm and wet winters in Europe and cold and dry winters in northern Canada and Greenland are experienced.

The Eastern USA and Southeastern Canada experience mild and wet winter conditions.

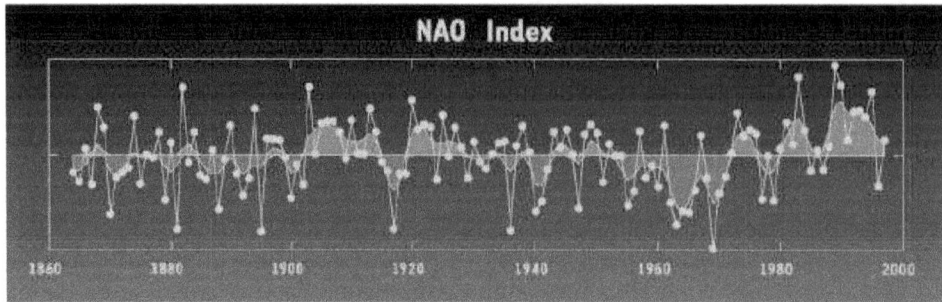

**Figure 7.6:** Variation of NAO Index

### *Characteristics of Negative North Atlantic Oscillation (NAO⁻)*

The negative NAO index phase shows a weak subtropical high and a weak Icelandic low.

The reduced pressure gradient results in fewer and weaker winter storms crossing on a more west-east pathway.

They bring moist air into the Mediterranean and cold air to northern Europe.

The US east coast experiences more cold air outbreaks and hence snowy weather conditions occur.

Greenland, however, will have milder winter temperatures.

### Pacific Decadal Oscillation

The concept of *Southern Oscillation,* proposed by Sir Gilbert Walker in 1925, led many scientists in later period to study the climatic variability in the Pacific Ocean. *The Pacific Decadal Oscillation*, or PDO, is often described as a long-lived El Niño-like pattern of Pacific climate variability (Zhang *et al.* 1997). In tune with the El Niño/Southern Oscillation (ENSO) phenomenon, the extremes in the PDO pattern are characterised by widespread variations in Pacific Basin and North American climate. In parallel with the ENSO phenomenon, the extreme phases of the PDO are classified into either *warm* or *cool*, as defined by ocean temperature anomalies in the northeast and tropical Pacific Ocean.

The *Pacific Decadal Oscillation* (PDO) is distinguished from ENSO by two main characteristics. First, typical PDO events show remarkable persistence relative to that attributed to ENSO events. Second, the climatic evidences of the PDO are most visible in the North Pacific/North American sector, while secondary signatures are marked in the tropics. It happens to be the opposite of ENSO. Mantua, Minobe and others (1997) found the cool PDO regimes occurring from 1890–1924 and again from 1947–1976 and the warm PDO regimes extending from 1925–1946 and from 1977 through the mid-1990's. Recent changes in Pacific climate suggest a possible reversal to cool PDO conditions in 1998.

### *Temperature and Pressure Variation in PDOs*

We can identify characteristic pressure, wind, temperature, and precipitation patterns related to the PDO. The SST pattern shows anomalously cool condition in the central North Pacific Ocean, whereas the SST pattern along the coast of North America indicates unusually warm condition. Intensifications of the *Aleutian Low* pressure cell generally coincide with periods of anomalously high Sea Level Pressure (SLPs) over western North America and the subtropical Pacific region.

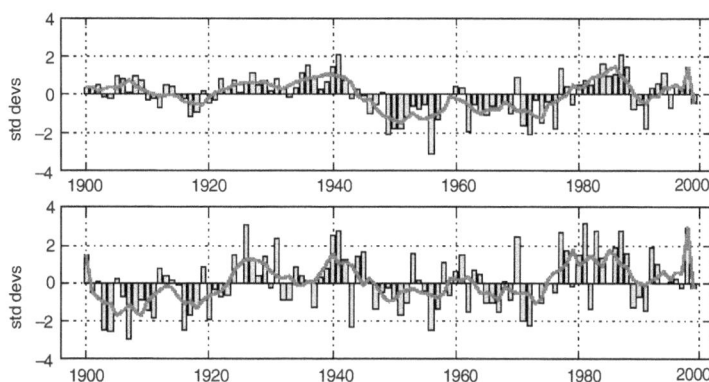

**Figure 7.7:** SST and SLP Variations in the Pacific Ocean over the Years

When SSTs are anomalously cool in the interior North Pacific and warm along the Pacific Coast, and when Sea Level Pressures (SLPs) are below average over the North Pacific, the respective indices indicate positive values. As the anomalous climatic patterns are reversed, with warm SST anomalies in the interior and cool SST anomalies along the North American coast, or above average Sea Level Pressures (SLPs) over the North Pacific, the respective indices indicate negative values. Winter/spring (October-March) average values for the PDO indices are presented in Figure 7.7 (SST in the top panel, SLP in the bottom panel). Probably the most important characteristic of these indices is the year-to-year persistence. It indicates much of their variability in the 20th century. Negative values in both indices correspond to the cool PDO periods, while positive values are indicative of the warm PDO periods. Within the 20 to 30 year regimes there are several short-lived signature alterations in the indices.

### *Impact of PDO on North American Climate*

The North American climate anomalies associated with PDO warm and cool phases closely resemble the phenomena related to El Niño and La Niña (Latif and Barnett 1995 and 1996). Warm phases of the PDO are correlated with North American temperature and precipitation anomalies similar to those correlated with El Niño: above average winter and spring temperatures in northwestern North America, below average temperatures in the southeastern USA, above average winter and spring rainfall in the southern USA and northern Mexico, and below average precipitation in the interior Pacific Northwest and Great Lakes regions. Cool phases of the PDO are simply correlated with the reverse climate anomaly patterns over North America, broadly resembling typical La Niña climate phenomenon. The following table indicates the impact of PDO on North American climate.

**Table 7.3:** North American Climate Anomalies Associated with Extreme Phases of PDO

| Climate Anomalies | Warm Phase PDO | Cool Phase PDO |
|---|---|---|
| Ocean surface temperatures in the northeastern and tropical Pacific | Above average | Below average |
| October-March northwestern North American air temperatures | Above average | Below average |
| October-March Southeastern USA air temperatures | Below average | Above average |
| October-March southern USA/ Northern Mexico precipitation | Above average | Below average |
| October-March Northwestern North America and Great Lakes precipitation | Below average | Above average |
| Northwestern North American spring time snow pack | Below average | Above average |
| Winter and spring time flood risk in the Pacific Northwest | Below average | Above average |

### Further Readings

Bitz, C.C., and D.S. Battisti (1999) Interannual to decadal variability in climate and the glacier mass balance in Washington, Western Canada, and Alaska. *Journal of Climate*, **12**.

Cayan, D.R. (1996) Interannual climate variability and snowpack in the western United States. *Journal of Climate*, **9**.

César N. Caviedes (2001) *El Niño in History: Storming through the Ages,* University Press of Florida, Florida.

Coghlan, C. (2002) El-Niño - causes, consequences and solutions. *Weather*, **57**.

Gershunov and Barnett (1998) Interdecadal modulation of ENSO teleconnections. *Bulletin of the American Meteorological Society*, **79**.

Gershunov, A., T. Barnett and D. Cayan (1999) 'North Pacific interdecadal oscillation seen as factor in ENSO-related North American climate anomalies'. *EOS,* **80**.

Graham, N.E. (1994) 'Decadal-scale climate variability in the 1970s and 1980s: observations and model results', *Climate Dynamics*, **10**.

Latif, M. and T.P. Barnett (1994) 'Causes of decadal climate variability over the north Pacific and North America', *Science,* **266**.

Latif, M. and T.P. Barnett (1996) 'Decadal climate variability over the North Pacific and North America: dynamics and predictability', *Journal of Climate*, **9**.

Mantua, N.J., S.R. Hare, Y. Zhang, J.M. Wallace, and R.C. Francis (1997) 'A Pacific decadal climate oscillation with impacts on salmon', *Bulletin of the American Meteorological Society*, **78**.

McCabe, G.J. and M.D. Dettinger (1999) 'Decadal variations in the strength of ENSO teleconnections with precipitation in the western United States', *International Journal of Climatology,* **9**.

National Research Council (1996) *Learning to Predict El Niño: Accomplishments and Legacies of the TOGA Program,* National Academy Press.

Oliver, J.E. and John J. Hidore (2002) *Climatology,* Pearson Education Inc., Delhi.

Philander, S. George (1990) *El Niño, La Niña and the Southern Oscillation,* Academic Press Inc., New York.

Ropelewski, C.F. and M. S. Halpert (1987) 'Global and Regional Precipitation patterns associated with the El-Niño/Southern Oscillation', *Monthly Weather Review*, **115**.

Saha, P.K. (2004) 'Impact of ENSO on global climate', in *Transact: Refresher Course in Geography,* University of Burdwan, Burdwan.

Trenberth, K.E. (1997) *The definition of El Niño*, Bulletin of the American Meteorological Society, **78**, **12**.

Trenberth, K.E. (1990) 'Recent observed interdecadal climate changes in the northern hemisphere', *Bulletin of the American Meteorological Society*, **71**.

Trenberth, K.E., and J. W. Hurrell (1994) 'Decadal atmosphere-ocean variations in the Pacific', *Climate Dynamics,* **9**.

WMO (1984) *The Global Climate System: A Critical Review of the Climate System During 1982– 1984,* World Meteorological Organization, Geneva.

Zhang, Y., J.M. Wallace and D.S. Battisti (1997) 'ENSO-like interdecadal variability: 1900–93', *Journal of Climate*, **10**.

## Chapter 8

## GLOBAL WARMING

### Understanding the Global Warming

*Global Warming* means the rise in the average temperature of the near-surface air and oceans since the mid-twentieth century and its projected continuation. The average global air temperature close to the earth's surface increased $0.74 \pm 0.18°$ C ($1.33 \pm 0.32°$ F) since 1905. The Intergovernmental Panel on Climate Change (IPCC) concludes that "most of the observed increase in globally averaged temperatures since the mid-twentieth century is very likely due to the observed increase in anthropogenic greenhouse gas concentrations" exerting the greenhouse effect. Natural reasons such as solar variation combined with volcanoes probably had a little warming effect from pre-industrial times to 1950 and a small cooling effect from 1950 onward.

### *Key Observations*

1.  Global mean surface temperatures have increased 0.5-1.0° F since the late 19th century. The 20th century's 10 warmest years all occurred in the last 15 years of the century. Of these, 1998 was the warmest year on record. The snow cover in the northern hemisphere and floating ice in the Arctic Ocean have decreased. Globally, sea level has risen 4-8 inches over the past century. Worldwide precipitation over land has increased by about one percent.
2.  Increasing concentrations of greenhouse gases are likely to accelerate the rate of climate change. Scientists expect that the average global surface temperature could rise 1-4.5° F (0.6-2.5° C) in the next fifty years, and 2.2-10° F (1.4-5.8° C) in the next century, with significant regional variation.
3.  Many palaeo-climate records from the North Atlantic region show a pattern of rapid climate oscillations, the so-called Dansgaard–Oeschger events, with a quasi-periodicity of 1,470 years for the late glacial period. Various hypotheses have been suggested to explain these rapid temperature shifts, including internal oscillations in the climate system and external forcing, possibly from the Sun. But whereas pronounced solar cycles of 87 and 210 years are well known, a 1,470-year solar cycle has not been

detected. Here we show that an intermediate-complexity climate model with glacial climate conditions simulates rapid climate shifts similar to the Dansgaard–Oeschger events with a spacing of 1,470 years when forced by periodic freshwater input into the North Atlantic Ocean in cycles of 87 and 210 years.

4. The glacial 1,470-year climate cycles could have been triggered by solar forcing despite the absence of a 1,470-year solar cycle.

## *A Few Questions*

The rise of temperature on a global scale was evidenced in different geological periods prior to human foothold on the earth's surface. It will be useful understanding the global warming phenomenon on our planet.

- Present state of global warming could be evidenced in the paleocene-eocene period.
- Onwards from the late-cretaceous period global temperature declined.
- Lowest temperature was recorded during the pliestocene period, when the continental ice sheet covered most of the earth's surface (13 m. sq.miles).
- The extinct of dinosaur during the glacial period drew attention of the scientists.
- Even the glacial age was interrupted with relatively warm inter-glacial periods, viz. Nebraska period (> 0.320 m. yrs), Kansan period (> 0.240 m. yrs), Illinois (> 0.115 m. yrs.) and Wisconsin period (60,000–25,000 yrs).
- Emilliani and others observed the changing temperatures for about 65 m. years, particularly during the Canozoic period.
- The Canozoic period was characterised by emergence of mammals, upheaval of mountains, intense tectonic movements, birth of homo-sapiens and beginning of glacial period.
- Emilliani and others observed the temperature and other conditions on the ocean floor, as the ocean was relatively less affected and residuals of earth's surface could be preserved on the ocean floor.
- Emilliani devised a geological thermometer (after Niggliani) based on isotope for the study of foraminifera, a species living at the bottom of Pacific Ocean.

## *Observations by Emilliani and Others*

Cezare Emilliani, a young Italian scientist in the laboratory of Professor Urey in the University of Chicago, was inquisitive to study the changing temperature conditions of the earth in the past geological periods. He observed in details the status of temperature with particular reference to the ocean bottom. Most of the signature could have been lost on the land surface, but could be found stored in the oceans.

- 32 million years ago the temperature of the bottom of Equatorial Pacific Ocean was 51° F (10.6° C) at a depth of 3000 m.

- 22 million years ago the temperature dropped to 44° F (6.7° C).
- 1 million years ago the temperature further dropped to 36° F (2.2° C). This record resembles nearly the present temperature of the bottom of Equatorial Pacific Ocean.
- They also recorded marginal changes of temperature during the last 0.5 million years.
- Research funded by the U.S. National Science Foundation (NSF) shows that global climate change quickly may have disrupted ancient ocean processes and could lead to drastic shifts in environments around the world.
- Researchers from the Scripps Institution of Oceanography observed that the events, unfolded millions of years ago and spanned thousands of years, historically were similar to current warming-induced changes in large-scale sea circulation and may help illuminate potential long-term effects of today's climate warming.
- The unique data set, based on the chemical make-up of tiny ancient sea creatures, single-celled animals called foraminifera, showed evidence of a monumental reversal in the circulation of deep-ocean patterns around the world.
- The researchers concluded that it was triggered by the global warming, the world experienced at the time, called the Paleocene/Eocene Thermal Maximum (PETM).
- Fifty-five million years ago, when the earth was in a period of global warmth, ocean currents rapidly changed direction, and this change did not reverse to original conditions for about 20,000 years.
- Fossil records indicated that the PETM caused a host of changes around the globe, ranging from a mass extinction of deep-sea bottom-dwelling marine life to key migrations of terrestrial mammal species, likely allowed by warm conditions that opened travel routes not possible in colder conditions.
- In this period the scientists found the earliest evidence for horses and primates in North America and Europe.
- The results indicated that deep-ocean circulation in the southern hemisphere abruptly stopped the conveyor-belt-like process called "overturning," in which cold and salty water in the depths exchanges with warm water on the surface.
- As this process virtually was shutting down in the south, overturning apparently became active in the Northern Hemisphere.
- This shift drove unusually warm water to the deep sea, releasing stores of methane gas that led to further global warming and a massive die-off of deep-sea marine life.
- Today, overturning in the modern North Atlantic Ocean is a primary way to draw heat into the far north Atlantic and keep temperatures in Europe relatively warmer than conditions in Canada.

### *A Few Other Observations on Changing Temperature Conditions*

- 32 million years ago the Antarctic Ocean was warmer like the Atlantic ocean near Rhodes Island.

- Advance and retreat of glaciers could be evidenced in different periods.
- Climatic condition was: 100 A.D. – similar to present, 350 A.D. – >wet, 500 A.D. – dry (Alpine passes used), 900 A.D. – again wet, 1000-1100 A.D. – again drier.
- Greenland was habitable around 1000 A.D. Vikings settled in 984 A.D., but abandoned in 1410 A.D.
- Europe evidenced 'Little Ice Age' during 1450–1850 A.D.
- 1860 A.D. onwards temperatures had been rising. Fossil fuels caused the rise as a result of industrialisation.
- The Green House Gases (GHG) play significant role in warming the earth's atmosphere. The terrestrial back-radiation travels as infrared radiation. This infrared radiation is absorbed in the atmosphere by several gases in the atmosphere and results in warming of the earth's atmosphere.
- The gases, which absorb infrared radiation, are known as Green House gases. These gases are carbon dioxide, methane, chloro-fluro carbon and nitrous oxide. These are called Green House Gases, as they behave like a green house in trapping the long wave radiation.
- Role of the Green House Gases are very important in global warming. If there be no Green House Gases in the atmosphere the average temperature on earth's surface would have been dropped from $15°$ C to $–18°$ C.
- So there is nearly $33°$ C rise of temperature by the absorption of long wave radiation by such green house gases. Most of the earth's surface would have been snow covered if there be no trapping of long wave radiation by such gases in the atmosphere. The role of different green house gases also varies in the extent of global warming.
- Of all green house gases, the carbon dioxide plays most significant role. Carbon Dioxide contributes 55 percent in global warming. The role of carbon dioxide in global warming was first precisely stated by John Tyndal, the noted British physicist, in 1861. He first proposed the theory that carbon dioxide controls temperature as carbon dioxide molecules in the air absorb infrared radiation.
- The carbon dioxide and other gases in the atmosphere are usually transparent to visible solar radiation, so that insolaton or solar rays can reach the earth's surface for heating it. But a lesser hot body radiates back much of the energy in the invisible infrared region of the spectrum. This raditation is most intense at wavelengths very close to the principal absorption band (13 to 17 microns) of the carbon dioxide spectrum. In case the concentration of carbon dioxide is high, its weaker bands also can absorb energy effectively and contribute in global warming.
- In general the carbon dioxide is only 0.03 percent of all the constituent gases in the atmosphere. This amount is also variable latitudinally and seasonally. The total amount of carbon dioxide in the atmosphere is estimated at 2300 billion tons. The status of carbon dioxide in the atmosphere depends on the interaction of oceans, rocks and living organisms.

- The oceans contain more than 50 times carbon dioxide than the atmosphere. The oceans as reservoir contain nearly $1.3 \times 10^{14}$ tons of carbon dioxide. Some of the gas is dissolved in water and but most is transformed into carbonate compounds. Interestingly the oceans also exchange carbon dioxide with the atmosphere. This amounts to 200 billion tons each year. When the equilibrium is disturbed, the oceans may absorb or release billion tons of additional carbon dioxide.
- The atmosphere and oceans also exchange carbon dioxide continually with rocks and living organisms. Carbon dioxide is added from the volcanic activity that releases gases from the earth's interior and from the respiration and decay of organisms.
- On the other hand carbon dioxide is lost to the weathering of rocks and the photosynthesis by plants. As these processes continue the content of carbon dioxide in the atmosphere varies, affecting the radiational balance and thus raising or lowering the earth's temperature.
- The balance is thus restored in the natural set up. But human interference has resulted in an imbalance in this set-up. From the Table 8.2 it can be understood that nearly 8 billion tons of carbon dioxide is added each year through human activities like burning of fossil fuels and cultivation of soils.

**Role of the Green House Gases (GHG)**

The greenhouse effect was discovered by Joseph Fourier in 1824 and was first investigated quantitatively by Svante Arrhenius in 1896. The atmosphere close to the earth's surface gains heat through the process of absorption and emission of infrared radiation. The Green House Gases (GHG) play significant role in warming the earth's atmosphere. The terrestrial back-radiation travels as infrared radiation. Several gases in the atmosphere absorb this infrared radiation causing warming of the earth's atmosphere. These gases, which absorb infrared radiation, are known as Green House Gases. These gases are carbon dioxide, methane, chloro-fluro carbon and nitrous oxide. These are called green house gases, as they behave like a green house in trapping the long wave radiation.

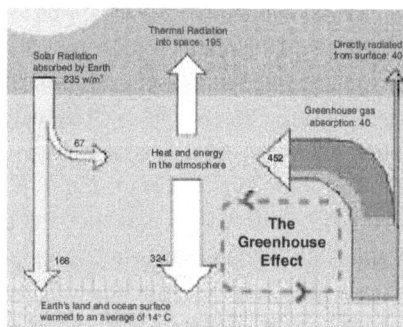

**Figure 8.1:** The Green House Effect

The role of the green house gases is very important in global warming. If there be no green house gases in the atmosphere the average temperature on earth's surface would have been dropped from 15° C to-18° C. So there is nearly 33° C rise of temperature by the absorption of long wave radiation by such green house gases. Most of the earth's surface would have been snow covered if there be no trapping of long wave radiation by such gases in the atmosphere. The role of different green house gases also varies in the extent of global warming. From the following table the role of different green house gases can be understood.

**Table 8.1:** Green House Gases

| Green House Gases | Present contribution | Present status & rate of increase | Projected rise in 2050 A.D. | Name of the GHG | Sink |
|---|---|---|---|---|---|
| Carbon Dioxide ($CO_2$) | 55% | 385 ppm. (0.5% rise annually) | 400 to 600 ppm. | Fossil fuels & deforestation | Oceans & biosphere |
| Methane ($CH_4$) | 15% | 1.6 ppm. (1% rise annualy) | 2.1 to 4.0 ppm. | Grazing lands, paddy lands, wetlands, gas & coal mines | Reaction with Hydrogen ion in the atmosphere |
| Chloro-Fluro Carbon ($CFC_{11}$ & $CFC_{12}$) | 24% | 0.2 ppb. ($CFC_{11}$) 0.4 ppb. ($CFC_{12}$) (4% rise annually) | 0.7 to 0.3 ppb ($CFC_{11}$) 2.0 to 4.8 ppb ($CFC_{12}$) | Freezing gas, aerosols and different industrial processes | Reaction with ultraviolet rays |
| Nitrous oxide ($N_2O$) | 6% | 300 ppb. (0.8% rise annually) | 350 to 450 ppb. | Fossil fuel, fertiliser and burning of organic materials | Reaction with ultraviolet and cosmic rays |

ppm. = parts per million and ppb= parts per billion
*Source:* Lagett, J. (1990)

### *Role of Carbon Dioxide*

Of all green house gases the carbon dioxide plays the most significant role. Carbon dioxide contributes 55 percent in global warming. John Tyndal, the noted British physicist first precisely stated the role of carbon dioxide in global warming in 1861. He first proposed the theory that carbon dioxide controls temperature as carbon dioxide molecules in the air absorb infrared radiation. The carbon dioxide and other gases in the atmosphere are usually transparent to visible solar radiation, so that solar rays can reach the earth's surface for heating it. But as a lesser hot body radiates back much of the energy in the invisible infrared region of the spectrum, this radiation is most intense at wavelengths very close to the principal absorption band (13 to 17 microns) of the carbon dioxide spectrum. In case the concentration of carbon dioxide is high, its weaker bands also can absorb energy effectively and contribute in global warming. In fact the envelope of carbon dioxide in the atmosphere acts as a blanket and prevents the escape of energy to space. Thus the entrapped infrared radiation from earth in the atmosphere warms up the earth-atmosphere system.

### *Status of Carbon Dioxide*

In general the carbon dioxide is only 0.03 percent of all the constituent gases in the atmosphere. This amount is also variable latitude-wise and seasonally. The total amount of carbon dioxide in the atmosphere is estimated at about 2300 billion tons. The status of carbon dioxide in the atmosphere depends on the interaction of oceans, rocks and living organisms.

The oceans contain more than 50 times carbon dioxide than the atmosphere. The oceans as reservoir contain nearly $1.3 \times 10^{14}$ tons of carbon dioxide. Some of the gas is dissolved in water and but most is transformed into carbonate compounds. Interestingly the oceans also exchange carbon dioxide with the atmosphere. This amounts to 200 billion tons each year. When the equilibrium is disturbed, the oceans may absorb or release billion tons of additional carbon dioxide. This acts in maintaining the equilibrium of carbon dioxide status in the atmosphere. As the atmospheric concentration of carbon dioxide rises, the oceans absorb large part of additional carbon dioxide. But when the concentration falls in the atmosphere, the oceans discharge additional carbon dioxide to the atmosphere in maintaining the balance.

The atmosphere and oceans also exchange carbon dioxide continually with rocks and living organisms. Carbon dioxide is added from the volcanic activity that releases gases from the earth's interior and from the respiration and decay of organisms. On the other hand carbon dioxide is lost to the weathering of rocks and the photosynthesis by plants. As these processes continue the content of carbon dioxide in the atmosphere varies, affecting the radiation balance and thus raising or lowering the earth's temperature. The balance is thus restored in the natural set up. But human interference has resulted in an imbalance in this set-up. From the following table it can be understood that nearly 8 billion tons of carbon dioxide is added each year through human activities like burning of fossil fuels and cultivation of soils.

### *Effect of Deforestation on the Status of Carbon Dioxide*

From Table 8.3 we find that the plants by photosynthesis can absorb 60 billion tons of carbon dioxide. The plants maintain a balance of carbon dioxide through absorption from the atmosphere and utilisation of carbon dioxide to produce carbohydrates in the plants. So the loss of plant life on earth's surface would have detrimental effect upon the status of carbon dioxide, resulting in rise of the gas in the atmosphere. Table 8.4 shows the relationship between the concentration of carbon dioxide in the atmosphere and loss of vegetation cover on earth's surface.

**Table 8.2:** Carbon Dioxide Balance

| | | |
|---|---|---|
| **Reserve:** | | |
| Atmosphere | : | $2300 \times 10^9$ tons |
| Oceans | : | $130,000 \times 10^9$ tons |
| Fossil beds | : | $40,000 \times 10^9$ tons (old) $+ < 0.1 \times 10^9$ tons (new) |
| Carbonates | : | $0.1 \times 10^9$ tons |
| **Gain by the atmosphere (+):** | | |
| Released from soil | : | $2 \times 10^9$ tons |
| Combustion of fossil fuels | : | $6 \times 10^9$ tons |
| Released from interior of earth | : | $0.1 \times 10^9$ tons |
| Respiration and decay | : | $60 \times 10^9$ tons |
| Released from oceans | : | $100 \times 10^9$ tons |
| **Loss by the atmosphere (–):** | | |
| Weathering of rocks | : | $0.1 \times 10^9$ tons |
| Photosynthesis by plants | : | $60 \times 10^9$ tons |
| Absorption by oceans | : | $100 \times 10^9$ tons |
| **Net gain by the atmosphere:** | | |
| Released from soil | | |
| + Combustion of fossil fuels | : | $2 \times 10^9$ tons $+ 6 \times 10^9$ tons |

**Table 8.3:** Average Annual Budget of $CO_2$ Perturbations for 1980-1989 (from Schimel 1995)

*Fluxes and reservoir changes of carbon are expressed in $10^{12}$ kg/yr*

| Carbon Dioxide Budget | $10^{12}$ kg / yr | Error |
|---|---|---|
| *Carbon Dioxide Sources* | | |
| Emissions from fossil fuel combustion and cement production | 5.5 | 0.5 |
| Net emissions from changes in tropical landuse | 1.6 | 1.0 |
| Total anthropogenic emissions | 7.1 | 1.1 |
| *Carbon Dioxide Sinks* | | |
| Storage in the atmosphere | 3.02 | 0.2 |
| Oceanic uptake | 2.0 | 0.8 |
| Uptake by northern hemisphere forest regrowth | 0.5 | 0.5 |
| Carbon Dioxide fertilization | 1.0 | 0.5 |
| Nitrogen deposition | 0.6 | 0.3 |
| Residual (Source) | (0.2) | (2.0) |

• *Errors are accumulated by quadrature*

Note: This analysis shows that the terrestrial biosphere was about in balance with regard to the emission and absorption of $CO_2$ for that period, a conclusion supported by recent measurements of atmospheric $CO_2$ concentrations. An estimated $1.6 \times 10^{12}$ kg/yr (1.6 billion metric tons) of carbon were released through land use change in the tropics, whereas about $2.1 \times 10^{12}$ kg/yr of carbon were absorbed by terrestrial ecosystems, through the combined effects of forest regeneration, $CO_2$ fertilization, and nitrogen deposition.

**Table 8.4:** Deforestation and Addition of Carbon Dioxide

| Country | Deforestation (sq. km) | Released Carbon Dioxide in m.tons | Released Carbon Dioxide in percentage |
|---|---|---|---|
| Brazil | 50,000 | 454 | 32.1 |
| Indonesia | 12,000 | 124 | 8.9 |
| Myanmar | 8,000 | 83 | 5.9 |
| Mexico | 7,000 | 64 | 4.6 |
| Thailand | 6,000 | 62 | 4.4 |
| Nigeria | 5,000 | 57 | 4.1 |
| Zaire | 5,000 | 57 | 4.1 |
| Malaysia | 4,800 | 50 | 3.6 |
| India | 4,000 | 41 | 2.9 |
| Vietnam | 3,500 | 36 | 2.6 |
| Papua | 3,500 | 36 | 2.6 |
| Ivory Coast | 3,500 | 36 | 2.6 |
| Peru | 3,400 | 32 | 2.3 |
| Central America | 3,300 | 30 | 2.1 |
| Philippines | 2,700 | 28 | 2 |
| Cameroon | 2,700 | 28 | 2 |
| Madegaskar | 2,700 | 28 | 2 |
| Ecuador | 2,600 | 27 | 1.9 |
| Bolivia | 1,500 | 14 | 1 |
| Venezuela | 1,500 | 14 | 1 |
| Others (5) | 3,300 | 38 | 2.7 |
| Total | 1,386,000 | 1,398 | 100 |

Source: Global Warming—The Green Peace Report, 1989.

*Recent Rise in Carbon Dioxide*

It is estimated that for the last hundred years there had been addition of 360 billion tons of carbon dioxide to the atmosphere due to human activities. The present concentration of $CO_2$ in the atmosphere is about 385 parts per million (ppm) by volume. Future $CO_2$ levels are expected to rise due to continued burning of fossil fuels and land-use change. The rate of rise will depend on uncertain economic, sociological, technological, and natural developments, but may be ultimately limited by the availability of fossil fuels.

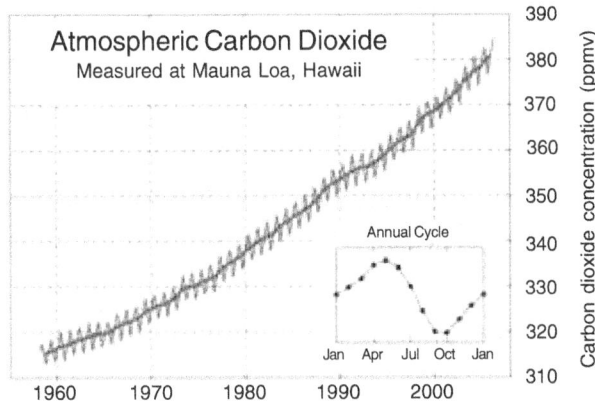

**Figure 8.2:** Rise in Carbon Dioxide Concentrations

The 'IPCC Special Report on Emissions Scenarios' has projected a wide range of future $CO_2$ scenarios, ranging from 541 to 970 ppm by the year 2100. Fossil fuel reserves are sufficient to reach this level and continue emissions past 2100, if coal, tar sands or methane clathrates are extensively used (IPCC). The concentration of $CO_2$ has increased by 31 percent in the atmosphere since the beginning of the industrial revolution in the middle of 18th Century. These levels are considerably higher than at any time during the last 650,000 years, the period for which reliable data has been extracted from ice cores. Such high concentration of carbon dioxide has led to the rise of global temperature in general. After the Little Ice Age in 1850 the global temperature had been steadily rising. The maximum rise was recorded in 1940, thereafter the rate of increase declined for a decade. Since 1950 the trend in the emissions of carbon dioxide and global rise of temperature appears to be disturbing. In 1990 the industrialised nations emitted about 4 billion tons of carbon dioxide (@ 3 tons per capita), of which USA alone contributed about 1.4 billion tons of carbon dioxide (@ 6 tons per capita). The developing nations released only 2 billion tons of carbon dioxide (@ 0.5 tons per capita). The Earth Summit, 1992 at Rio de Janeiro cautioned the global community and called for the stabilisation of global temperature in high latitudes, which had increased by 4 to 7 degree Celsius than in the middle of the last century compared to 1900. Particularly since 1985 the rate had accelerated. The rise has been estimated at about 0.8 degree Celsius per decade. So the global community has become concerned with such alarming rise in global temperature. The Earth Summit observed that the green house gas concentrations in the atmosphere at existing level would have dangerous interference with the climate systems. Unfortunately since that time, the condition had not improved and emissions of green house gases continued to rise, even as the community of climate scientists has moved towards a consensus that already human activities are having a discernible impact on global climate (IPCC, 1996). For equity considerations it is essential for climate protection that the industrialised nations shall undertake immediate action to drive down their emissions towards the level of 20 percent below the 1990 emissions by 2005. The goal of reducing carbon dioxide had been 7 percent for USA, 8 percent for European Unions and 6 percent for Japan.

A study by Tellus Institute and SEI Boston has shown that, "protecting global climate will require that annual emissions of green house gases be reduced by at least 60 percent below the present rates". This study also finds that the United States alone can reduce its annual emissions to 10 percent below 1990 levels by 2005 and to 22 percent below 1990 levels by 2010, at net savings to its economy relative to the present energy-intensive and fossil-based path. This is true for all developed nations emitting large concentration of carbon dioxide gases. The industrialised world must reduce its annual carbon dioxide emissions from about 3 tons per capita to about 0.3 tons per capita and the industrialised world must achieve economic growth along a path to low carbon intensity.

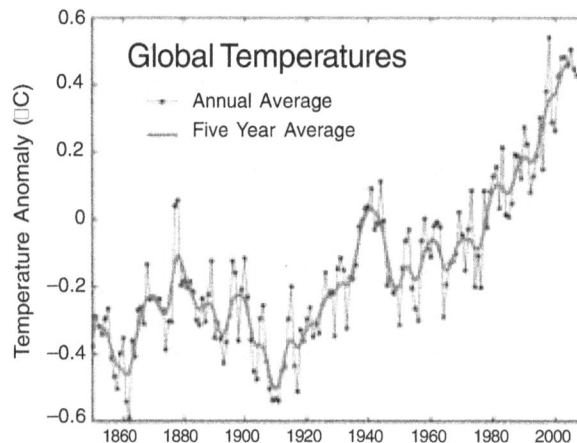

**Figure 8.3:** Global Temperatures—Annual Average and Five Year Average

## Trends in Global Rise of Temperatures

- IPCC concluded in 1995 that global warming is real, serious and accelerating.
- The most likely cause is primarily from burning of fossil fuels—coal, oil and gasoline.
- Increasing the amount of carbon dioxide and other GHGs trapped in the earth's atmosphere leads to rising global temperatures.
- The burning of fossil fuels has increased nearly five times since 1950.
- 1998 was the hottest year in history. 2005 was a close second. The five hottest years had occurred in the 1990s.

Climatologists in general have the consensus that the global average surface temperature had risen over the last century. Within this general agreement, some individual scientists disagree with the scientific consensus that human activities induce most of this warming.

The scientific consensus was summarised in the 2001 Third Assessment Report of the Intergovernmental Panel on Climate Change (IPCC) as follows:

1.  The global average surface temperature has increased by $0.6 \pm 0.2°$ C since the late 19th century, and $0.17°$ C per decade in the last 30 years.
2.  "There is new and stronger evidence that most of the warming observed over the last 50 years is attributable to human activities", in particular emissions of the greenhouse gases—carbon dioxide and methane.
3.  If greenhouse gas emissions continue the warming will also continue, with temperatures projected to increase by $1.4°$ C to $5.8°$ C between 1990 and 2100. Accompanying this temperature increase will be increases in some types of extreme weather and a projected sea level rise of 9 cm to 88 cm, excluding "uncertainty relating to ice dynamical changes in the West Antarctic ice sheet". On balance the impacts of global warming will be significantly negative, especially for larger values of warming.

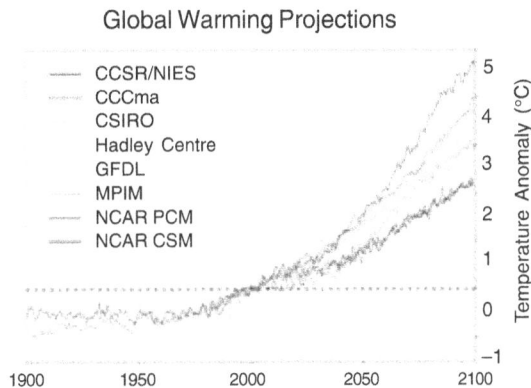

**Figure 8.4:** Global Warming Projections by Different Agencies

## *Impacts of Global Warming*

*   Rising temperature is estimated at $0.8°$ C per decade.
*   Melting of the ice cap over the poles—lowering and diminishing icebergs are common over the polar seas.
*   Sea levels are rising. On average sea levels are 10 to 25 cm. higher than a century ago.
*   Some 80 percent of sea beaches are eroding, often at the rate of many metres per year.
*   Rising tides threaten the survival of low lying coral islands—Marshall Islands, Anguilla, Tocklau and Maldives. Even the cities like Venice and Alexandria are threatened.
*   SST has been rising, particularly in the Pacific Ocean. $1°$ C rise has been recorded in the west of Californea. SST is related to ENSO causing climatic uncertainity.

The global warming has serious impact upon global climate. Not only the temperature of the earth's surface and lower atmosphere has been rising, but also it affects the thermal structure of the stratosphere thus causing an uncertainty in global climatic pattern. Due to formation of ozone holes the chloflurocarbons can increasingly penetrate the stratosphere. If this happens, then the ozone concentration in the stratosphere will be largely affected. It may result in cooling of the stratospheric zone and lapse rate will be reversed. It will impute an uncertainty of weather processes in the troposphere. The observed trend since 1960 has been a cooling of the lower stratosphere. Reduction of stratospheric ozone also has a cooling effect, but substantial ozone depletion was not evidenced until the late 1970s. It is not possible at this stage to predict the change. But the change is certain. Moreover, the ozone hole will allow the penetration of UV rays to the earth's surface causing serious health hazards of carcinogenic diseases. Even the plants and biotic lifes on earth's surface will be seriously perturbed. Ills are many—melting of the ice caps raising the sea level and inundating many coastal lowlands. Even the ENSO phenomena are related to global warming, which largely affects the monsoon rainfall, some years wet and some years dry. The climate protection scenario demands effective control of GHG emissions on earth's surface.

Though it is difficult to relate specific weather events to global warming, an increase in global temperatures may initiate macro changes, including glacial retreat, Arctic shrinkage, and sea level rise. Variation in the amount and pattern of precipitation may cause flooding in one region and drought in other region. There may also be shift in the frequency and intensity of extreme weather events. Other effects may induce changes in agricultural yields, reduced summer streamflows, species extinctions, and rise in the range of disease vectors.

### *Climate Model Predictions*

*Temperature Prediction*

Climate models predict that the effect of rising greenhouse gases will result in a warmer climate. However, even when the same assumptions of future greenhouse gas levels are used, there still exists a considerable range of climate sensitivity.

Climate model projections by the IPCC indicate that average global surface temperature will likely rise a further 1.1 to 6.4° C (2.0 to 11.5° F) during the twenty-first century. The range of values results from the use of differing scenarios of future greenhouse gas emissions as well as models with differing climate sensitivity. Although most studies referred to the period up to 2100, the warming and sea level rise may continue for more than a thousand years even if greenhouse gas levels are stabilised. The delay in reaching equilibrium may be caused by the large heat capacity of the oceans.

Climate models suggest a good match to observations of global temperature changes over the last century, but do not simulate all aspects of climate. These models do not unambiguously attribute the warming that occurred from approximately 1910 to 1945 to either natural variation or human effects. The models emphatically indicate that the warming since 1975 is dominated by anthropogenic greenhouse gas emissions.

Based on the estimates made by the World Meteorological Organization (WMO) and the Climatic Research Unit it may be concluded that 1998 was the warmest year followed by 2005 as the second warmest year. Temperatures in 1998 were unusually warm due to the occurrence of strongest El Nino in that year.

*Attributed and Expected Effects*

An increase in global temperatures may in turn attribute remarkable changes: melting of the glaciers in polar and high altitude region, shrinkage of the Arctic sea, sea level rise, submergence of coral and mid oceanic islands, inundation of deltas and mangroves in the littoral areas and worldwide sea level rise. Changes in the amount and pattern of precipitation may result in flooding and drought. There may also be shifts in the frequency and intensity of extreme weather events. Other effects may also manifest changes in agricultural yields, reduction of summer stream-flows, extinction of species, and increases in the range of disease vectors.

A 2001 report by the IPCC suggests that glacier retreat, ice shelf disruption such as that of the Larsen Ice Shelf, sea level rise, changes in rainfall patterns, and increased intensity and frequency of extreme weather events, are being attributed in part to global warming. From the satellite imageries the diminishing sizes of Larsen Ice Shelf could be seen (Figure 8.7).

The IPCC Fourth Assessment Report states that there is observational evidence for an increase in intense tropical cyclone activity in the North Atlantic Ocean since 1970, relating to the rise in sea surface temperature. Increased intensity of storms and extreme weather events may also be evidenced in other oceanic basins, particularly in the tropics.

Additional probable effects would be the rise of sea level 110 to 770 millimetres (0.36 to 2.5 ft) between 1990 and 2100. This estimate is high above the normal change in eustatic level. Another important aspect of change will be the lowering of pH of ocean water. This may lead to possible slowing of thermohaline circulation in the ocean. The shift in oceanic circulation may result in uncertainty of climate, as the oceanic circulation is closely linked to atmospheric circulation.

Uncertainty and changing pattern of climate along with extreme weather events may cause certain repercussions to agriculture. This would result in reduction in crop production in many parts of the world resulting in the problem of food security. Countries having the dense population will suffer badly in coming future. Signs of such adversity have already emerged in recent years in the less developed countries. Increasing deaths, displacements, and economic losses are expected in the regions of growing population densities due to extreme weather conditions, attributed to global warming. However, the temperate regions are supposed to experience some benefits, such as fewer deaths due to cold exposure.

Global warming along with the variability of climate would affect the biotic environment in general. Living conditions for different species of flora and fauna could be either lost or reduced in many natural habitats. One study predicts 18 percent to 35 percent of a sample of 1,103 animal and plant species would be extinct by 2050, based on future climate projections. Loss of biodiversity in many parts of the world would be evidenced in near future. It may result in the rise of disease vectors and help the spread of diseases like malaria, dengue fever and avian flu.

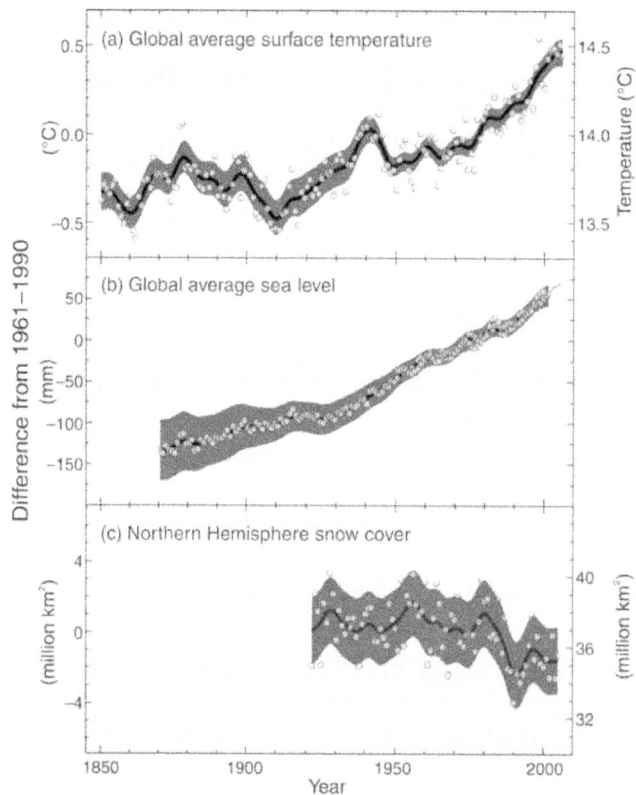

**Figure 8.5:** Global Status: Average Surface Temperature (*Top*), Average Sea Level (*Middle*) and Northern Hemisphere Snow Cover (*Bottom*).

*Source:* IPCC

### *Present Warning Signs for the Effects of Global Warming*

*Temperature Rise*

1.  Many parts of the world now record high temperatures in the summer months deviating from the normal, in some areas by as much as 4 degrees Fahrenheit. In fact, in USA

all states experienced either "above normal" or "much above normal" average temperatures in 2006.

2. The National Oceanic and Atmospheric Administration (NOAA) declared 2006 to be the second warmest year on record for the United States, with an annual average temperature of 55 degrees Fahrenheit, close to the highest record in 1998. Every year from 1998 through 2006 ranks among the top 25 warmest years on record for the United States, an unprecedented occurrence, according to NOAA.

## *Drought and Wildfire*

1. Extensive droughts are occurring. In many parts of the world drought frequency has increased.
2. Wild fire is common in many parts of the world, starting from Indonesia to USA. Close to 100,000 fires were reported and nearly 10 million acres burnt, 125 percent above the 10-year average.
3. With continued rise of temperature and related occurrence of wildfire fire-fighting expenditures have consistently exceeded $1 billion per year.

## *Intensification of Rain Storms*

1. Annual precipitation has increased by 5 to 10 percent during the 20th century, with heavy downpours in some areas.
2. The IPCC reports the increased frequency of intense rain events during the last 50 years. Global warming more likely contributed to the trend.
3. According to NOAA statistics, the Northeast region of USA recorded its wettest summer in 2006, exceeding the previous record by more than 1 inch.

## *Health Hazards*

1. In 2003, extreme heat waves claimed an estimated 35,000 lives in Europe. In France alone, nearly 15,000 people succumbed to death due to soaring temperatures, which was recorded as high as 104 degrees Fahrenheit and remained extreme for two weeks.
2. Most parts of North America experienced a severe heat wave in July 2006, which caused deaths of at least 225 people.
3. Studies show that a higher level of carbon dioxide results in an increase in the growth of weeds whose pollen triggers allergies and exacerbates asthma.
4. Mosquitoes as vectors of diseases are favoured with the warm climate. Mosquitoes as vectors of dengue fever viruses were previously restricted to elevations of 1000 metre but now found at an elevation even exceeding 2000 metre in the Andes Mountains

of Colombia. Malaria has been detected in new higher-elevation areas in Indonesia and Kenya.

*Warming of Ocean Water*

1. Occurrences of categories 3 and 4 storms have markedly increased during the last 35 years as a consequence of rising Sea Surface Temperature (SST).
2. The hurricane season in 2005 over the Atlantic Ocean recorded 27 storms of which 15 turned to be hurricanes. It was unprecedented. Of these 5 became category IV and 4 developed into category V hurricanes.
3. Hurricane Katrina in the month of August, 2005 was the costliest and one of the deadliest hurricanes in U.S. history.

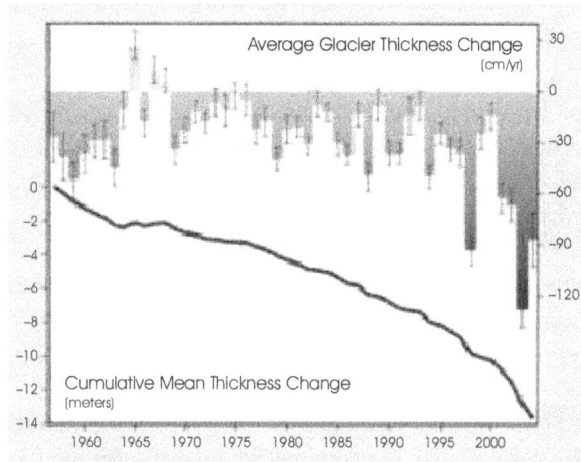

**Figure 8.6:** Lowering of Glacier Thickness due to Melting

*Retreat and Melting of Glaciers*

1. Glaciers over the high altitudes are retreating. One estimate shows that at the present rate of retreat all the glaciers over the Glacier National Park in USA will be lost by 2070. The glaciers over the Himalayas, Alps and Andese are also retreating.
2. Satellite imageries show progressive disintegration of the Larsen B ice shelf over the Antarctic since 1995. The ice shelf virtually collapsed between January and March, 2002.
3. NASA estimate shows the melting of polar ice cap at an alarming rate of 9 percent. Since 1960 the thickness of Arctic ice has shrunk by 40 percent.
4. According to the National Snow and Ice Data Centre the Arctic ice extent showed nearly half a million square miles less ice in September, 2007 than observed in

September, 2005. Over the past 3 decades, more than a million square miles of perennial sea ice, an area of the size of Norway, Denmark and Sweden combined, has disappeared.

5. The climate models indicate that sea ice over the Arctic region will be lost at the current rate of increasing temperature. By the summer of 2040 the Arctic Sea will be ice-free.

**Figure 8.7:** Melting and Collapse of Larsen B Glaciers over the Antarctic

## Rise of Sea Level

1. Sea level rise has been recorded at 4 to 8 inches in the last century (20th century) due to both thermal expansion of the oceans (eustatic level) and melting of the glaciers and ice caps. The pace of sea level rise appears to be alarming. IPCC estimates show that the sea level could rise by 10 to 23 inches by 2100.

2. In 1990 though the Greenland Ice Mass remained stable, the ice sheet has increasingly melted. Such melting contributes nearly one hundredth of an inch rise of the global sea level per year.

3. Greenland has 10 percent of the total global icemass. If melting occurs the sea level could rise to height of 21 feet.

## Disruption of Ecosystem

1. Nearly 2000 plant and animal species are shifting towards the poles at an average rate of 3.8 miles per decade to survive the excess heat due to global warming. Another study indicated the movements of species to higher altitudes in Alpine areas at a rate of 20 feet per decade in the later half of the 20th century.

2. The latest IPCC report indicated that approximately 20 to 30 percent of plant and animal species are threatened with extinction if there be rise of global temperature exceeding 2.7 to 4.5° F.

3. Polar bears are confronting the hazards of drowning as they have to swim more from one ice shelf to other. Their habitat has been threatened. By 2050 nearly half of the total number of polar bear will become extinct.

4. Due to increased acidification as a result of warming coral reefs are denuded at a faster rate. With the rise of 3.6° F temperature nearly 97 percent of the coral reef will be wiped out.

5.   Due to loss of habitat nearly 33 percent of penguin population declined in the Antarctic region.

### Global Cooling

*Global cooling* in general refers to an overall cooling of the earth's surface. It refers primarily to a conjecture during the 1970s of imminent cooling of the earth's surface and atmosphere along with a posited commencement of glaciation. This hypothesis never had significant scientific support, but gained temporary popular attention due to press reports that did not accurately reflect the scientific understandings of glacial cycles and a slight downward trend of temperatures from the 1940s to the early 1970s.

In the 1970s, there was rising awareness that estimates of global temperatures indicated cooling since 1945. The climate scientists were divided in two groups—majority advocating the concept of global warming and the minority (about 10 percent) held a fractured view of global cooling. The general public had little awareness on the effects of carbon dioxide on climate, although Paul R. Ehrlich advocated climate change from the greenhouse gases in 1968. By the time the idea of global cooling reached the public press in the mid-1970s, the temperature trend had stopped going down, and there was concern in the climatologic community about the effects of carbon dioxide. It was accepted that both natural and man-made effects caused variations in global climate.

On April 28, 1975 an article in *Newsweek* magazine, titled '*The Cooling World*' was published. The contents referred to 'ominous signs that the earth's weather patterns have begun to change' and pointed to 'a drop of half a degree [Fahrenheit] in average ground temperatures in the Northern Hemisphere between 1945 and 1968'. The *Newsweek* article did not state the cause of cooling; it stated that 'what causes the onset of major and minor ice ages remains a mystery' and cited the NAS conclusion that 'not only are the basic scientific questions largely unanswered, but in many cases we do not yet know enough to pose the key questions'.

What were just conjectures in the late seventies, now evidences and causes of global cooling are piling up. The concept of global cooling in one way is linked with the geologists' view on glacial cycle and interpretation of changing paleo-climate. We will discuss the issue in the next chapter on 'Climate Change' in detail. However, an assessment of this minority view of global cooling has been made in this chapter for comprehensive knowledge on the subject.

The concept of global cooling has its root in the changing climate of the world, alternately warm and cool at an interval exceeding 10,000 years. Even the evidences of Little Ice Age in the recent past strengthen the idea of global cooling. The drop of temperature from 1940 onwards up to 1980 indicated that the global warming was not continuous and carbon dioxide could not be the immediate cause. Dissenting scientists considered the global warming as a temporal phase to be replaced by cooling at the end of cycle. Milankovich's *Astronomical*

*Theory* for climate change boosted the ideas for global cooling. Orbital changes occur over thousands of years, and the climate system may also take thousands of years to respond to orbital forcing. Theory suggests that the primary driver of ice ages is the total summer radiation received in northern latitude zones where major ice sheets have formed in the past, near 65° N latitude. Past ice ages correlate well to 65° N summer insolation (Imbrie 1982). Astronomical calculations show that 65° N summer insolation should increase gradually over the next 25,000 years, and that no 65° N summer insolation declines sufficient to cause an ice age are expected in the next 50,000–100,000 years.

As for the prospects of the end of the current interglacial period (Note: only valid in the absence of human perturbations), it is not certain that the interglacial periods previously only continued for about 10,000 years. Milankovitch-type computations indicate that the present interglacial period would probably continue for tens of thousands of years naturally. Other estimates (Loutre and Berger, based on orbital calculations) put the unperturbed length of the present interglacial at 50,000 years. Berger assumed that the present $CO_2$ perturbation would last long enough to suppress the next glacial cycle entirely.

Many of the scientists who are opposed to the doctrine of global warming now consider, judging from the record of the past interglacial ages, that the present time of high temperatures should be drawing to an end, leading into the next glacial age. However, it is possible, or even likely, than human interference has already altered the environment so much that the climatic pattern of the near future will follow a different path. In 1972 Emiliani cautioned, "Man's activity may either precipitate this new ice age or lead to substantial or even total melting of the ice caps". By 1972 a group of glacial-epoch experts at a conference agreed that "the natural end of our warm epoch is undoubtedly near".

Apart from the phenomena of glacial cycles and its explanation by Milankovitch, reknowned cosmologist Carl Sagan had predicted cooling due to increased albedo on earth's surface. In the science series 'Cosmos: A Personal Voyage', Carl Sagan warned of catastrophic cooling through the burning and clear cutting of forests. He postulated that the increased albedo of the earth's surface might draw a new ice age nearer.

Concerns on nuclear winter emerged in the early 1980s from several reports. Many scientists speculated the effects due to catastrophes such as asteroid impacts and massive volcanic eruptions. A suggestion that massive oil well fires in Kuwait and other large oil fields would cause significant effects on climate was, however, quite incorrect. Volcanic eruptions in the past certainly played important role in global cooling. As the sky had been overcast with clouds for a longer period, incoming solar radiation decreased and cooling was initiated.

### *Recommended Policies for Carbon Dioxide Reduction*

Debate on global cooling *vis-à-vis* global warming still continues. Global warming has become an established fact that would affect us continuously throughout the next centuries, and even millennium. No expert now doubted that $CO_2$ and other greenhouse gases were at least partly responsible for the unprecedented warming all around the globe since the 1980s. In 2005 a team compared computer calculations with long-term measurements of temperatures in the world's ocean basins. In each separate ocean basin, they showed a close match between observations of rising temperatures at particular depths, and calculations of where the greenhouse effect should appear. Nothing but greenhouse gases could produce the observed ocean warming—and other changes that were now showing up in many parts of the world, as predicted.

Recent computations also indicated an imbalance. The earth now gets from insolation nearly a watt per square metre more than it was radiating back into space, averaged over the earth's surface. That was enough energy to cause truly serious effects if it continued. It is called the 'smoking gun' and considered as a strong evidence for greenhouse effect on global warming.

So the global community now seeks remedial measures to resolve the adverse impact of global warming. IPCC and many organisations including NGOs are involved in critical scientific studies to achieve the conservation of habitats and goal of sustenance. We need action to cut the GHG emissions, particularly the Carbon Dioxide in the global sphere to a level prior to 1990. The climate protection policies would help in the growth of an economically and environmentally sustainable future through technical innovations and diffusions. These policies would guide the economy toward more efficient, lower cost, less polluting, more secure, and more sustainable production and consumption of energy. These are principally:

i) *Renewable content standards* would demand an increasing share of renewable resources for electricity and automobile fuels, implemented through credit trading market mechanisms.

ii) *Pollutant emissions caps* would limit $SO_2$, $NO_x$, particulates and $CO_2$ emissions in the electric power sector through allowance trading systems.

iii) *Biomass co-firing expansion* would quickly establish a large amount of renewable energy in the electric supply sector without demanding substantial investment in new power plants.

iv) *Advanced vehicles initiative* would introduce stronger fuel economy and emissions standard along with pricing reform, incentives and require management, to support commercialisation of clean and efficient vehicles and better travel patterns.

v) *Investment tax credits* would advance the adoption of modern and energy efficient technologies for new manufacturing equipment through an incentive that could be paid for fees on energy purchases.

vi) *Support for research, development and technology* would help in the advancement of innovation at a pace that makes improved energy efficient and low carbon technologies available more rapidly.

vii) *Market transformation initiatives* would reduce the cost of transaction and help move products from prototype to commercial production lowering the key hurdles between potential and realised energy savings.

viii) *Initiative to accelerate adoption of industrial Combined-Heat-and-Power (CHP)* would help in refining citing protocols and ensuring market access, to fairly represent its economic and market benefits.

ix) *Appliance and building standards* would set up norms for equipment, design and performance that, through purchases and practices, would reduce energy used in providing services in homes and offices.

---

## Further Readings

Barry, R.G. and R.G. Chorley (1968) *Atmosphere, Weather and Climate,* Methuen & Co., London.

Brasseur, G.P., J.J. Orlando and S.G. Tyndall (1999) *Atmospheric Chemistry and Global Change,* Oxford University Press, New York.

Critchfield, H.J. (1975) *General Climatology,* Prentice Hall India Ltd., New Delhi.

Emilliani, C. (1955) 'Pliestocene temperatures', *Journal of geology,* **53**.

Fairbridge, R.N. (1975) 'The changing level of the sea', *Earth Sciences,* **805**.

Hiddore, J.J. (1996) *Global Environmental Change,* Prentice Hall, Upper Saddle River, N.J.

Intergovernmental Panel on Climate Change (2007) *Fourth Assessment Report,* on line *www.ipcc.ch/*

Intergovernmental Panel on Climate Change (2001) *Summary for Policy Makers*, on line *www.ipcc.ch/*

Keppler, Frank; Marc Brass, Jack Hamilton, Thomas Röckmann (2006) *Global Warming—The Blame is not with the Plants,* Max Planck Society.

Lagett, J. (1990) *Global Warming,* The Green Peace Report, Oxford University Press, Oxford.

Lamb, H.H. (1970) 'Climatic variation and our environment today and in the coming years', *Weather,* **45**.

Oliver, J.E. and J.J. Hiddore (2002) *Climatology,* Pearson Education, Delhi.

Panel on Climate Change Feedbacks (2004) *Understanding Climate Change Feedbacks,* Climate Research Committee, National Research Council, Washington.

Plass, G.N. (1956) 'The carbon dioxide theory of climatic change', *Tellus,* **8**.

Plass, G.N. (1962) 'Carbon dioxide and the climate', *in White (ed.) 'The Study of the Earth'*.

Reynolds, R.W. (1988) 'A real-time global sea surface temperature analysis', *Journal of Geology,* **1**.

Royal Society (2001) *The Science of Climate Change,* Royal Society, London.

Saha, P.K. and P.K. Bhattacharya (1994) *Adhunik Jalavayu Vidya,* West Bengal State Book Board, Government of West Bengal, Calcutta.

Sherwood, B. Idso (1989) *Carbon Dioxide and Global Change,* Institute of Biospheric Research Press, Arizona.

Tans, Pieter (2008) *Trends in Atmospheric Carbon Dioxide,* National Oceanic and Atmospheric Administration, Mauna Loa.

Turekian, K.K. (1996) *Global Environmental Change,* Prentice Hall, Upper Saddle River, N.J.

Warrick, R.A., T.M. Barrow and T.M. Wigley (1993) *Climate and Sea Level Change,* Cambridge University Press, New York.

# Chapter 9

## CLIMATE CHANGE

### Introduction

Weather is the day-to-day state of the atmosphere, and is a chaotic non-linear dynamical system. On the other hand, *climate* is the average state of weather and fairly stable and predictable. Climate includes the average temperature, amount of precipitation, days of sunlight, and other variables that could be measured at any given site. Climate change states any long-term significant change in the average weather in a given region time specific. Average weather indicates diurnal average temperature, precipitation and wind patterns. It involves changes in the variability or average state of the atmosphere over durations ranging from decades to millions of years. These changes might be due to dynamic process, external forces including variations in sunlight intensity, and more recently by anthropogenic interferences.

We can find that the climate had been changing since the inception of the earth relating to dynamic change in the atmosphere. We know from the past records that initially the climate on earth's surface was extremely hot with extensive cloud cover and incessant rainfall. The formations of oceans, creations and growth of life shaped the climatic conditions to a great extent. It was difficult to trace the climatic situation prior to the Cambrian period because no fossil records were available due to non-existence of life in that period on our planet. In later periods the nature of climate could be outlined from the fossil records.

With the advancement of science and applications of new tools we can preview the past climate, ranging the period 350 million to 500 million years back. Extremely high temperatures associated with very intense rainfall caused the growth of rain forests over wide areas on earth's surface in the carboniferous period. In this geological era of reptiles, about 150 million years ago, the dianosaurs reigned over the earth. The climatic conditions continually changed since the cretaceous period with the progressive fall of temperatures. Reasons were many, of which the orogenic theory by Lowenstem and Epstein was the most probable. In the quaternary period extremely cold climate prevailed on the earth's surface with the initiation of glacial period. In this period glaciers extended over wide areas on earth's surface. It covered large parts of Europe, North America and other continents extending up to present sub-tropical latitudes. The depth of the glaciers extended for about 1,500 metres. The glaciers over the Himalayas, Alps, Rockies and Andese invaded the low lands.

At present glaciers cover nearly 6 million square miles area of the continents. These are found in the Antarctic, Arctic and highland regions of the world. In the past (Pliestocene Ice Age) the glaciers were extended over 13 million square miles area of the present glacier-free regions of the world. The waning of glaciers after the pliestocene age was the key pointer to a changing climate, returning from cold to warm conditions. Glaciers are considered to be the most sensitive indicators of climate change. With modern tools we can gauge the advance of the glaciers during the cooling period and similarly the retreat during the warming period on moderate time scales. Glaciers grow and collapse, both contributing to natural variability and effectively signalling externally forced changes. The advance and retreat, growth and collapse of the glaciers on a global scale could only occur due to swinging of climate from cold to warm in different geological periods in the past.

**Inter-glacial Cycles**

The glacial period in the pliestocene age was not continuous. There had been breaks of intervening warm periods resulting in the retreat of glaciers on a large scale. Again with the advent of cold conditions glaciers reemerged. A cycle could be evidenced in the glacial period, cold—warm—cold. It is probable we are traversing amidst the interglacial period. After the warm phase the cold phase may again emerge. So, the climate change deems to be natural phenomena.

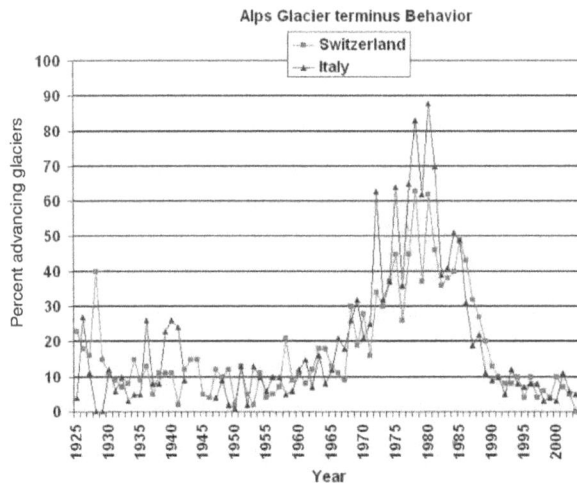

**Figure 9.1:** Advance and Retreat of Alpine Glaciers during the Last Century

In North America 4 glacial periods could be recognised with interglacial periods. These are: 1. Nebraska period–320,000 years ago; 2. Kansan period–240,000 years ago; 3. Illinois period–115,000 years ago and early and late Wisconsin period–60,000 to 25,000 years ago.

Similarly in the Alpine region of Europe 4 glacial periods could be distinguished. These are: 1. Gunj period, 2. Mindel period, 3. Riss period and 4. Wurm period. Penck and Brookner named these glacial periods after the names of 4 tributaries of the River Daneube.

The study of the glacial periods came to the minds of natural scientists exploring the changing climatic conditions. It was a big question why there had been swinging thermal conditions throughout the entire glacial periods. Even in the recent past a few centuries back in the early 15th century (1400 A.D.) to mid-19th century (1850 A.D.) Europe experienced a very cold climate. This period is known as *Little Ice Age.* During this period all the Alpine glaciers advanced. Arctic glaciers were also very much extensive. After 1850 the temperature conditions began to change, temperature rising corresponding to the industrial age and burning of fossil fuels.

The most significant climate processes of the last several million years are the glacial and interglacial cycles of the pliestocene ice age. It is difficult to explain how these are caused. It is of great concern to understand the reasons for such changes for better understanding of climate change in the present context. The Milankovich theory of orbital variations may explain the initiation of cool conditions, the internal responses involving continental ice sheets and 130 m sea-level change certainly played a key role in deciding what climate response would be observed in most regions. Other changes, including Heinrich events, Dansgaard–Oeschger events and the Younger Dryas show the potential for glacial variations to influence climate even in the absence of specific orbital changes. Many other hypotheses for such enormous cooling could be postulated. These include orogenic movement theory, volcanic dust theory, plate tectonics, solar radiation variation and sun spot cycle concepts. The role of carbon dioxide in the atmosphere is also highlighted in understanding the global cooling due to deficit in the atmosphere under certain conditions.

### *Orbital Variations*

Variations in the Earth's orbit, in essence, an extension of solar variability. The slight variations in the Earth's orbit can lead to changes in the distribution and abundance of sunlight reaching the earth's surface. Such orbital variations, known as Milankovitch cycles, could be caused due to mutual interactions of the earth, its moon, and the other planets. These variations are considered the driving factors underlying the glacial and interglacial cycles of the present ice age. Subtler variations are also present, such as the repeated advance and retreat of the Sahara desert in response to orbital precession. Changes in the tilt of the earth in the orbit can result in the severity of the seasons, more tilt causing more severe seasons—warmer summers and colder winters; less tilt causing less severe seasons—cooler summers and milder winters. Wobbling of the earth continues in space so that its tilt changes between about 22 and 25 degrees on a cycle of about 41,000 years. The cool summers and moderate winters can persist for longer periods than the hot summers and cool winter. It is the cool summers which are considered to deposit snow and ice to last from year to year in high

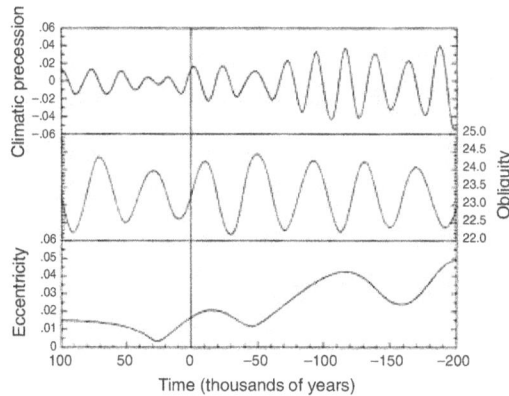

**Figure 9.2:** Milankovitch Cycle Showing the Predictability of Variations

latitudes, eventually forming the massive ice sheets. The albedo factor also plays, as well, because an earth covered with more snow reflects more of the sun's energy into space, causing additional cooling. In the 1960s, Mikhail Budyko, a Russian climatologist, developed a simple energy-balance climate-model to study the effect of ice cover on global climate. Budyko observed that if ice sheets advanced far enough out of the polar regions the increased reflectiveness (albedo) of the ice could lead to further cooling and the formation of more ice until the entire earth was covered in ice and stabilised in a new ice-covered equilibrium. Moreover it appears that the amount of carbon dioxide in the atmosphere consequently drops as ice sheets grow, also enhancing the cooling of the climate. The intervening period for orbital variations was calculated at 41,000 years by Milancovch. Emilliani confirmed the calculation made by Milancovich by his study on glacial cycles.

### *Volcano Eruptions*

When the volcanoes erupt a great mass of emitted smoke and dust particles cover the canopy of sky. We have learnt that the overcast sky can affect the insolation. On an overcast day the solar radiation will be much less than in the cloud-free days. If such cloud cover with ash and dust particles persist for longer periods due to volcanic eruption on a large scale it would certainly affect radiational balance of the earth. If the radiational balance becomes tilted with more outgoing energy than less incoming energy the earth would continue to cool and in turn ice age may emerge.

Large eruptions, known as large igneous provinces, occur only a few times every hundred million years, but can reshape climate for millions of years and cause mass extinctions of living species. There had been many occasions of volcanic eruptions in the geological past and even in the recent times. During the cretaceous period the earth was probably very unstable charging the volcanoes to be active. We can even locate once very active volcanoes in the *suture zone* during the formation of the Himalayas in the upper cretaceous period. The volcanic belt even today extends along the structurally weak belt of the young fold mountains extending from Iceland to Phillipines and also penetrating the Central Pacific basin. Many volcanoes are still active, some dormant and can erupt any time. In 1883, the eruption of Cracatoa in the East Indies occurred. The explosion was so massive that the sound of explosion was heard from a distance of 5,000 kilometre. The emitted dust and ash particles covered the sky for about two years. Though instrumental recordings were not possible globally in all the areas, cool temperature condition was perceived in many parts of the world. Even in the recent years the effects of volcanic eruptions in the East Asian countries in 1991 and 1992 had been felt in the climatic conditions. For example, the eruption of Mount Pinatubo in 1991 affected climate substantially.

A single eruption of the kind that occurs several times per century can affect climate, causing cooling for a period of a few years. Initially, scientists thought that the dust emitted into the atmosphere from large volcanic eruptions was responsible for the cooling by partially blocking the transmission of insolation on the earth's surface. However, measurements indicate that most of the dust emitted to the atmosphere returns to the earth's surface within six months. Exception may occur in case of large explosion as in the geological past as evidenced from the variation of climate in different periods. Over very long (geological) time periods, volcanoes release carbon dioxide from the earth's interior, counteracting the uptake by sedimentary rocks and other geological carbon dioxide sinks. So the volcanoes appear to be the extended part of the carbon cycle. However, this contribution is insignificant compared to the current status of emissions by human activities. The US Geological Survey estimates show that human activities generate more than 130 times the amount of carbon dioxide emitted by volcanoes.

### *Tectonic Theory*

Lowenstem and Epstein analysed the fossils of the cretaceous period and showed that the temperature on earth's surface was extremely high. The *Tethys,* the shallow seas were extensive over the earth's surface. Then there existed mountain ranges of lower heights. It is assumed that in the late cretaceous period subduction occurred over the large part of Pacific Ocean basin. As a result orogenic movements started along the weak structural belts of the earth's surface. There was great upheaval of landmasses on the earth's surface due to such intense instability. All the great mountains of the world, the Himalayas, Alps, Rockies and Andese, were formed. Hundreds of millions square kilometres area, earlier submerged under water, rose above as landmasses. We know that the landmass radiates heat more than the watermass and becomes cooler at a faster rate than the watermass. So the earth's surface tended to be cooler due to rise of landmass and shrinkage of large watermass. Due to continuous cooling large tracts of Antarctic, Arctic and Greenland became ice covered. The albedo factor, as stated earlier, enhanced the rate of cooling.

### Continental Drift and Plate Tectonics Concept

The continental drift theory by Wegner suggested the disintegration of the large continental mass *pangea* and the disintegrated blocks of landmass started to move about 300 million years ago. Due to such movements the relative position of the poles and equator changed. On the longest time scales, plate tectonics could shift the position of the continents, shape the oceans, form and tear down the mountains and generally serve to define the stage upon which climate exists. More recently, plate motions have been implicated in the intensification of the present ice age when, approximately 3 million years ago, the North and South American plates collided to form the Isthmus of Panama and shut off direct mixing between the Atlantic and Pacific Oceans.

### Solar Variation and Sun Spot Cycle

The sun is the ultimate source of essentially all heat energy in the climate system. The energy output of the sun, converted to heat energy at the earth's surface, determines shaping the earth's climate. On the longest time scales, the sun itself is getting brighter with higher energy output; as it continues its main sequence. This slow change or evolution has effect on the earth's atmosphere. It is supposed that, early in earth's history, the sun was too cold to support liquid water at the Earth's surface, leading to what is known as the *Faint young sun paradox*.

Presently we know that there are different forms of solar variation. Also there exist the 11-year solar cycle and longer-term modulations. However, the 11-year sunspot cycle does not manifest itself clearly in the climatological data. But the sun spot cycles are related to the solar intensity on sun's surface. It is suggested that the variations of solar intensity could have been influential in triggering the Little Ice Age, and for some of the warming observed from 1900 to 1950. The cyclical nature of the sun's energy output is not yet fully understood; it differs from the very slow change that is happening within the sun.

### Role of the Green House Gases

We know that the Green House Gases play significant role in global warming (*see also Chapter 8*). Greenhouse gases are also important in understanding the earth's climate history. The green house effect plays a key role in regulating earth's temperature.

Over the last 600 million years the concentrations of carbon dioxide varied from more than 5000 ppm to less than 200 ppm. It was caused primarily due to the effect of geological processes and biological innovations. The variations in greenhouse gas concentrations over tens of millions of years, however, are not well correlated to climate change, with plate tectonics perhaps playing a more dominant role. More recently the climate sensitivity indices have been used to derive a value for the $CO_2$–climate correlation. There are several illustrations of marked changes in the concentrations of greenhouse gases in the earth's atmosphere that deem to correlate to strong warming, including the Paleocene–Eocene thermal maximum, the Permian–Triassic extinction event, and the end of the Varangian snowball earth event.

## Paleo-climatic Evidences of Climate Change

The study of paleo-climatic conditions helps understanding of climate change in the present context. Different features of paleo-climate could be observed in the natural archives even today. Various methods and tools are applied to predict the nature of past climate.

### *Geological Records*

Petrographic characteristics, particularly of sedimentary rocks, indicate the nature of rocks under which climatic condition they were formed in various geological periods. For example, formation of sandstone indicates a dry climate compared to mud stone, which was formed under relatively wet condition. Formations of limestone and other carbonate rocks on landmass refer to the marine transgression and retreat of sea in later periods. Thus from the sequence of rock formations we can outline the trends of past climates in distant geological periods. Due to continued weathering and erosion many features on land surface had been lost. But the deposits in the deep ocean basins still exhibit the paleo-climatic features to construct the past climate. The lysocline marks the depth at which carbonate spontaneously dissolves in the oceans. Presently the lysocline level is around 4000 metre comparable to the median depth of the oceans. This depth depends on (among other things) temperature and the amount of $CO_2$ dissolved in the ocean. The lysocline rises with the addition of carbon dioxide, resulting in the dissolution of deep water carbonates and acidification. This deep-water acidification can be observed in ocean cores, which exhibit an abrupt change from grey carbonate ooze to red clays (followed by a gradual grading black to grey). It is suggested that acidification was more concentrated in the North Atlantic core, related to a greater rise in the level of the lysocline. It is estimated that the lysocline level rose by around 2000 metre in just a few thousand years.

### *Glacial Records*

The advance and retreat of glaciers on a time scale display the thermal conditions of the earth's surface during their movement. The atmospheric moisture conditions in the past could also be traced by the study of glaciers. Alman and others painstakingly studied the advance and retreat of glaciers in Sweden and determined positive correlation between the movement of glaciers and climate conditions. When the climate was cold and wet the glaciers advanced south extending over wide regions of Western and Central Europe. To the contrary with the advent of warm and dry conditions the glaciers retreated significantly. Now the radio-isotope technology is applied to study the glaciers. With this application it has been possible to analyse the changing temperature in Greenland and Antarctica for the past 100,000 years. About 60,000 years ago Antarctica was cooler by 5° C. Thenafter the temperature began to rise slightly. But the temperature again started to fall about 30,000 years back coinciding with the glacial age. It is vividly evidenced that the glacial cycles continued in different periods.

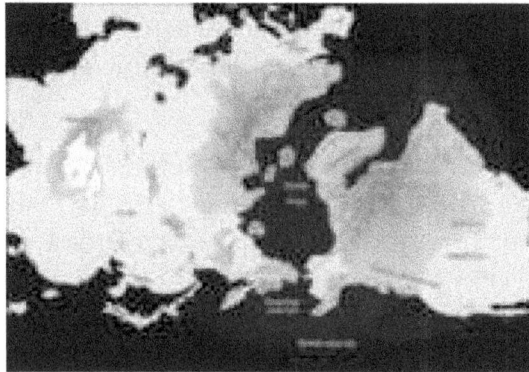

**Figure 9.3:** Northern Hemisphere Glaciation during the last Ice Ages. The Set up of 3 to 4 km Thick Ice Sheets Caused a Sea Level Lowering of about 120 m. Shaded Areas Indicate the Extent of Glaciation.

### *Biotic Records*

We know that there exists a correlation between the vegetation growth and climate. Vegetation characteristics indicate the nature of climate prevailing. Fossil records show luxuriant growth of rainforest coinciding with the Paleo-Eocene Thermal Maximum (PETM). Many signals of past climate could also be studied from the fossils of flora and fauna. One of the important methods is the study of tree-ring. The growth of tree-ring depends on climatic conditions. So it could indicate the nature of climate in different periods of its life cycle. Douglas found a correlation between annual precipitation and tree-ring growth. The study of pollens may also indicate the climate change in the past. We can record the phenomena of extinction of 35-50 percent of *benthic foramanifera* (particularly in deeper waters) over the course of ~1000 years. Emilliani studied in detail on *benthic foraminifera* to trace the ice age. The extinction of species, *benthic foraminifera*, was more marked than the extinction of dinosaur. To the contrary, planktonic foramanifera diversified, and dinoflagellates bloomed.

The emergence of mammals occurred radiating profusely around the Paleo-Eocene Thermal Maximum. The increase in mammalian abundance remains to be curious. There is no evidence of any increased extinction rate among the terrestrial biota. Increased $CO_2$ levels could probably promote dwarfing, which might encourage speciation. Many major mammalian orders, including the Artiodactyla, horses and primates, appeared as if from nowhere, and spread across the globe, 13,000 to 22,000 years after the initiation of the PETM.

The study of paleo-soils may also indicate the climate change. The soils, which are formed *in situ*, display their characteristics as influenced greatly by climate-vegetation complex. The deposited soils in the flood plains, lakes and littoral areas also indicate the changing pattern of climate during the period of their deposition.

### *Historical Records*

From the available records and documents in the form of literatures, paintings, sculptures, inscriptions etc. we can portray the climate change in the recent past. Most of the records and documents are available in Europe. In Europe and Mediterranean region during the first century the nature of the climate was almost the same as of today. By 350 A.D. the climate became wetter. In the fifth century hot and dry climate dominated. During this period water of many lakes dried in North America. Europe also experienced a hot and dry climate in the seventh century. Many of the passes in the Alps, which are now covered by ice, were open for traverse of cart movement. Same climatic conditions prevailed in North America. During this period in hot and dry climate the flow of water in the Nile was lean. In the ninth century Europe became wetter, but hot and dry conditions reemerged in the tenth and eleventh centuries. The Vikings settled in Greenland in 984 A.D. under relatively warm comfortable climate. The climate turned to be harsh and cold later. The Vikings were compelled to abandon their settlement over Greenland by 1410 A.D. In the thirteenth century the North Sea and North Atlantic Ocean were stormy. Thereafter cold conditions prevailed over Europe from the middle of the fifteenth century. The cold conditions remained harsh till the middle of the nineteenth century. This period extending from 1450 A.D. to 1850 A.D. in Europe and midlatitude region, was known as Little Ice Age. The glaciers in the Alps advanced. The ice sheets of the Arctic extended over wide areas in Europe. After 1850 A.D. the climatic conditions were reversed and returned to the sixteenth century status.

From the late half of the nineteenth century the temperature began to rise. Coincidently the rise was activated by the burning of fossil fuels with the advent of industrial revolution. The global temperature continued to rise, with a break of temperature fall from 1945 to 1975, till to date. (We have discussed the global warming in the previous chapter. So, repetition is not desired.)

### Tracing the Ice Age

Cezare Emilianni started his work in the laboratory of Prof. Harold C. Urey in the Chicago University. Prof. Urey devised a geological thermometer after Paul Nigli. Paul Nigli found that isotope analysis of the carbonate deposits would indicate whether the carbonates had been deposited in the fresh water environment or saline environment. Urey applied the isotope method to distinguish between the fresh water carbonate and saline carbonate. He further found that the availability of oxygen isotope in the carbonate would depend on the temperature condition of water during deposition. By application of this method prof. Urey attempted to measure the fossil temperature. For this he examined the remains of a cigar type species, known as *belemnite*. Belemnites were the ancestor of squids, and found 150 million years ago in the sea near Scotland. Emilliani, a student of Prof. Urey, applied this technique on the fossils of *foraminiferra* and measured the temperature conditions of the past 65 million years.

Emilliani and others observed the variations of temperature during the Cenozoic period in particular. It was a point of concern that the dinosaurs, large dominating reptiles, began to become extinct in this period due to marked fall of temperature. They studied in detail the temperature of the bottom of the Pacific Ocean with the application of isotope method. It was a remarkable finding in the study of natural history. It was found that 32 million years ago the temperature at the bottom (3000 meter deep) of Pacific Ocean was 10.6° C (51° F). The temperature dropped to 6.7° C (44° F) about 22 million years ago. About 1 million years back the temperature was recorded at 2.2° C (36° F). For the last 500,000 years there had been marginal variations of this temperature.

### Evidences of Past Glaciation

Evidences of past glaciation could be traced from various records—geological, chemical and palynological. We can specify different periods of glaciation in the earth's history and record different ice ages. Such evidences indicate the fluctuations of climate on the earth's surface in distant geological past. All the processes are natural without any anthropogenic interference, as the ice ages (with the exception of little ice age) occurred before the advent of man. Ice ages emerged and later lost. In between inter-glacial periods were warm and most of the earth's surface was ice-free. We can identify the following ice ages broadly:

### Huronian Ice Age

The earliest ice age, called the Huronian, probably prevailed around 2.7 to 2.3 billion years ago during the early Proterozoic Eon.

### Cryogenian Ice Age

The earliest well-documented ice age, and probably the most severe of the last 1 billion years, occurred from 850 to 630 million years ago during the Cryogenian period. During this period permanent ice covered the entire globe and a Snowball Earth was produced. This period was terminated very fast as water vapour returned to the earth's atmosphere. It is assumed that the end of this ice age was responsible for the subsequent Ediacaran and Cambrian Explosion, though this theory is recent and controversial.

*Andean-Saharan Ice Age*

A minor ice age, the Andean-Saharan, was evidenced from 460 to 430 million years ago, during the Late Ordovician and the Silurian periods. Ice covered the Andean and Saharan regions for a brief period.

*Permo-carboniferous Ice Age*

The *Permo-Carboniferous* period includes the latter parts of the Carboniferous and early part of the Permian period. During the Permo-Carboniferous period, about 300 million years ago, a great glaciation occurred on the earth's surface. It was probably the first instance of glaciation after a long warm period. It could be caused due to continental drift, as proposed by Wegner. The widespread distribution of Permo-Carboniferous glacial sediments in South America, Africa, Madagascar, Arabia, India, Antarctica and Australia indicates the evidence of glaciation. Glacial activity spanned virtually the whole of Carboniferous and Early Permian time. The Gondwana continent, the southern part of Pangea, was located near the south-pole about 290 million years ago at the end of carboniferous period. Glacial centres extended across the continents, forming glacial tillites and striations in pre-existing rocks. The Permo-Carboniferous ice sheet was so extensive that it would occupy a circle spanning 50 degrees of latitude centred on the pole. There had been extensive polar ice caps at intervals from 350 to 260 million years ago, during the Carboniferous and early Permian Periods, associated with the Karoo Ice Age.

*Karoo Ice Age*

The glacial deposits had been found in the Carboniferous strata in tropical continental landmass such as India and South America. It was earlier considered that the Karoo Ice Age glaciation reached the tropical landmass. Later it was realised that this tropical landmass was the *Gondowana* land, part of the *Pangea*. However, a continental reconstruction showed that ice was in fact limited to the polar parts of the supercontinent Gondwanaland.

*Pliestocene Ice Age*

The present ice age, known as *pliestocene* or *recent ice age,* started 40 million years ago with the growth of an ice sheet in Antarctica. It intensified during the late Pliocene, about 3 million years ago. The ice sheets spread over the Northern Hemisphere, and continued in the Pleistocene period. Since then, the world had been witnessing cycles of glaciation with ice sheets advancing and retreating on 40,000- and 100,000-year time scales. The most recent glacial period was terminated around ten thousand years ago.

Ice ages or glacial periods can be further specified by location and time. In North America 4 glacial periods could be recognised with interglacial periods. These are: 1. Nebraska period

–320,000 years ago; 2. Kansan period–240,000 years ago; 3. Illinois period–115,000 years ago and early and late Wisconsin period–60,000 to 25,000 years ago. Similarly the *Riss* (180,000–130,000 years ago) and *Würm* (70,000–10,000 years ago) refer specifically to glaciation in the Alpine region. Two others Alpine glaciation—Gunj and Mindel are also recognised.

Some scientists believe that the Himalayas play major role in the current ice age. It is suggested that the Himalayas have increased earth's total rainfall and therefore the rate at which $CO_2$ is washed out of the atmosphere, lowering the greenhouse effect. The Himalayas' formation began in the late cretaceous period, about 70 million years ago, when the Indo-Australian Plate collided with the Eurasian Plate. The Himalayas are still rising by about 5 mm per year as the Indo-Australian plate still traverses @ 67 mm/year. The history of the Himalayas, in general, is analogous with the long-term decrease in earth's average temperature since the mid-Eocene period, 40 million years ago.

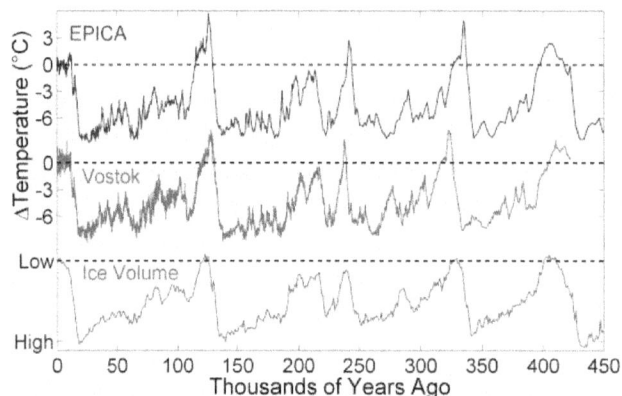

**Figure 9.4:** Ice Age Temperature Changes

However, it must be noted that the maximum spread of ice could not be maintained for the full interval. Unfortunately, the scouring action of each glaciation removed most of the evidences of prior ice sheets almost completely, except in areas where the later sheet failed to cover whole areas.

Ice age conditions were not uniform. Within the ice ages (or at least within the last one), more temperate and more severe conditions occurred. The colder periods are called *glacial periods* and the warmer are *interglacial periods*. Glacial periods are characterised by cooler and drier climates over most of the earth's surface. Large land and sea ice masses moved outward from the poles. Mountain glaciers or the valley glaciers in otherwise unglaciated areas moved downwards to lower elevations due to a lower snow line. The height of the snow line dropped due to lowering of temperature. Sea levels also dropped due to the removal of large volumes of water above sea level in the icecaps. There are reasons that ocean circulation patterns were also disrupted by glaciations. Since the earth has significant continental glaciation even today in the Arctic and Antarctic landmass, it may be said that we are currently in a glacial minimum of a glaciation. Such a period between glacial maxima is known as the *interglacial period.*

It is assumed that the earth has been passing through an interglacial period, known as the *Holocene* period, for more than 11,000 years. It was conventionally accepted that the typical interglacial period could last for about 12,000 years, after glacial epoch could step in. Recently it has been suggested by some scientists that the current interglacial period might continue like a previous interglacial period that prevailed for about 28,000 years. Predicted changes in orbital forcing predict that the next glacial period would not begin before about 50,000 years from now, irrespective of man-made global warming. In addition, anthropogenic forcing from increased greenhouse gases might outweigh orbital forcing for as long as intensive use of fossil fuels continues and the present interglacial cycle would continue further.

### Climate of Recent Glaciations

*Dansgaard-Oeschger Events*

*Dansgaard-Oeschger* events display rapid fluctuations of climate occurring every ≈ 1470 (± 532) years throughout the last glacial period. Twenty-three such events have been identified between 110,000 and 23,000 years ago. The Dansgaard-Oeschger events could be evidenced in the Greenland cores as traced back to the end of the last interglacial, the Eemian interglacial. Somewhat less direct evidence from Antarctic cores (the pattern of warmings; and the methane record) suggests that they were present in previous glacial periods as well.

In the Northern Hemisphere, they take the form of rapid warming episodes, even in terms of decades, each followed by gradual cooling over a longer period. Around 11,500 years ago, averaged annual temperatures on the Greenland icepack caused warming, which recorded a rise of about 8° C over 40 years, in three steps of five years. A variation of 5° C over 30-40 years was more common.

*Heinrich Events*

Heinrich events preceded the Dransgaard-Oeschger warmings and emerged as cold spells. These events are intimately linked with the Dransgaard-Oeschger events. It is suggested that D-O cycles might cause the events, or at least constrain their timing. So the course of a D-O event experienced a rapid warming of temperature, followed by a cool period lasting a few hundred years. This cold period resulted in the expansion of polar front, with ice floating further south across the North Atlantic Ocean.

The processes responsible for the timing and amplitude of these events (DO/Heinrich) are still unexplained. The pattern in the southern hemisphere appears to be different, with slow warming and much smaller temperature fluctuations. Indeed, the Vostok ice core was drilled before the Greenland cores, and the existence of Dansgaard-Oeschger events was not widely known until the Greenland cores were drilled. The events appear to reflect changes in the North Atlantic Ocean circulation, perhaps triggered by an influx of fresh water.

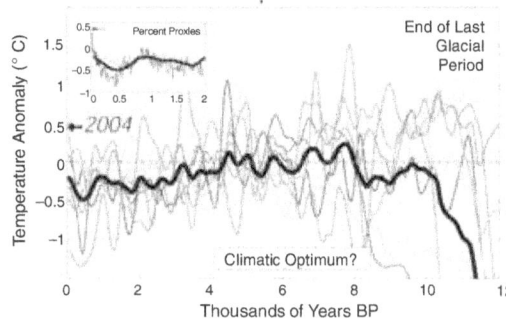

**Figure 9.5:** Holocene Temperature Variations

The events may be caused by an amplification of solar forcings, or by a cause internal to the earth system—either a "binge-purge" cycle of ice sheets accumulating so much mass they become unstable, as postulated for Heinrich events, or an oscillation in deep ocean currents. Maslin *et al.* (2001) suggested that each ice sheet had its own conditions of stability, but that on melting, the influx of freshwater was enough to reconfigure ocean currents, causing melting elsewhere. Due to associated influx of melt-water, the strength of the North Atlantic Deep Water current (NADW) was probably reduced, weakening the northern hemisphere circulation and therefore resulting in an increased transfer of heat polewards in the southern hemisphere. This warmer water caused melting of Antarctic ice, thereby reducing density stratification and the strength of the Antarctic Bottom Water current (AABW). This initiated the return of NADW to its previous strength, driving northern hemisphere melting. So another D-O cold event emerged. This theory also explains the Heinrich events' apparent connection to the D-O cycle. Even the Little Ice Age, a few centuries back (400 to 200 years ago) could be interpreted as the cold part of a D-O cycle.

*Younger Dryas*

The term *Younger Dryas* stadial was coined after the alpine/tundra wildflower *Dryas octopetala.* It was a brief (approximately 1300 ± 70 years) period of cold climate, also known as the *Big Freeze.* It occurred at the end of the Pleistocene between approximately 12,800 to 11,500 years ago following the Bölling/Allerød interstadial and preceding the Preboreal of the early Holocene. The Younger Dryas also resembles the Blytt-Sernander climate period, detected from the layers in North European bog peat. Blytt-Sernander climate period occurred approximately 12,900-11,500 years ago (calibrated), or 11,000-10,000 years ago (uncalibrated). The Younger Dryas experienced a rapid return to glacial conditions in the higher latitudes of the northern hemisphere between 12,900-11,500 years before present, in sharp contrast to the warming of the preceding interstadial deglaciation.

In Western Europe and Greenland, the Younger Dryas appears to be a well-defined synchronous cool climate period. But cooling in the tropical North Atlantic might have preceded this by a few hundred years; South America shows a less well defined initiation but a sharp termination. The Antarctic Cold Reversal appears to have started a thousand years before the Younger Dryas. In Western North America the Younger Dryas were probably less intense than in Europe. Younger Dryas cooling also occurred in the Pacific Northwest.

The Younger Dryas resulted due to a significant reduction or shutdown of the North Atlantic thermohaline circulation in response to a sudden influx of fresh water from Lake Agassiz and de-glaciation in North America. The South America cooled first, but the reasons remain still unexplained.

*Bond Events–1500 Years' Climate Cycle*

The climate cycle during the Holocene period is known as Bond events. The North Atlantic climate fluctuations occurred every ≈ 1470 years throughout the Holocene period as the *Bond events.* Eight such events had been identified. Bond events during the Holocene period were similar to the Pleistocene Dansgaard-Oeschger events. The hypothesis of 1500-year climate cycles in the Holocene was postulated by Gerard C. Bond of the Lamont-Doherty Earth Observatory at Columbia University, based principally on petrologic tracers of drift ice in the North Atlantic.

The existence of climatic changes, possibly on a quasi-1500 year cycle, is well established for the last glacial period, as studied from the ice cores. But the continuation of these cycles into the Holocene period is not well explained. Bond *et al.* (1997) proposed a cycle of climate around 1470 ± 500 years ago in the North Atlantic region. They suggested that many if not most of the Dansgaard-Oeschger events of the last ice age, conform to a 1500-year pattern, as like the climate events of later periods, such as the Little Ice Age, the 8.2 kiloyear event, and the start of the Younger Dryas. Even the occurrence of most weak events of the Asian monsoon had been correlated with the North Atlantic ice-rafting events over the past 9000 years.

## Recent Climate Changes

Interglacial warm periods dominated as changing phases in the Holocene period. The *Holocene Climate Optimum* was a warm period that occurred roughly during the interval 9,000 to 5,000 years ago. This event is also known by many other names. These are *Hypsithermal, Altithermal, Climatic Optimum, Holocene Optimum, Holocene Thermal Maximum*, and *Holocene Megathermal*. This warm period gradually declined until about 2,000 years ago.

The Holocene Climate Optimum warm event recorded the rise of temperature up to 4° C near the North Pole. Northwestern Europe also experienced warming, while cooling occurred in the south. The warm conditions changed gradually to cool conditions in the south with the drop of temperature rapidly with the lowering of latitude. Global average temperatures were probably less than 0.5° to 2° C warmer than the mid-20th century. Of 140 sites across the Western Arctic, there is clear evidence for warmer-than-present conditions at 120 sites. Northwestern North America had peak warmth first, from 11,000 to 9,000 years ago, while the Laurentide ice sheet still chilled the continent. Northeastern North America experienced peak warming 4,000 years later.

The 'African Humid Period' was evidenced at an interval between 16,000 and 6,000 years ago when Africa was much wetter due to a strengthening of the African monsoon by changes in summer radiation resulting from long-term variations in the earth's orbit around the sun. During this period, numerous lakes existed over the Sahara arid zone. The existence of such water bodies in Sahara had been confirmed from the fossils of typical African lake crocodile and hippopotamus fauna. In the far southern hemisphere (e.g. New Zealand and Antarctica), the warmest period during the Holocene continued around 8,000 to 10,500 years ago, immediately following the termination of the last ice age. About 6,000 years ago, associated with the Holocene Climatic Optimum in the northern hemisphere, these regions recorded the temperatures similar to those existing in modern times.

### Little Ice Age

After the warmer era, known as the *Medieval Climate Optimum,* cooling started. This period within the precept of history, a few centuries back, was known to be the *Little Ice Age.* Climatologists and historians find it difficult to agree on either the beginning or withdrawal dates of this period. However, it has been accepted in general that the Little Ice Age existed around the 16th century to the mid 19th century. It is also generally agreed that there were three minima, beginning about 1650, about 1770, and 1850, each separated by slight warming intervals.

It was initially believed that the LIA was a global phenomenon; but not substantiated to be true. The Intergovernmental Panel on Climate Change (IPCC) characterised the LIA as "a modest cooling of the Northern Hemisphere during this period of less than 1° C." It was further stated that the "current evidence does not support globally synchronous periods of anomalous cold or warmth over this timeframe, and the conventional terms of 'Little Ice Age' and Medieval Warm Period appear to have limited utility in describing trends in hemispheric or global mean temperature changes in past centuries." However, recent evidences confirmed that the Little Ice Age also affected the Southern Hemisphere.

The scientists did not agree on the start year of the Little Ice Age. Any of several dates ranging over 400 years may indicate the beginning of the Little Ice Age:

- • 1250 for when Atlantic pack ice began to grow.
- • 1300 for when warm summers stopped being dependable in Northern Europe.
- • 1315 for the rains and Great Famine of 1315-1317.
- • 1550 for the re-emergence of worldwide glacial expansion.
- • 1650 for the first climatic minimum.

However, there is a consensus that the Little Ice Age was terminated in the mid-19th century.

During the Little Ice Age, extremely severe and bitter cold winters were experienced in many parts of the world. In the mid-17th century, glaciers in the Swiss Alps advanced like avalanche, gradually engulfing farms and destroying entire villages. The River Thames and the canals and rivers of the Netherlands often became frozen during the winter. The historical records show that people skated and even held frost fairs on the ice. The first Thames frost fair was in 1607; the last in 1814. In the winter of 1780, New York Harbour froze, allowing people to walk from Manhattan to Staten Island. Compare the scenario of New York as portrayed in the film 'Day after tomorrow'. Sea ice surrounding Iceland extended for miles in every direction, closing that island's harbours to shipping.

The extent of mountain glaciers had been mapped in the late 19th century. In both the north and the south temperate zones of the earth, the height of the snowlines dropped 100 m lower than they were in 1975. In Glacier National Park, the last episode of glacier advance was recorded in the late 18th and early 19th century. Even in Africa (in Ethiopia and Mauritania) permanent snow was reportedly found on mountain peaks at levels where it is absent today.

There is limited evidence of Little Ice Age in Australia, though lake records in Victoria suggest that conditions at least in the south of the state were wet and/or unusually cool. In the north of the continent the sparse evidence indicate that fairly dry conditions prevailed. The coral cores from the Great Barrier Reef show no effect of Little Ice Age and indicate similar rainfall as of today but with less variability. Tropical Pacific coral records also show that frequent and intense El Niño-Southern Oscillation activity occurred in the mid 17th century, during the Little Ice Age.

### Temperature Records of Past 1000 Years

The temperature record of the past 1,000 years shows the reconstruction of temperature for the last 1000 years over the northern hemisphere. A reconstruction will be useful because a reliable surface temperature record exists only since about 1850. The past climate study is now growing concern for the climate scientists in order to improve the understanding of current climate variability and, relatedly, providing a scientific basis for future climate projections. In particular, if the nature and magnitude of natural climate variability is established, it will help to identify and quantify human induced climate variability (anthropogenic global warming). Although temperature reconstructions from proxy data help us understand the character of natural climate variability, attribution of recent climate change largely depends on a broad range of methodologies of which the proxy reconstructions are only a small part.

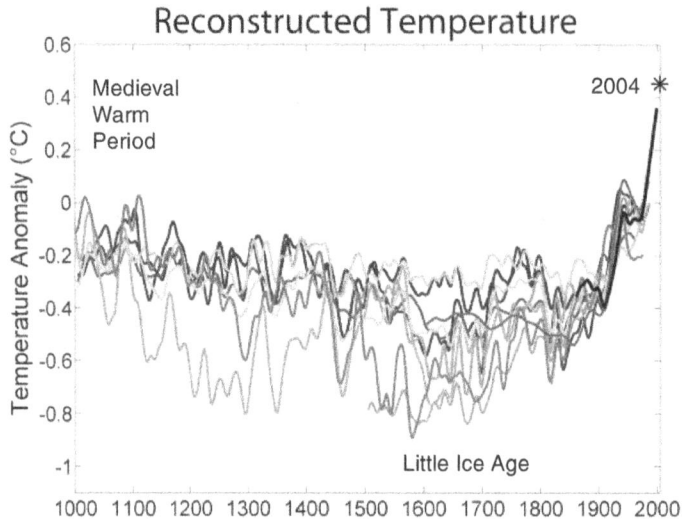

**Figure 9.6:** Development of Global Average Temperatures during the Last Thousand Years.

All major temperature reconstructions indicate the increase in temperature in the 20th century. The temperature in the late 20th century appears to be the highest in the record.

**Review of Climate Change**

Climate change is certain. Climate change in terms of IPCC refers to any change of climate over time, whether due to natural variability or as result of human activity. Climate changed markedly in different geological past even prior to the footprints of human on the earth's surface. Climate change, million years or billion years ago, was supposed to be caused by natural processes. But in recent history the anthropogenic interference has put complex variables within the natural system. It is not certain whether the natural system will persist, or the man induced changes will lead to more uncertainty.

In the human history the Industrial Society may be considered the culmination of all out hostilities against the nature. The primitive societies interacted with the nature, marginally changing some of its attributes, because it could only use his muscle power. The agricultural societies effectively utilised the animal power to derive more energy from the nature and thus caused deprivation of the nature. It resulted into loss of many forests and grasslands, and degradation of fertile alluvial lands. Many ecosystems suffered badly and many species became extinct. The industrial societies have transcended all limits in the exploitation and consumption of energy from the nature. So in this industrial age we find increasing burning of fossil fuels and emissions of green house gases. Temperature change is considered to be the controlling factor of climate change. Unwise action of human, particularly in the last century, has led to concentration of carbon dioxide in the atmosphere and rise of global temperature.

Climate change appears to be one of the most critical global challenges of our time. Recent events have emphatically demonstrated our growing vulnerability to climate change. Climate change impacts are many and will be disastrous even for our survival with endangered food security, inundation due to sea-level rise, accelerated erosion of coastal zones, increasing intensity of natural disasters, species extinction and the spread of vector-borne diseases.

IPCC in its Third Assessment Report on Climate Change has specified warming and other changes in the climate systems.

### Changes in the Physical Systems

*Sea Ice:* In the Arctic there had been decline of sea ice, 10 to 15 percent, since 1950. No significant trends were apparent in the Antarctic sea ice extent.

*Glaciers and Permafrost:* Mountain glaciers were receding on all continents and northern hemisphere permafrost had been thawing.

*Snow Cover:* The extent of snow cover in the northern hemisphere had decreased by 10 percent since late 1960 and 1970.

*Snowmelt and Runoff:* Snowmelt and runoff had occurred increasingly in Europe and Western North America since 1940s.

*Lake and River Ice:* Duration of lake and river ice had been reduced by two weeks in the mid-latitude and high latitude rivers and lakes in northern hemisphere and become more variable.

### Changes in the Biological Systems

*Range:* Plant and animal ranges had shifted polewards and towards higher elevation.

*Abundance:* Within the ranges in some plants and animals, population size had changed, increasing in some areas and declining in others.

*Phenology:* Timing of many life-cycle events, such as blooming, migration and emergence of insects, had shifted earlier in the spring and later in the autumn.

*Differential Change:* Species changed in different speeds and variable directions that decoupling of species interactions occurred (e.g. prey-predator relationships).

### Preliminary Evidences for Changes in the Human Systems

*Damages due to Droughts and Floods:* Changes in some economic systems had been related to persistent low rainfall in Sahelian region of Africa and to increased precipitation extreme in North America. Most of the increase in damages is due to increased wealth and exposure. However, part of the increase in losses was attributed to climate change, in particular to more frequent and extreme weather events in some regions.

### Impact of Climate Change in India

The Climate Change would have greater impact on India with its growing population, rapid industrialisation and urbanisation. The Climate Change could stress an additional pressure on its overall ecology and socio-economic systems. Increasing global warming has caused various climate-related disasters thereby adversely affecting agriculture, health, food security, water resources, and biodiversity as a whole.

Extreme temperature, heat spells are now commonly experienced in many parts of the country, particularly in the northern and eastern part of the country. Climate change would also have a greater impact on agricultural production due to adverse monsoon condition, which remains the primary source of water for agriculture, drinking water and also crucial for its rich biodiversity. Frequent change in weather pattern almost every year has been noted with concern. Fluctuations in monsoon circulation cause major impact on agricultural productivity and shifting crop pattern.

Impact of climate change will be pervasive in various sectors in India. We can specify the following sectors.

i)   *Impact of climate change on water resources of India*: It is probable that quantity of surface run off due to climate change would vary across the river basins as well as sub-basins of the principal rivers. There would be a general reduction in the quantity of the available run off due to receding glaciers and lower rainfall.

ii)  *Impact of climate change on agriculture in India:* In the context of changing rainfall and temperature conditions there would be large impact on crop pattern and yield. Global warming, particularly changes in rainfall and temperature, was likely to enhance the current stresses and increase vulnerability of food production and livelihoods of the agricultural community. The recent Inter-Governmental Panel on Climate Change and a few other studies indicate a probability of 10 to 40 per cent loss in crop production with increases in temperatures by the end of the century due to global warming. Already lower yield in wheat and other grain crops have been noticed. India being a vast country has 35 climate subdivisions. So the impact will be rationalised, while in one region there might be crop failure, in others crops would be abundant. No definite trends have yet been established.

iii) *Impact of climate change on forestry and natural ecosystem in India*: Forests are projected to be vulnerable and biodiversity would be adversely affected due to climate change. Due to rising temperature wildfires would damage the forest cover. The loss of biodiversity would have a greater impact of serious concern. Global estimate shows that the number of large (more than 500 acres) wildfires have increased from an average of 6 per year in the 1970s to 21 per year in the first decade of the 21st century.

iv) *Impact of climate change on coastal zones in India*: The coastal zones of India are highly vulnerable to climate change. Due to rise in sea level low lying coastal areas of Gulf of Kutch and West Bengal could be inundated. The sea level reveals a high variability along the Indian coastline with rise along the eastern coast and fall along the western coast in general. So the eastern coast, particularly the deltaic tracts, would be susceptible to inundation with the rise of sea level. IPCC marked the Bengal Delta in West Bengal and adjoining Bangladesh with a rise of about 5 cm. per decade in sea level.

v) *Impact of climate change on human health in India*: With global warming there could be increase in vector borne diseases such as Malaria and Dengue. Diseases would spread in areas where earlier not reported. Due to depletion of ozone resulting in the penetration of UV rays more likely there would be increased incidences of carcinogenic diseases.

vi) *Impact of climate change on infrastructure in India*: Increased extreme weather events are associated with the climate change. Such extreme events like cyclones, heavy rains, land slides and floods would cause damages to large infrastructure, such as dams, roads and bridges incurring high costs of repair and constructions.

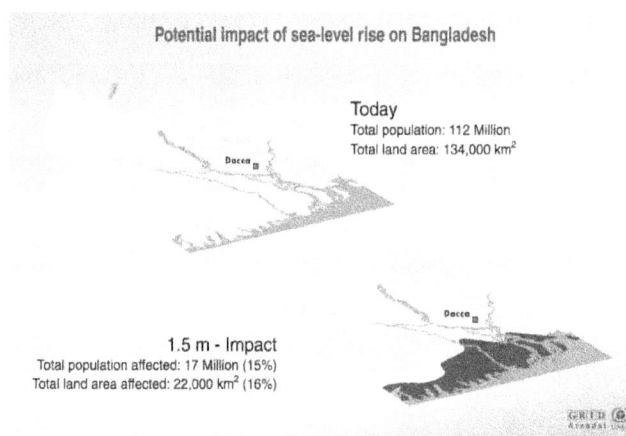

**Figure 9.7:** Inundation of Large Tracts in the Delta of Bangladesh and West Bengal

*Source:* UNEP, Geneva

The observation indicated a long term average rising trend of 1mm/year in sea level and a projection of rise in sea level in the range of 46-59 cm by the end of twenty first century.

## Policies for Adaptation and Mitigation for Climate Change

It is now universally accepted that global temperature will tend to rise further. Global warming remains and will remain to cause the change in climate all over the world. Human interferences play the key role in global warming with increased burning of fossil fuels and emission of green house gases resulting in the concentration of carbon dioxide in the earth's atmosphere. The stabilisation of carbon dioxide through the reduction of green house gas will be the policy action in the coming days.

The global community now seeks remedial measures to resolve the adverse impact of global warming. IPCC and many organisations including NGOs are involved in critical scientific studies to achieve the conservation of habitats and goal of sustenance. We need action to cut the GHG emissions, particularly the carbon dioxide in the global sphere to a level prior to 1990. The climate protection policies would help in the growth of an economically and environmentally sustainable future through technical innovations and diffusions. These policies would guide the economy toward more efficient, lower cost, less polluting, more secure, and more sustainable production and consumption of energy.

## Further Readings

Barry, R.G. and R.G. Chorley (1968) *Atmosphere, Weather and Climate,* Methuen & Co., London.

Brasseur, G.P., J.J. Orlando and S.G. Tyndall (1999) *Atmospheric Chemistry and Global Change,* Oxford University Press, New York.

Critchfield, H.J. (1975) *General Climatology,* Prentice Hall India Ltd., New Delhi.

Emilliani, C. (1955) 'Pliestocene temperatures', *Journal of Geology,* **53**.

Fairbridge, R.N. (1975) 'The changing level of the sea', *Earth Sciences,* **805**.

Hiddore, J.J. (1996) *Global Environmental Change,* Prentice Hall, Upper Saddle River, N.J.

Intergovernmental Panel on Climate Change (2007) *Fourth Assessment Report,* on line *www.ipcc.ch/*

Intergovernmental Panel on Climate Change (2001) *Summary for Policy Makers*, on line *www.ipcc.ch/*

Keppler, Frank; Marc Brass, Jack Hamilton, Thomas Röckmann (2006) *Global Warming—The Blame is not with the Plants,* Max Planck Society.

Lagett, J. (1990) *Global Warming,* The Green Peace Report, Oxford University Press, Oxford.

Lamb, H.H. (1970) 'Climatic variation and our environment today and in the coming years', *Weather,* **45**.

Oliver, J.E. and J.J. Hidore (2002) *Climatology,* Pearson Education, Delhi.

Panel on Climate Change Feedbacks (2004) *Understanding Climate Change Feedbacks,* Climate Research Committee, National Research Council, Washington.

Plass, G.N. (1956) 'The carbon dioxide theory of climatic change', *Tellus,* **8**.

Plass, N. (1962) 'Carbon dioxide and the climate', *in White (ed.) 'The study of the Earth'.*

Reynolds, R.W. (1988) 'A real-time global sea surface temperature analysis', *Journal of Geology,* **1**.

Royal Society (2001) *The Science of Climate Change,* Royal Society, London.

Saha, P.K. and P.K. Bhattacharya (1994) *Adhunik Jalavayu Vidya,* West Bengal State Book Board, Government of West Bengal, Calcutta.

Sherwood, B. Idso (1989) *Carbon Dioxide and Global Change,* Institute of Biospheric Research Press, Arizona.

Tans, Pieter (2008) *Trends in Atmospheric Carbon Dioxide,* National Oceanic and Atmospheric Administration, Mauna Loa.

Turekian, K.K. (1996) *Global Environmental Change,* Prentice Hall, Upper Saddle River, N.J.

Warrick, R.A., T.M. Barrow and T.M. Wigley (1993) *Climate and Sea Level Change,* Cambridge University Press, New York.

# Chapter 10

## CLIMATE AND ECOSYSTEM

### Understanding the Relationship

Ecosystem appears to be an interacting component in the earth system. It is the wheels within the wheels in a system. There is now general recognition that anthropogenic alterations of the earth system and especially of atmospheric composition and the character of the land surface, have led to changes in global and regional climatic conditions and will continue to do so at least throughout the present century. We are now all concerned with the effects of these changes on the living environment.

Ecologists find it difficult to decide any one of two approaches for studying the effects of climate on a particular ecosystem. The two approaches are: 1. Traditional approach using local weather measures, and 2. Modern approach using large scale climatic indices. Local weather gains the advantage of being directly linked to mechanisms actually affecting the particular system. On the other hand the use of large-scale climate indices also indicates some advantages as noted below:

I.   *Spatial variation:* Large-scale climate indices show spatial correlation in weather patterns. So this approach will be ecologically useful, particularly for marine ecosystem.

II.  *Predictability:* Application of large-scale indices helps predicting the future of the ecosystems. As some of the climate systems could be predicted, large scale indices with broad data-base would facilitate the successful prediction of ecological effects.

III. *Biological impact:* The most effect of the climate on individuals happens to be the result of interacting variables. Large-scale indices may represent the climatic effects better than any single local variable.

### Biodiversity and Climatic Change

Biodiversity is intimately linked with the climate control. Biodiversity moderates the ability of terrestrial ecosystems absorbing carbon dioxide, determines the rates of evapotranspiration of plants and affects the temperature conditions. All these are related to climate at local and global levels. Biodiversity influences the removal of carbon dioxide from the atmosphere primarily through its effects on species' characteristics. The appropriate choice of species is important for maximising carbon absorption from the atmosphere. The loss of biodiversity as a result of the clearing and burning of vegetation also leads to global warming through the release of greenhouse gases to the atmosphere. Degradation and loss of biodiversity due to human activities will increasingly affect the global climate. There is evidence that reductions in biodiversity restrict ecosystem resilience, or its ability to recover to its original state after natural or human-induced disturbances.

Climate change on any scale will have impact on biodiversity. The Intergovernmental Panel on Climate Change (IPCC) calculated a rise of the global mean surface temperature by 0.6° Celsius over the last century. The decade of the 1990s was the warmest of all centuries on record in recent times. Change in precipitation patterns was also evidenced both spatially and temporally. Global sea level has recorded a rise of 0.1-0.2 m. It is assumed that impacts of climate change will be more pronounced in biodiversity loss and changes in ecosystem services at the global level. It is projected by IPCC that there will be a further increase in global mean surface temperature of two to six degrees celsius above pre-industrial levels by 2100, increased incidence of floods and droughts, and further rises in sea level of several centimetres.

Climate changes demonstrate significant impacts on biodiversity and ecosystems, including changes in species distribution, population sizes, the timing of reproduction or migration events, and increases in the frequency of pest and disease outbreaks. It has been studied that the spatial distribution of many species has migrated 6.1 km per decade towards the poles or 1 m in elevation per decade. Flowering and leaf flushing phenomena are found to occur on average 2.3 days earlier per decade. It causes the seasonal shifting of species. Coral reef ecosystem has been affected as a major event. Such of the changes are sometimes partially reversible. Rise of surface temperatures above the average for the warmest months has affected the bleaching episodes of coral reefs.

Adaptation strategies for the conservation of biodiversity are urgently needed in the context of changing climate on a global scale. The basic understanding of dynamic interactions of biotic and abiotic factors will help formulating the strategies for the protection of biodiversity.

    I.   The extent to which current networks of protected areas will remain effective in the face of climatic change.

   II.   The extent to which protected area networks are functional in the sense that they form connected networks.

  III.   The characteristics that a landscape must have if it is to be habitable to species adjusting to climatic change.

The resilience of ecological systems, the pattern and extent of the genetic variance of species may facilitate or hinder their adaptation to climatic change. The climate change may lead the components of ecological systems at greatest risk. As a consequence of climatic change the regions having such threatened systems will be at maximum risk.

**Significance of Biomes**

A biome is a climatically and geographically defined area of ecologically similar climatic conditions such as communities of plants, animals, and soil organisms. Biomes are often referred to as ecosystems and defined based on factors such as plant structures (such as trees, shrubs, and grasses), leaf types (such as broadleaf and needle-leaf), plant spacing (forest, woodland, savanna), and climate. The climatic relationship with the ecosystem is more pronounced in the characteristics and distribution of biomes. To enumerate the impact of climate on ecosystem it is urgently needed to study the biomes on earth's surface. The climate change would induce variation in the characteristics of biomes in a specific ecological region.

**Biomes of the World**

Biomes are the major living communities of the world, classified according to the predominant vegetation and characterised by adaptations of organisms to that particular environment (Campbell, 1996). Biomes are not ever lasting. As controlled by several abiotic factors biomes changed and moved many times during the history of life on earth. Changes in natural constituents in different geological periods caused marked variations and even resulted in extinction. Anthropogenic interferences have also been drastically affecting these communities in recent times. Thus, conservation and preservation of biomes have become a major concern to all.

We can observe six major types of biomes:

1. Freshwater, 2. Marine, 3. Desert, 4. Forest, 5. Grassland and 6. Tundra.

**Freshwater Biome**

Freshwater means the water having a low salt concentration—usually less than 1%. Plants and animals in freshwater regions are adapted to the low salt content and would not be able to survive in areas of high salt concentration (i.e., ocean). We can identify different types of freshwater regions:

- Ponds and lakes
- Streams and rivers
- Wetlands.

*Ponds and Lakes*

Ponds and lakes vary in size from just a few square metres to thousands of square kilometres. Many of these lakes are remnants from the Pleistocene glaciation. Many ponds are seasonal, lasting just a couple of months while lakes may exist for hundreds of years or more. Limited species diversity is observed since they are usually isolated from one another and from other water sources like rivers and oceans. One can distinguish three different "zones" in ponds and lakes which are usually determined by depth and distance from the shoreline.

The topmost zone near the shore of a lake or pond is the *littoral zone*. This zone happens to be the warmest since it is shallow and can absorb more of the insolation. A fairly diverse community can be found in this zone sustaining several species of algae (like diatoms), rooted and floating aquatic plants, grazing snails, clams, insects, crustaceans, fishes, and amphibians. In the case of the insects, such as dragonflies and midges, only the egg and larvae stages are found in this zone. The vegetation and animals living in the littoral zone become food for other creatures such as turtles, snakes, and ducks.

The *limnetic zone* appears to be the near-surface open water surrounded by the littoral zone. The limnetic zone is well-lighted (like the littoral zone) and dominated by plankton, both phytoplankton and zooplankton. Planktons are small organisms having a crucial role in the food chain. Most of the living organisms depend on aquatic planktons. A variety of freshwater fish is also the habitant in this zone.

The deep-water part of the lake/pond is called the *profundal zone*. This zone is much colder and denser. Little light penetrates all the way through the limnetic zone into the profundal zone. The fauna are heterotrophs, meaning that they eat dead organisms like dead planktons and use oxygen for cellular respiration.

Temperature varies in ponds and lakes seasonally. During the summer, the temperature can range from 4° C near the bottom to 22° C at the top. During the winter, the temperature at the bottom can be 4° C while the top is 0° C (ice). In between the two layers lies a narrow zone, called the *thermocline*, where the temperature of the water changes rapidly. During the spring and fall seasons, there is a mixing of the top and bottom layers by wind circulation that results in a uniform temperature distribution in water around 4° C. This mixing also circulates oxygen throughout the lake. Of course there are many lakes and ponds that do not freeze during the winter, thus the top layer would be a little warmer.

### Streams and Rivers

Streams and rivers are bodies of flowing water moving in one direction. Streams and rivers have their sources at springs, glaciers and snowfields or even lakes. They flow along the courses towards their mouths, meeting another water channel or the ocean. The characteristics of a river or stream change during the journey from its source to mouth. The temperature of the flowing water usually drops from source to mouth. The water is also clearer, having higher oxygen levels closer to its source. Freshwater fish such as trout and heterotrophs are found. Towards the middle part of the stream/river, the width increases. Species diversity occurs with numerous aquatic green plants and algae. Toward the mouth of the river/stream, the water becomes murky with increased sediment's load decreasing the amount of light that can penetrate through the water. Since there is less light, there is less diversity of flora, and because of the lower oxygen levels, fish that require less oxygen, such as catfish and carp are normally found.

## *Wetlands*

Wetlands are areas of accumulated water that support aquatic plants. Water standing more than three months is considered as the wetlands. Marshes, swamps, and bogs are all designated wetlands. Plant species grow under very moist and humid conditions. These are called hydrophytes. Pond lilies, cattails, sedges, tamarack, and black spruce are the important species. Species like cypress and gum are known to be other marsh flora. The biodiversity of wetlands is considered to be the richest having the highest species diversity of all ecosystems. Many species of amphibians, reptiles, birds (such as ducks and waders), and furbearers have their habitat in the wetlands. Wetlands appear to be transitional between the terrestrial and marine ecosystems. Wetlands also include salt marshes which support different species of animals, such as shrimp, shellfish, and various grasses.

## Marine Biome

Marine biome stretches over three-fourths area on our earth. It includes oceans, coral reefs, and estuaries. Marine algae contribute much of the world's oxygen supply and receive back huge amount of atmospheric carbon dioxide. The evaporation of seawater is the source of rainfall.

- Oceans
- Coral reefs
- Estuaries

## *Oceans*

Of all the ecosystems, oceans appear to be the largest with huge water bodies on earth's surface. The ocean regions are classified into separate zones: inter-tidal, pelagic, abyssal, and benthic. All four zones contain a great diversity of species. In fine, the ocean contains the richest diversity of species even though it contains lesser species compared to land.

The *inter-tidal zone* is meeting point of land with ocean. It is alternately submerged and exposed, as waves and tides come in and out. As a result the communities are constantly changing. Where only the highest tides can reach, there remain only a few species of algae and mollusks. In those areas usually submerged during high tide, there is a more diverse array of algae and small animals, such as herbivorous snails, crabs, sea stars, and small fishes. The bottom of the inter-tidal zone is only exposed during the lowest tides. Many invertebrates, fishes, and seaweed are found in this zone.

The *pelagic zone* is basically the open ocean. In the pelagic zone constant mixing of warm and cold ocean currents occur. The surface seaweeds remain to be common flora. The fauna include many species of fish and some mammals, such as whales and dolphins. Many feed on the abundant plankton.

The *benthic zone* is the area down the pelagic zone. However, this is not the deepest parts of the ocean. The bottom of the zone consists of sand, silt, and/or dead organisms. In this zone temperature decreases with increasing depth toward the abyssal zone due to lesser penetration of light through the deeper water. Flora are represented primarily by seaweeds while the fauna, being very nutrient-rich, include all sorts of bacteria, fungi, sponges, sea anemones, worms, sea stars, and fishes.

The deep ocean is known as the *abyssal zone*. The water in this region is very cold (around 3° C), heavy and rich in oxygen content, but low in nutritional status. The abyssal zone supports many species of invertebrates and fishes.

### *Coral Reefs*

Coral reefs are widely found in warm waters of shallow seas. They are usually found as barriers along the continents (e.g., the Great Barrier Reef off Australia), fringing islands, and atolls. The dominant organisms in coral reefs are corals. The reef waters tend to be poor in nutrition. So, corals obtain nutrients through the algae *via* photosynthesis and also by extending tentacles to obtain plankton from the water. Besides corals, the other fauna include several species of microorganisms, invertebrates, fishes, sea urchins, octopuses, and sea stars.

### *Estuaries*

Estuaries are the fringe areas where freshwater river-flow merges with ocean. This mixing of fresh waters with higher salt concentrations generates a very interesting and unique ecosystem. It is considered to be one of the most productive ecosystems. Microflora like algae, and macroflora, such as seaweeds, marsh grasses, and mangrove trees (only in the tropics) are found. Estuaries support a diverse fauna, including a variety of worms, oysters, crabs, and waterfowl.

### Desert Biome

Nearly 20% of the earth's surface remains to be deserts, where rainfall is less than 50 cm/year. Most of the deserts stretch over the low latitude region. These include such as the Sahara of North Africa, deserts in Arab and Asia Minor and the deserts of the southwestern U.S., Mexico, and Australia. We can also find cold deserts in Central and Eastern Asia and also in the basin and range area of Utah and Nevada. Usually deserts have a considerable amount of specialised vegetation, as well as specialised vertebrate and invertebrate animals.

Soils are rich in nutrients, but saline in many cases. However, water stress is common restricting productive agriculture. A few large mammals are found in the deserts due to paucity of water and extreme hot climate. The dominant animals of warm deserts are non-mammalian vertebrates, such as reptiles. Mammals are usually small, like the kangaroo mice of North American deserts.

Four major types of deserts are identified:

- Hot and dry
- Semiarid
- Coastal
- Cold

### Hot and Dry Desert

Hot and Dry deserts include the deserts in Western and South Asia, North American deserts of Chihuahuan, Sonoran, Mojave and Great Basin, deserts of South and Central America, deserts of Africa and Australian deserts.

The seasons are generally warm throughout the year and very hot in the summer. Extreme temperature conditions are recorded in these regions, sometimes indicating the maximum up to 49° C. Minimum temperature during night hours may drop even up to –18° C. Scarce rainfall occurs in winter months. On many occasions rainfall turns to be of high intensity for a brief period causing flash floods. Total rainfall is lower than 1.5 cm in a year in Sahara and Atacama deserts. However, more than 10 cm rainfall may be recorded in many of the deserts.

In deserts plants are mainly ground-hugging shrubs and short woody trees. Leaves are rich with nutrients possessing water-conserving characteristics. They tend to be small, thick and covered with a thick cuticle (outer layer). In the cacti, the leaves are much-reduced (to spines) and photosynthetic activity is restricted to the stems. Some plants open their stomata (microscopic openings in the epidermis of leaves that allow for gas exchange) only at night when evaporation rates are lowest. These plants include: yuccas, ocotillo, turpentine bush, prickly pears, false mesquite, sotol, ephedras, agaves and brittlebush.

Small nocturnal (active at night) carnivores dominate the animal kingdom in the deserts. These are mostly the burrowers and kangaroo rats. The insects, arachnids, reptiles and birds are also found. The animals stay idle in protected shelter in hot day-hours and appear at dusk, dawn or at night, when the desert turns cooler.

### Semi-arid Desert

Semi-arid deserts lie in between the arid and forest climate. Summers are moderately long and dry. In the winter months there is low concentration of rainfall. Summer temperatures usually range between 21°-27° C. It rarely exceeds 38° C. Evening turns to be cooler recording temperature around 10° C. Cool nights prevail and help both plants and animals by reducing moisture loss from transpiration, sweating and breathing. Moreover, condensation helps formation of dew due to night cooling. Such precipitation may equal or exceed the rainfall received by some deserts. As in the hot desert, rainfall is often very low and/or concentrated. The average rainfall ranges from 2 to 4 cm annually. It has a fairly low salt concentration, compared to deserts which receive a lot of rain (acquiring higher salt concentrations as a result).

Many plants in semi-arid deserts have large number of spines. Such large number of spines help the plant surface enough to reduce transpiration effectively. The hairs on the woolly desert plants also have the same effects. Many plants possess silvery or glossy leaves, allowing them to reflect more radiant energy. Important semi-arid plants are Creosote bush, bur sage (*Franseria dumosa* or *F. deltoidea*), white thorn, cat claw, mesquite, brittle bushes (*Encelia farinosa*), lyciums, and jujube.

In day-light, insects remain around twigs to seek shelter on the shady side; jack rabbits traverse the moving shadow of a cactus or shrub. Usually, many animals seek shelter in underground burrows where they are insulated from both heat and aridity. These animals include mammals such as the kangaroo rats, rabbits, and skunks; insects like grasshoppers and ants; reptiles like lizards and snakes; and birds such as burrowing owls and the California thrasher.

### *Coastal Desert*

Coastal deserts are found in moderately cool to warm areas in the littoral zone. A good example is the Atacama of Chile.

The average summer temperature is recorded between 13°-24° C. Winter temperature remains 5° C or below. The maximum annual temperature is about 35° C and the minimum is about –4° C. In Chile, the temperature varies from –2° to 5° C in July and 21°-25° C in January.

The average rainfall records are 8-13 cm in many areas. The maximum annual rainfall may be as high as 37 cm with a minimum of 5 cm.

The plants have thick and fleshy leaves or stems which can take in large quantities of water as available and store it for future use. In some plants, the surfaces are corrugated with longitudinal ridges and grooves. With available water the stem swells so that the grooves are shallow and the ridges far apart. After the use of water the stem shrinks so that the grooves are deep and ridges close together. The plant types include the salt bush, buckwheat bush, black bush, rice grass, little leaf horsebrush, black sage, and chrysothamnus.

A few animals adapt themselves for sustenance under the desert heat and lack of water. Some toads keep themselves in burrows with gelatinous secretions and remain inactive for eight or nine months until a heavy rain occurs. Amphibians going through larval stages have accelerated life cycles with chances of reaching maturity before the waters evaporate. Laid eggs of some insects remain dormant until the environmental conditions are suitable for hatching. The fairy shrimps also lay dormant eggs. Other animals include: insects, mammals (coyote and badger), amphibians (toads), birds (great horned owl, golden eagle and the bald eagle), and reptiles (lizards and snakes).

## *Cold Desert*

Cold deserts record cold winters with snowfall and relatively high rainfall throughout the winter and occasionally over the summer. Short, moist, and moderately warm summers with fairly long, cold winters indicate climatic characteristics. The mean winter temperature ranges between –2 to 4° C and the mean summer temperature is 21°-26° C.

The winter months record appreciable snowfall. The mean annual precipitation ranges from 15-26 cm. Annual precipitation may be as high as 46 cm with a minimum record of 9 cm. The heaviest rainfall of the spring usually occurs in April or May. In some areas, rainfall may be heavy in autumn.

The plants are widely scattered. Plant heights range between 15 cm and 122 cm. The plants are mostly deciduous, having spiny leaves. Important animals are jack rabbits, kangaroo rats, kangaroo mice, pocket mice, grasshopper mice, and antelope ground squirrels. All animals are burrowers with the exception of *jack rabbits*. The carnivores like the badger, kit fox, and coyote also have the burrowing habit. Many lizards can do some burrowing and move soil. Deers are found only in winter.

## Forest Biome

Forests occupy nearly one-third of land surface on earth. Forests play important role in the ecosystem absorbing about 70% of carbon present in living things. The forest biomes are dominated by trees and other woody vegetation. Three principal types of forests may be classified in terms of latitudinal extension:
- Tropical
- Temperate
- Boreal forests (taiga)

## *Tropical Forests*

Tropical forests possess the greatest diversity of species. They are mostly found near the equator, within the latitudinal extent of 23.5 degrees N and 23.5 degrees S. Tropical forests show their distinct seasonality. Winter is absent, and only two seasons are distinct (rainy and dry). The length of daylight extends for 12 hours and varies marginally.
- Average temperature ranges between 20°-25° C and varies little throughout the year. The average temperatures of the three warmest and three coldest months rarely differ by more than 5 degrees.

- Precipitation is evenly distributed throughout the year, with annual rainfall exceeding 2000 mm.
- Soil is acidic and poor in nutrients. Decomposition rate is high. Intense leaching occurs due to high rainfall.
- Canopy in tropical forests is multilayered and continuous. Light penetration is low due to dense canopy.
- Flora appears to be highly diverse. Nearly 100 different tree species may be found in one square kilometre area. Trees are usually tall reaching the height of 25 m, with buttressed trunks and shallow roots. The trees are mostly evergreen, with large dark green leaves. Plants such as orchids, bromeliads, vines (lianas), ferns, mosses, and palms remain to be the natural vegetation in tropical forests.
- Fauna are numerous including many types of birds, bats, small mammals, and insects.

Other specific distinctions are identified in different types of tropical forests:

- *Evergreen rainforest*—no dry season.
- *Seasonal rainforest*—short dry period in a very wet tropical region (the forest exhibits definite seasonal changes as trees undergo developmental changes simultaneously, but the general character of vegetation remains the same as in evergreen rainforests).
- *Semi-evergreen forest*—longer dry season (the upper tree story consists of deciduous trees, while the lower story is still evergreen).
- *Moist/dry deciduous forest (monsoon)*—the length of the dry season increases further as rainfall decreases (all trees are deciduous).

The tropical rainforests remain to be the sink of carbon dioxide. Unfortunately more than one half of tropical forests have already been destroyed causing a great concern for the degradation of ecosystem.

### Temperate Forest

Temperate forests grow in the mid-latitude region under temperate climatic conditions. Such forests extend over eastern north America, northeastern Asia, and western and central Europe. The seasons are well marked with a distinct winter specifying this forest biome. Moderate climate and a growing season of 140-200 days during 4-6 frost-free months characterise the temperate forests. The characteristics of temperate forests are:

- Temperature ranges from –30° C to 30° C.
- Precipitation (75-150 cm) occurs evenly throughout the year.
- Canopy is moderately dense and allows light to penetrate, helps in well-developed and richly diversified undergrowth vegetation and stratification of animals.
- Flora is not much diversified. Only 3-4 tree species per square kilometer are found. Trees are deciduous and have broad leaves that are lost annually during fall. The trees include such species as oak, hickory, beech, hemlock, maple, basswood,

cottonwood, elm, willow, and spring-flowering herbs. Conifer trees are found in colder region and remain evergreen.

- Fauna includes squirrels, rabbits, skunks, birds, deer, mountain lion, bobcat, timber wolf, fox, and black bear.

These forests have specific characteristics based on seasonal distribution of rainfall:

- *Moist conifer and evergreen broad-leaved forests*—wet winters and dry summers (rainfall is concentrated in the winter months and winters are relatively mild).
- *Dry conifer forests*—dominate higher elevation zones; low precipitation.
- *Mediterranean forests*—precipitation falls in winter recording less than 1000 mm per year.
- *Temperate coniferous forests*—mild winters, high annual precipitation (greater than 2000 mm).
- *Temperate broad-leaved rainforests*—mild, frost-free winters, high precipitation (more than 1500 mm) evenly distributed throughout the year.

Most of the temperate forest are now lost for excessive pressure on land. Only scattered remnants of original temperate forests stand today.

### Boreal Forest/Taiga

Boreal forests or the Taiga appear to be the largest terrestrial biome. These are found in high latitude ranging between 50 and 60 degrees north latitudes. Boreal forests occupy the broad belt of Eurasia and North America—mostly extending two-thirds in Siberia with the rest in Scandinavia, Alaska, and Canada. Seasons are divided into short, moist, and moderately warm summers and long, cold, and dry winters. The length of the growing season stands to be 130 days. These forests show specific characteristics:

- Cold temperatures prevail.
- Precipitation is primarily in the form of snow, 40-100 cm annually.
- Soil is thin, acidic and poor in nutrients.
- Canopy allows low light penetration. So, undergrowth vegetation is limited.
- Flora displays mostly the cold-tolerant evergreen conifers with needle-like leaves, such as pine, fir, and spruce.
- Fauna include woodpeckers, hawks, moose, bear, weasel, lynx, fox, wolf, deer, hares, chipmunks, shrews, and bats.

The boreal forests are exploited for extensive logging. Current extensive logging in boreal forests may soon result in their disappearance.

### Grassland Biome

Grasslands are the lands dominated by grasses rather than large shrubs or trees. Two principal types of grasslands are:

- Tropical grasslands or savannas
- Temperate grasslands

### *Savanna*

Savanna is characterised by grassland with scattered isolated trees. Savanna type grasslands cover nearly half of Africa and large areas of Australia, South America, and India.

Savannas are found in warm or hot climates where the annual rainfall ranges from about 50.8 to 127 cm (20-50 inches) per year. The rainfall usually occurs in six or eight months of the year, followed by a long period of drought.

Different types of savannas support different grasses due to variation in rainfall and soil conditions. The savanna provides habitat to a large number of species competing for living. In drier savannas such as those on the Serengeti plains or Kenya's Laikipia plateau, the dominant grasses on well-drained soils are Rhodes grass and red oat grass. In East African savannas star grasses are dominant. The lemon grasses are common in many western Uganda savannas. Deciduous trees and shrubs are also found scattered in the landscape. One type of savanna commonly is identified as grouped-tree grassland, where the trees grow only on termite mounds, the intervening soil being too thin or poorly drained to support the growth of trees at all. This type of savanna is prevalent in southwestern Kenya, Tanzania, and Uganda. Frequent fires and large grazing mammals kill seedlings resulting in lower density of trees and shrubs. Savannas record an average annual rainfall of 76.2-101.6 cm (30-40 inches). A few savannas even record lower rainfall ranging from 15.24 cm (6 inches) to 25.4 cm (10 inches) per year.

Alternate dry and rainy season prevail in the savanna region. Seasonal fires play a significant role in the savanna's biodiversity. In October, a series of violent thunderstorms, followed by a strong dry wind, indicates the beginning of dry season. Fire appears to be common around January, at the height of the dry season. Fires are often caused by human interference. The poachers clear away dead grass for hunting their prey. Most of the animals killed by the fires are insects with short life spans. Underground holes and crevices provide a safe refuge for small creatures. Larger animals can escape the fire. Unlike grasses and shrubs, trees can survive a fire by retaining some moisture in all their above-ground parts throughout the dry season. Sometimes they have a corky bark or semi-succulent trunk covered with smooth resinous bark, both being fire resistant. When the rains come again, savanna bunch grasses grow vigorously. Some of the larger grasses grow an inch or more in 24 hours. There is a surge of new life at this time in the savannas. Many antelope calves are born. With so much grass to feed on, mothers have plenty of milk. Other animals include giraffes, zebras, buffaloes, kangaroos, mice, moles, gophers, ground squirrels, snakes, worms, termites, beetles, lions, leopards, hyenas, and elephants.

Poaching, overgrazing, and clearing of the land for crops appear to be a critical concern for the denudation of the savannas.

## Temperate Grassland

Temperate grasslands are found in cool climate with dominant vegetation of grasses. Trees and large shrubs are commonly absent. Temperature range is higher in summer than in winter. The amount of rainfall is less in temperate grasslands than in savannas.

Temperate grasslands are commonly known as the *veldts* in South Africa, the *puszta* in Hungary, the *pampas* in Argentina and Uruguay, the *steppes* in the former Soviet Union, and the *plains* and *prairies* in central North America. Temperate grasslands usually manifest hot summers and cold winters. Rainfall is moderate. The amount of annual rainfall signifies the height of grassland vegetation, with taller grasses in wetter regions. Like the savanna, seasonal drought and occasional fires play important role for biodiversity, but their effects are less significant in temperate grasslands than in savannas. Edaphic conditions are better in the temperate grasslands with deep and dark soils having fertile upper layers. It is nutrient-rich from the growth and decay of deep, many-branched grass roots. The rotted roots hold the soil together and provide a food source for living plants.

The seasonal drought, occasional fires, and grazing by large mammals all restrict woody shrubs and trees from growing. However, a few trees, such as cottonwoods, oaks, and willows grow in river valleys, and some non-woody plants, specifically a few hundred species of flowers, grow among the grasses. The various species of grasses include purple needle-grass, blue grama, buffalo grass, and galleta. Different types of flowers blossom including asters, blazing stars, coneflowers, goldenrods, sunflowers, clovers, psoraleas, and wild indigos. Precipitation in the temperate grasslands usually falls in the late spring and early summer. Total annual precipitation in average comes to about 50.8 to 88.9 cm (20-35 inches). Annual temperature range appears to be very high, ranging from summer temperatures over 38° C to winter temperatures as low as –40° C.

The fauna include gazelles, zebras, rhinoceroses, wild horses, lions, wolves, prairie dogs, jack rabbits, deer, mice, coyotes, foxes, skunks, badgers, blackbirds, grouses, meadowlarks, quails, sparrows, hawks, owls, snakes, grass-hoppers, leafhoppers, and spiders.

There are also environmental concerns regarding the temperate grasslands. Most of the natural prairie regions have been turned into farms or grazing land. Temperate grasslands on a global scale had been denuded to make it a granary, but affecting adversely the ecosystem.

*Prairies* are grasslands with tall grasses while *steppes* are grasslands with short grasses. However, they possess the similar characteristics. Prairies/Steppes are dry areas of grassland with hot summers and cold winters. They receive 25.4-50.8 cm (10-20 inches) of rainfall a year. They are usually found in the interiors of North America and Europe. Plants growing in prairies/ steppes exceed 1 foot height. These include blue grama and buffalo grass, cacti, sagebrush, speargrass, and small relatives of the sunflower. The fauna includes badgers, hawks, owls, and snakes. Today, people use steppes to graze livestock and to grow wheat and other crops. Overgrazing, plowing, and excess salts left behind by irrigation waters have caused deterioration of many prairies/steppes.

## Tundra Biome

Tundra is the coldest region of all the biomes. Tundra means a treeless plain. Tundra is characterised by:

1.  Very cold climate
2.  Lower biodiversity
3.  Simple vegetation structure
4.  Restrictive drainage
5.  Short growing season and period of reproduction
6.  Dead organic material forms the pool of energy and nutrients.
7.  Large population oscillations

Tundra is divided into two types:
*   Arctic tundra
*   Alpine tundra

### *Arctic Tundra*

Arctic tundra is located in the North Polar Region and even extending south to the coniferous forests of the taiga. The arctic is characterised by its cold, desert-like conditions. The growing season extends from 50 to 60 days only. The average winter temperature is below freezing as low as –34° C (–30° F). However, the average summer temperature of 3°-12° C (37°-54° F) enables this biome to sustain life. Rainfall may vary in different regions of the arctic. Annual precipitation, including melting snow, is 15 to 25 cm (6 to 10 inches). A layer of permanently frozen subsoil, known as *permafrost,* exists. It comprises mostly of gravel and finer material. When water saturates the upper surface, bogs and ponds may form, providing moisture for plants. No deep root systems in the vegetation of the arctic tundra exist. There are about 1,700 kinds of plants in the arctic and sub-arctic regions. These are mostly:

*   Low shrubs, sedges, reindeer mosses, liverworts, and grasses
*   400 varieties of flowers
*   Crustose and foliose lichen

All these plants are adapted to sweeping winds and conditions of the soil. Plants are short and grouped together to resist the cold temperatures and protect themselves from the snow during winter. They possess the properties to carry out photosynthesis at low temperatures and low light intensities. The growing seasons are short and most plants reproduce by budding and division rather than sexually by flowering.

The fauna in the arctic region remains diversified. It includes:

*   *Herbivorous mammals:* lemmings, voles, caribou, arctic hares and squirrels
*   *Carnivorous mammals:* arctic foxes, wolves, and polar bears
*   *Migratory birds:* ravens, snow buntings, falcons, loons, ravens, sandpipers, terns, snow birds, and various species of gulls

- *Insects:* mosquitoes, flies, moths, grasshoppers, blackflies and arctic bumble bees
- *Fish:* cod, flatfish, salmon, and trout.

Animals are adapted to the conditions of long cold winters and can breed and raise young quickly in the summer. Animals such as mammals and birds also have additional insulation from fat. Many animals hibernate during the winter because food is not abundant or migrate south in the winter, like birds do. Reptiles and amphibians are few or absent because of the extremely cold temperatures. Because of constant immigration and emigration, the population continually oscillates.

### Alpine Tundra

Alpine tundra is located on high mountains, above the tree-line. The growing season is around 180 days. The night-time temperature is usually below freezing. Unlike the arctic tundra, the soil in the alpine is well drained. The plants are very similar to those of the arctic ones and include:

- Tussock grasses, dwarf trees, small-leafed shrubs, and heaths.

Animals living in the alpine tundra commonly include:

- *Mammals:* pikas, marmots, mountain goats, sheep and elk
- *Birds:* grouselike birds
- *Insects:* springtails, beetles, grasshoppers and butterflies.

### Ecosystems as Interacting Components of the Earth System

Climate changes occur due to interactions between the land surface and the climate system. The potential for feedbacks, whether positive or negative, as a result of changes in ecological systems deems to be the consequence of climatic change. In many parts of the world, for example the humid tropics, direct and local human actions mainly cause changes in land-surface characteristics, even though their importance may become secondary in coming decades as the impacts of climatic change become greater. It will be an area of great interest and importance, how to reduce uncertainty in risk assessments and how to engage with uncertainty when conveying the outcomes of such assessments to policymakers and the public. We must consider securing a low entropy situation for reduction in uncertainty.

### Consequences of Climate Change on Coral Reefs and Tidal Marshes

The potential consequences of climate change on coral reefs and tidal marshes appear to be the result of climate variability, changes in temperature and precipitation patterns, and sea level rise. Coral reefs provide a variety of important services including: maintenance of biodiversity, shoreline protection, tourism, fisheries, trade, and aesthetic and cultural value. Climate related stressors affecting coral reefs include: temperature, solar radiation, changes in $CO_2$, change of frequency and severity of storms, sea level rise, changes in water quality, and salinity fluctuations.

Tidal marshes are wetlands that are frequently or continually inundated with water, characterised by emergent soft-stemmed vegetation adapted to saturated soil conditions, and influenced by the motion of the tides. Tidal marshes may be freshwater, brackish or saline. Tidal marshes provide many ecosystem services. They buffer stormy seas, slow shoreline erosion, absorb excess nutrients before they reach oceans and estuaries, and provide vital food and habitat for aquatic biota. Climate related stressors affecting tidal marshes include: climate variability, changes in precipitation and temperature patterns, changes in $CO_2$ concentrations, and sea level rise.

For coral reefs or tidal marshes, we must study how the relationships among ecological processes and their associated services will be quantified. Changes in ecological processes due to climate change stressors can be estimated based on existing data and literature where possible or original research where necessary.

## Climate Change and Oceans

Many recent studies on climate change have indicated that the Sea Surface Temperature (SST) and also the temperature of ocean water are rising resulting in a change in the ocean life. Change in the ocean chemistry, particularly the acidification, has been impeding the ability of organisms to build shells. Due to rise of Sea Surface Temperature ocean circulations have been affected. El Nino/La Nina episodes appear to be the direct consequence of changing climate. El Nino significantly affects the marine ecosystem. Moreover, the change in cloud cover and sea ice retard the incoming light to the ocean surface, which in turn perturbs the ecosystem.

The most important example of climate change's impact is the swimming sea snail. The tiny pteropod has difficulty growing a shell in a warmer acidified ocean waters. Given the snails' role at the base of the cold-water food chain, its struggle threatens the entire polar ecosystem, through salmon to seals and whales.

The problem is one of many associated with ocean acidification. That change is the consequence of warming that has already happened and fossil-fuel emissions that have long since been dumped into the atmosphere. Today's ocean chemistry is already hostile for many creatures fundamental to the marine food web. The world's oceans, for so long a neat and invisible sink for humanity's carbon dioxide emissions, are about to pay a price for all that waste.

The rate of change is alarming. The transformation has happened over 250 years, faster than anything in the historical record. And if emissions remain unchecked, the oceans in 40 years will be more acidic than anything experienced in the past 20 million years.

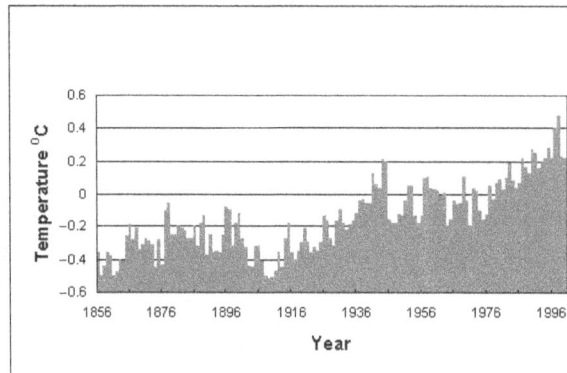

**Figure 10.1:** Changes in Global Sea Surface Temperature

Over the next several centuries the pH changes may be larger than any inferred from the geologic record of the past 300 million years, with the exception of a few rare extreme events.

The process is fairly simple. For eons prior to the industrial revolution, oceans were at equilibrium with the atmosphere, absorbing as much carbon dioxide as they released.

The oceans so far have absorbed some 30 percent of the carbon dioxide that humans have added to the atmosphere since the beginning of the industrial revolution and nearly 80 percent of the heat generated by those gases.

Today the world's oceans absorb some 30 million metric tons of extra carbon dioxide every day, roughly twice the amount of carbon dioxide emitted each day by the United States.

With increasing acidity of the oceans reefs are struggling in many parts of the world, shell growth rates are slowing, life phases, particularly reproductive maturity, are being affected.

So we find that the marine ecosystem has largely been affected with the changing climate. We can discuss the effect of changing climate under the following heads:

1. Physical and chemical effects of climate change leading to variations of – *Deep-ocean Circulation Patterns, Wind Patterns, Ocean Stratification and Chemical Change.*
2. Effects on bio-diversity affecting the life of – *Plankton, Fish, Sea Birds and Marine Mammals.*

**Physical and Chemical Effects of Climate Change**

Physical and chemical effects of climate change on oceans are difficult to assess. The changes are not evidenced uniformly in all the oceans. However, the temperature of ocean water has been rising worldwide. The rising sea surface and ocean water temperature cause changes in the following conditions.

### *Deep-ocean Circulation Patterns*

Wind drives the surface currents, which cause upwelling and down-welling, as well as by thermohaline circulation. Wind movement is part of the thermal and dynamic general circulation in the atmosphere and controlled by the global pressure belts. Wind circulation drives ocean water around the earth as controlled by the pressure belts. Cold, hypersaline and dense water from Polar Regions sink and move towards equator across the ocean floor. Warmer waters from lower (tropical) latitudes alternately move towards the poles as surface flow. This continuous circulation oxygenates the deep oceans and redistributes heat from equator to the poles. If there be rise of temperature, sea ice would be melted reducing the salinity. As a result density of polar waters would decrease, and in turn the strength of thermohaline circulation would be weakened. This may restrict the transport of oceanic heat from warm tropical oceans to cold oceans of the world. On other hand warmer waters in the tropics would contain less dissolved carbon dioxide further releasing more in the atmosphere.

### *Wind Patterns*

Sea surface temperatures influence the patterns of atmospheric pressure, which in turn cause the generation of wind. Accelerated warming of the oceans may cause stronger wind circulation in certain areas, and increase the frequency of extreme events such as storms and hurricanes.

Tropical storms are formed when the sea surface temperature exceeds 27° C and heated air rises to form low pressure area with dense cloud formation and rainfall. Air from high pressure areas rushes in at high speed to replace this warm air, and large amounts of moisture are absorbed in a spiraling effect finally precipitating heavily. The threshold temperature for tropical storms could be attained more readily in a changing climate and such storms could move to higher latitudes.

Changes in wind generated surface currents would not only affect the weather conditions but an alteration of the upwelling process would have critical effects on the marine ecosystem. Upwelling may affect the nutrient-rich deeper water, vital for primary production. If reduced it could seriously affect species distribution and abundance.

### *Ocean Stratification*

The oceanic water column is stratified with warmer water on the surface. In warm, calm conditions this stratification is intensified and becomes more resistant to mixing by surface winds. This mixing helps bringing more nutrients from deeper water to replenish those being used in the upper layers. In more temperate areas stratification of the water column occurs during the warmer summer months. In winter due to cooling of water and together with the action of winter winds, stratification breaks down. This seasonal cycle is important as winter mixing brings nutrients from deeper waters to the surface. Phytoplankton can have the nutrients in spring and summer, once stratification forms to trap them in the warm waters near the surface. Extremely rising summer temperatures will cause greater stratification and winter winds will hinder the mixing of these two layers.

## *Chemical Composition*

Increase in temperature of water will cause lesser concentrations of dissolved gases that the oceans can hold. Both dissolved oxygen and carbon dioxide are important for all stages of food production and breakdown of organic matter. Lesser concentration of these gases would affect the food production and breakdown of organic matter. The reduced ability of seawater to hold carbon dioxide would mean more of the gas present in the atmosphere further causing the temperature rise. Such an impact may be mitigated by an opposing effect. Seawater of lower salinity holds higher concentrations of carbon dioxide than more saline water and, therefore, a greater input of freshwater from the melting of polar ice, could allow more carbon dioxide to be dissolved at higher latitudes.

The ability of the oceans to contain more dissolved carbon dioxide would change the chemical composition of surface waters. Increases in carbon dioxide are considered to reduce concentrations of aragonite. Aragonite is a form of calcium carbonate and an important component of the coral skeleton. Growth rates and skeletal strength may be reduced on coral reefs. Other marine organisms having calcium carbonate into their skeletal structures may be similarly affected.

Seawater appears to be a complex, but dynamic fluid. It is pure water having dissolved inorganic salts and gases, dissolved organic substances, and various microscopic living organisms. Circulation patterns help transporting all these components to areas where they are required for maintenance of life systems. Climate change may induce a change or cessation in this circulation severely altering the distribution of these elements. As a result damages for food production in the sea could be caused.

## Impact of El Nino Southern Oscillations (ENSO) and North Atlantic Oscillations (NAO)

El Nino Southern Oscillation is known to be a continuous or seasonal pattern in ocean-atmosphere dynamics which drives natural variation in climate over longer and less predictable time-scales. ENSO is caused by a naturally occurring oscillation of atmospheric pressures in the Pacific Ocean that results in weakening of the trade winds. These trade winds normally carry warm water away from the eastern Pacific. As a result an upwelling occurs. As the trade winds weakens, the warm water shift in that region is leass consequently lowering the nutrient concentrations. ENSO events may occur on average every 2-8 years. Many of these effects on the oceans are known to be similar to those predicted for climate change.

The frequency and duration of these events have risen over the last few decades. It is now predicted that this trend may prevail with year-to-year variations and become more extreme. An additional factor to consider is that the "natural" fluctuations associated with ENSO events may aggravate with rising base-line temperature.

The North Atlantic Oscillation (NAO) causes climate variability over the Atlantic Ocean. The two phases, negative and positive, are related to varying strengths of the subtropical high to the Icelandic low pressure system. The negative phase shows a weak subtropical high and weak Icelandic low with weaker pressure gradient resulting in lesser and weaker storms over the Atlantic. Over the last 30 years there has been a trend towards a more positive phase which shows a stronger than usual high pressure and deeper than normal Icelandic low. These positive phases have caused stronger winter storms across the Atlantic. The rise in frequency of these storms is associated with climate change on a large scale.

## Animal and Plant Responses to Climate Change

It is universally accepted that animals are capable of moving to migrate in response to climate change. The movements of commercially important species, such as cod, in response to changes in ocean temperature, have been documented for centuries. A variety of animal species have migrated in response to the warming in the 20th century. Individual plants cannot migrate, but plant species can and do migrate in response to changes in climate. All plants have seed dispersal mechanisms and therefore are constantly trying to establish seedlings in new areas. Seedlings thrive in a more desirable climate, but fail in a less desirable climate, shifting the range of the plant as climate changes. The total change in range can be dramatic. Fossil evidences indicate that since the end of the last Ice Age, the balsam fir migrated from the southeastern U.S. to northern Canada, while the black spruce migrated from the central plains to Alaska. It is understood that plants and animals could migrate in response to climate change, but we are concerned with at least four actions. It will result in a sufficient degree to prevent large scale species extinction:

1. Human activities, particularly habitat disruption, will block potential migration routes;
2. Even if they can migrate, the members of a given ecosystem will migrate at different rates leading to imbalances that will result in species extinctions;
3. Plants may not be able to migrate fast enough to keep up with projected rates of climate change; and
4. Plants and animals that live in restricted niches, e.g., near mountain tops, will have no place to migrate.

These concerns assume no human intervention to help wild species to adapt to climate change. In light of the growing and successful effort to reintroduce species such as beaver and wolf to their former habitats, to replant native plants, and to remove invasive plant species, this assumption appears to be conservative.

## Effects on Species

### *Plankton*

Microscopic plants, known as phytoplankton, are found in the illuminated surface layers of the ocean. There are tens of thousands of different species of phytoplankton. Importance of phytoplankton has now been well understood. We can enumerate the significance of phytoplankton as follows:

1.  Phytoplankton plays the central role in the global carbon, oxygen and nutrient cycles and produces half the oxygen on the planet.
2.  Changes in ocean temperature and circulation will compel species to migrate southward on the east coast of U.S.A. Abundance is already changing and climate change is likely to alter the environmental conditions that affect the occurrence of toxic or harmful algal blooms that can accumulate in food webs.
3.  Warming as an event of climate change can induce changes in phytoplankton community composition and, for example, the productive south-eastern temperate phytoplankton region may be reduced considerably in area.

Phytoplankton appears to be the most important biomass producers in the oceans, removing carbon dioxide from the atmosphere and transferring the carbon to other trophic levels. They have a vital link to deep ocean organisms when they die and organic material deposits on sea bottom. Phytoplankton species are concentrated in the top layer (euphotic zone or sunlit zone) where they get enough solar radiation for their photosynthetic requirements as well as nutrients that are transported by the upwelling process.

Rise in sea surface temperature causes alternate wind patterns and increased stratification of the water column. As a result there has been significant reduction of nutrients reaching the euphotic zone. So a decrease in primary production is evidenced. This means lowering of overall productivity as well as less carbon dioxide absorption from the atmosphere. Reduced carbon input will consequently affect the deep sea ecosystem. The species composition of phytoplankton populations in different regions will be affected by alterations in nutrient levels and temperature causing serious repercussions for the marine environment. A shift in species in one area could lead to less carbon dioxide absorption from the atmosphere as carbon absorbing species are replaced by others.

As phytoplankton happens to be responsible for carbon fixation, bacterioplankton plays the important role in the mineralisation of nutrients and provides a trophic link to higher organisms. The breakdown products of organic matter are consumed by bacterioplankton. So an indirect result of climate change would be a reduction in nutrients available to these microorganisms, impeding their function in the carbon flux of aquatic ecosystems.

The reduction in numbers or change in species composition of phytoplankton could lead to a reduction in the abundance of zooplankton.

Apart from temperature changes increased penetration of UV rays cause a threat to their survival. UV radiation actually hinders photosynthesis in phytoplankton. It is assumed that rapid vertical mixing in the water column further restricts photosynthesis whereas a greater cloud cover reduces the harmful effects of UV radiation. Moreover, UV rays can affect growth and reproduction as well as the functioning of enzymes and cellular proteins in phytoplankton. Bacterioplankton appears to have less UV-screening pigments. So UV radiation affects the enzymes responsible for breaking down their food. It is also known that UV radiation causes damage to zooplankton. Even small rise in UV exposure would result in significant reductions in the consumer community.

**Effects on Animals**

*Mammals*

Mammals are primarily terrestrial (land-dwelling) animals. Responses of mammals to rising temperatures and other climate changes appear to be diverse. Due to rising temperature and climate change many small mammals breed earlier in the year than they did several decades ago. Other mammals are known to shift to higher altitudes abating the rise of temperature. A few mammals indicate to have larger body sizes, probably due to increasing food availability and higher temperatures.

Melting of polar sea ice has caused lower reproductive success in polar bears. Polar bears, seals, migratory birds, caribou and reindeer are all affected by climatic changes causing dramatic effects on their species and habitats. Polar bears are unlikely to survive as a species if there is an almost complete loss of summer sea-ice cover. The seals that polar bears hunt fail to adapt in the absence of summer sea ice, as they give birth to and nurse their pups on the ice. Summer ice has been the place for their resting. Moreover, temperature increases in the Arctic cause sea ice to break up earlier in the summer. So the polar bears have less time to build up vital fat reserves to last them through the winter months. Sea ice supports other marine mammals like walruses and ringed seals by providing critical breeding and resting-places. Due to rising temperature tundra vegetation zone has been shifting further north. Caribou and reindeer populations so decline for their dependence on tundra vegetation.

*Invertebrates and Insects*

Invertebrates include 97 percent of all animal species. Most invertebrates are tiny in size, but their influence on their surroundings appears to be very high. Bees, moths, ants and other insects play a critical role in the life of seed plants by transferring pollen. We know the importance of insect pollination for production of certain fruits, nuts and vegetables.

Climate change indicates both positive and negative impacts on invertebrates and insects. In Alaska due to warming spruce budworms (insect pest) has shifted further north to reproduce.

South-central Alaska also experienced a massive outbreak of spruce bark beetles (insect pest) in the 1990s, resulting in 10-20 percent mortality of trees. In addition many invertebrates that are considered pests or disease organisms are found to shift toward the poles (northward in the Northern Hemisphere) or to higher elevations. Butterflies' habitat ranges in North America are also found to have migrated northward and in elevation as a result of rising temperatures. A few butterfly species, such as the Edith's Checkerspot Butterfly, have become extinct in the southern location of their range.

### Birds

Birds play many important roles in seed dispersal, pollination, and as both predator and prey in the ecosystem. It is now observed that birds are breeding and laying their eggs earlier. The migratory species have altered their wintering and/or critical stopover habitats. The Arctic Climate Impact Assessment has shown that the timing of bird arrival in the Arctic may no longer coincide with the availability of their insect food sources. Important breeding and nesting areas have declined due to shifting of trees northward, penetrating deeper into tundra area. With the rise of sea level due to melting of polar ice, more tundra area, and thus more habitat for birds and their prey, will be lost. This could eventually cause serious consequence of the breeding of several hundred million birds that migrate to the Arctic each summer from lower latitudes. This will in turn affect the population sizes of birds at lower latitudes.

The changing climate has affected the life cycles of birds as are related with its food supply, warming temperatures and other ecological processes that are essential to ecosystem health. Pollination, seed dispersal, and pest control by birds depend on careful timing of birds' arrival, atmospheric temperature and other climate-related factors. So the events of climate change would disrupt the ecosystem in general. Sea level rise can result in the loss of wetlands in coastal areas, thus affecting the habitats of some waterfowls in the winter months. Sea level is rising around the world, and is projected to continue rising throughout this century. So this important habitat may be lost affecting the ecosystems due to climate change.

### Fish

Fish population in marine environment depends on plankton production in sea. We have noted the decline of plankton production due to rise of sea surface temperature resulting in the loss of nutrients. A decrease in plankton production, resulting from a changing climate, would affect fish populations with lower availability of foods. Moreover, temperature rise would compel many species to migrate poleward. Because poleward the climate is cooler and hence metabolic rates could be kept low. Warm-water species may emerge in abundance with more favourable conditions at lower latitudes. The changing scenario would have significant impact on commercial fisheries such as that of Atlantic cod as this species feeds on temperature-sensitive ones like mackerel and herring. The decline of important salmon stocks on the west coast of North America in high latitudes suggests that climate change is already causing impact.

ENSO events in the Eastern Pacific Ocean off Peruvian coast in some years cause havoc loss of fish species such as Peruvian anchovies and sardines. Further large-scale disaster could result if the unusual temperatures experienced during an ENSO period become more marked associated with climate change.

Incoming UV radiation also affect the eggs and larvae of many fish species which are sensitive to ultraviolet (UV) radiation. Depletion of ozone layer may be one of the important causes for high mortality of fish larvae. Other causes of population decline are predation, poor food supply for larvae, exhaustive fishing of adults, water temperature, pollution and disease. UV radiation could be harmful at both primary and secondary production levels.

IPCC Report states that certain fish species are becoming less abundant worldwide. Fish populations and other aquatic resources are likely to be affected by warmer water temperatures, changes in seasonal flow regimes, total flows, lake levels, and water quality. These are to a great extent associated with climate change. In turn it will have impacts on the health of aquatic ecosystems, particularly with impacts on productivity, species diversity, and species distribution (IPCC, 2007).

Changes in the geographic distribution of ocean fish stocks appear to be the impact of climate-ocean system variations such as the *El Niño* events. Fluctuations in fish availability are considered to be the biological response to climate-ocean variations, and not just as a result of over-fishing and other human factors. Climate change would have compound impact of natural variation and fishing activity. It is also observed that increased temperatures would accelerate mortality of winter flounder eggs and larvae and lead to later spawning migrations. With rise in sea surface temperature tuna populations are likely to migrate toward cooler water in temperate regions. A recent study has indicated that climate change will be most severe at high latitudes. Many indicators confirm the changing environmental conditions in these regions.

### *Reptiles and Amphibians*

Reptiles and amphibians can adapt to changes in climate depending in part on their ability to migrate to more suitable habitat. A European study indicated that most reptile and amphibian species could stretch their ranges in a warmer climate if dispersal were unlimited, but if they were unable to disperse then the ranges of nearly all species (more than 97 percent) would become smaller (IPCC, 2007).

We have already noted the impact of climate change on coral reefs and mangrove. Coral reefs and mangroves happen to be the maximum productive region of all ecosystems. Impacts of climate change on coral reefs and mangroves may cause adverse effect on sea turtles and crocodiles. Tropical storms with higher intensity and frequency may also affect sea turtle populations adversely. Hurricane Emily in 2005 caused the destruction of 1,500 sea turtle nests along the Mexican coast. Orisssa super-cyclone in 1998 similarly caused destruction of 'Ridley turtle' nests along the Orissa coast (Bhitor Kanika) in India.

In North America, many amphibians, such as some species of frogs and salamanders, lay their eggs in temporary pools that form in early spring after snowmelt. In a warmer climate such ponds would be dry earlier in the season. As a result the amphibian populations would suffer.

## Adaptations of Plants and Animals to Climate Change

Plants and animals have been adapting to climate change for billions of years. However, not all plant and animal species will be successful in adapting. It is true that natural climate change has been a major factor in past species extinctions. Any change in climate, whether natural or human-induced, will increase the risk that some marginal species will become extinct. Despite the concern about climate change, habitat disruption will continue to be the largest threat posed by human activities to the survival of plant and animal species.

Migration is the major response that plants and animals can do under the climate change. Many concerns have been raised about the ability of plants and animals to migrate given habitat disruption under the situation involving high rates of climate change during the 21st century. Societal efforts to counter adverse effects by relocating endangered plant and animal species to more favourable habitats could reduce the impact of these changes.

Protection of endangered species enjoys widespread support in many countries, but the understanding needed to implement realistic programmes that are in balance with other priorities is inadequate. Many fundamental questions remain unresolved. We do not know yet how many species are becoming extinct. It remains unanswered. It is desirable to spot two areas specifically on the potential impacts of climate change on ecosystems. Efforts are needed to develop:

1. Improved models which project the combined transient effect of multiple stresses, including climate change, on ecosystems; both the IPCC and NAS agree that current models are inadequate; and
2. Better techniques to help endangered species migrate in response to climate change.

'Many innovative programs are now being undertaken to help plants and animals survive counter the adverse effects of habitat disruption, and these programs will help make these species more resilient to climate change' (IPCC, 2007).

**Box A: Climate Change Stressors**

- Climate variability
- Altered temperature
- Altered precipitation
- Altered solar radiation
- Sea level rise
- Increased atmospheric $CO_2$

**Box B: Impacts on Ecological Components**

- Loss of hard coral and/or key species
- Loss of tidal marsh area and/or key species

**Box C: Impacts on Ecological Processes**

Examples

- Coral reefs: Bioerosion, algaii overgrowth
- Tidal Marshes: Altered chemical uptake and filtrartion capacity

**Box D: Changes in Ecosyatem Service Flows**

Examples

Coral Reefs:
- Loss of coral diversity and associated potential pharmaceuticals
- Changes in coral reef refineries
- Reduced shoreline protection

Tidal Marshes
- Loss of tidal marsh plant biodiversity and associated harvestable products
- Reduced water purification
- Reduced flood protection

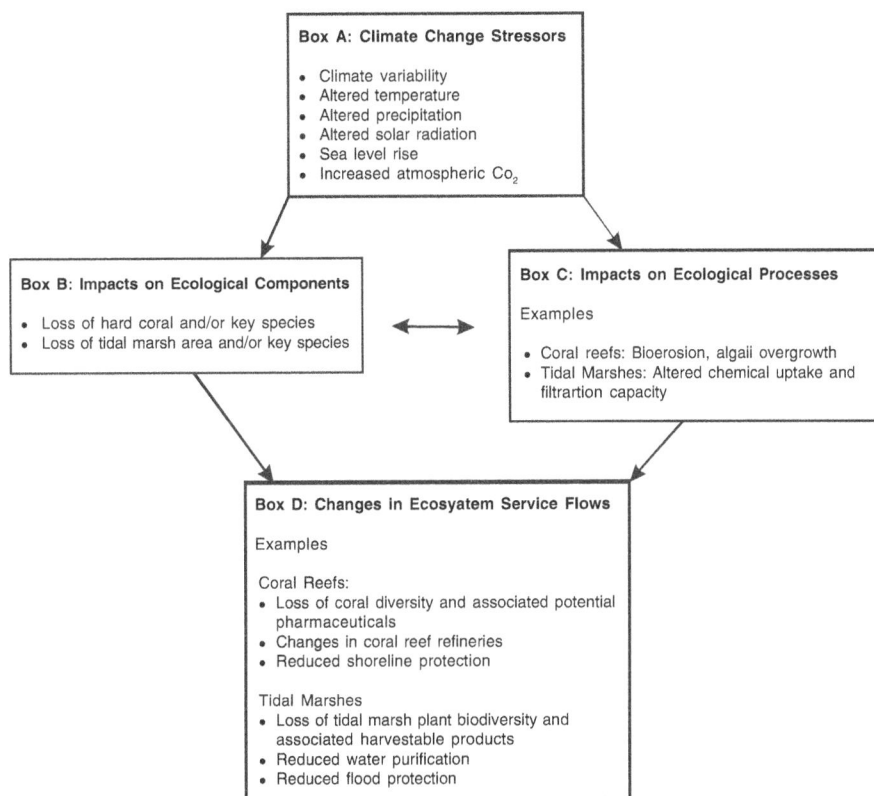

**Figure 10.2:** Climate Change Effects on Ecosystems

## Further Readings

Arctic Climate Impact Assessment (ACIA), (2004) *Impacts of a Warming Arctic: Arctic Climate Impact Assessment*, Cambridge University Press, Cambridge, United Kingdom and New York, NY, USA.

Barnett, T.P.; J.C. Adam, D.P. Lettenmaier (2005) "Potential impacts of a warming climate on water availability in snow-dominated regions". *Nature* **438**.

Board on Natural Disasters (1999) "Mitigation Emerges as Major Strategy for Reducing Losses Caused by Natural Disasters", *Science* **284**.

Botkin, Daniel B.; *et al.* (2007) "Forecasting the Effects of Global Warming on Biodiversity" (PDF). *BioScience* **57**.

Margules, C.R. and R.L. Pressey. (2000) "Systematic Conservation Planning." *Nature,* **405**.

Emanuel, K. (2003) "Tropical cyclones", *Annual Review of Earth and Planetary Sciences,* **31**.

EPA (2000) *Global Warming: Impacts: Forests.* United States Environmental Protection Agency, Washington.

Hunter, M.L. (1996) *Fundamentals of Conservation Biology.* Blackwell Science Inc., Cambridge, Massachusetts.

IPCC (2007) *Climate Change 2007:* Impacts, Adaptation, and Vulnerability. *Contribution of Working Group II to the Third Assessment Report of the Intergovernmental Panel on Climate Change* [Parry, Martin L., Canziani, Osvaldo F., Palutikof, Jean P., van der Linden, Paul J., and Hanson, Clair E. (eds.)]. Cambridge University Press, Cambridge, United Kingdom.

IPCC (2002) Climate Change and Biodiversity (PDF, p.86) [*Eds.* Gitay, Habiba, Suarez, Avelino, Watson, Robert T., and Dokken, David Jon]

IPCC (2002) Climate Change and Biodiversity, In-*Technical Paper 5*, April 2002.

Intergovernmental Panel on Climate Change (2007) M.L. Parry, O.F. Canziani, J.P. Palutikof, P.J. van der Linden and C.E.Hanson. ed. Summary for Policymakers. Climate Change 2007: Impacts, Adaptation and Vulnerability. *Contribution of Working Group II to the Fourth Assessment Report of the Intergovernmental Panel on Climate Change.* Cambridge University Press, Cambridge, UK.

Intergovernmental Panel on Climate Change (2008) *Fourth Assessment Report, Working Group II Report* "Impacts, Adaptation and Vulnerability". Chapter 6: Coastal Systems and Low Lying Areas. http://www.ipcc.ch/pdf/assessment-report/ar4/wg2/ar4-wg2-chapter6.pdf

IUCN (1993) *Impacts of Climate Change on Ecosystems and Species: Environmental Context,* Geneva, Switzerland.

Kevin J. Gaston and John I. Spicer (2004), *"Biodiversity: an Introduction"*, Blackwell.

Larry O'Hanlon (2006) *"Rising Ocean Acidity Threatens Reefs",* Discovery News. http://dsc.discovery.com/news/2006/07/05/acidocean_pla.html.

Le Bohec, Céline; Joël M. Durant, Michel Gauthier-Clerc, Nils C. Stenseth, Young-Hyang Park, Roger Pradel, David Grémillet, Jean-Paul Gendner, and Yvon Le Maho (2008), "King penguin population threatened by Southern Ocean warming" (abstract). *Proceedings of the National Academy of Sciences* **105**.

Leveque, C. and J. Mounolou (2003) *Biodiversity*, New York: John Wiley.

Mayhew, Peter J; Gareth B. Jenkins, Timothy G. Benton (2007) "A long-term association between global temperature and biodiversity, origination and extinction in the fossil record". *Proceedings of the Royal Society,* Royal Society Publishing **275**.

M.E. Soulé and B.A. Wilcox. (1980) *Conservation Biology: An Evolutionary-Ecological Perspective,* Sinauer Associates. Sunderland, Massachusetts.

Millennium Ecosystem Assessment (2005) *Ecosystems and Human Well-being: Biodiversity Synthesis,* World Resources Institute, Washington, DC.

Myers N. (1988) "Threatened biotas: 'hot spots' in tropical forests", *Environmentalist,* **8**.

Myers N. (1990) "The biodiversity challenge: expanded hot-spots analysis", *Environmentalist,* **10**.

Permesan, Camille (2006) "Ecological and Evolutionary Responses to Recent Climate Change" (PDF). *Annual Review of Ecology, Evolution, and Systematics* **37**.

Root, Terry L.; Jeff T. Price, Kimberly R. Hall, Stephen H. Schneider, Cynthia Rosenzweig & Alan Pounds (2003) "Fingerprints of global warming on animals and plants ", *Nature* **421**.

*The Independent*, April 27, 2005 "Climate change poses threat to food supply, scientists say" - Report

Walther, Gian-Reto; *et al.* (2002) *"Ecological responses to recent climate change"*, *Nature* **416**.

Whittaker, R.H. (1972) "Evolution and measurement of species diversity", *Taxon*, **21**.

# Chapter 11

## CLIMATE CHANGE AND HUMAN HEALTH

### Recognising Climate Change

Globally 'Climate Change' is now a conclusive assumption. There is now widespread consensus that the earth is warming at a rate unprecedented during post hunter-gatherer human existence. The last decade was the warmest since instrumental records began in the nineteenth century, and contained 9 of the 10 warmest years ever recorded. The Third Assessment Report of the Intergovernmental Panel on Climate Change states that "*There is new and stronger evidence that most of the warming observed over the last 50 years is likely to be attributable to human activities*", most importantly the release of greenhouse gases from fossil fuels.

The report of the Intergovernmental Panel on Climate Change also confirmed that there is overwhelming evidence that humans are affecting the global climate, and highlighted a wide range of implications for human health. Climate variability and change have become instrumental to cause increasing mortality and disease through natural disasters, such as heat-waves, floods and droughts. It is also known that many important diseases are highly sensitive to changing temperatures and precipitation. These include common vector-borne diseases such as malaria and dengue; as well as other major fatal diseases such as malnutrition and diarrhoea. Climate change already causes the global burden of disease, and it is expected to rise further.

### Potential Health Impacts of Climate Change

Prior to 1990, there was little knowledge and understanding of the risks posed to human health by global climate change. It was due to epidemiologists' limited conventional approach to environmental health. The environment was considered predominantly as a repository for specific human-made pollutants: in air, water, soil and food. Such pollutants were supposed to cause particular risks to human communities and individual consumers. There was lack of awareness among physical and natural scientists that changes in climatic conditions, biodiversity stocks, ecosystem productivity, and so on happen to be the potential risks to human health. It was IPCC first to suggest the possibility that global climate change might affect human health. The Third Assessment Report IPCC reviewed the impact of climate change on the health of global populations.

Global climate change causes an impact on human health *via* complex pathways with varying scale, direction and timing. Impacts also vary geographically as being a function comprising environment, topography and the vulnerability of the local population. Impacts are both positive and negative, though dominantly negative. The reason is that climate change would disrupt or otherwise modify a large range of natural ecological and physical systems which remain to be the integral part of earth's life-support system. The weather extremes like heat-waves, winter cold-waves, floods, cyclones, storm-surges, droughts exert more direct impacts on health. Enhanced generation of certain air pollutants and aeroallergens (spores and moulds) supplement the impacts.

Climate change may be the indirect actor transmitting many infectious diseases (e.g. water, food and vector-borne diseases) and affecting regional food productivity (especially cereal grains). These indirect impacts could be more vulnerable and are likely to have greater magnitude than the more direct in longer term. Vector-borne infections *vis-à-vis*, the distribution and abundance of vector organisms and intermediate hosts are related to various physical (temperature, precipitation, humidity, surface water and wind) and biotic factors (vegetation, host species, predators, competitors, parasites and human interventions). It is now known that an increase in ambient temperature would result in net increases in the geographical distribution of particular vector organisms (e.g. malarial mosquitoes) worldwide. Moreover, temperature related changes in the life-cycle dynamics of both the vector species and the pathogenic organisms (flukes, protozoa, bacteria and viruses) are likely to accelerate the potential transmission of many vector-borne diseases such as malaria (mosquito), dengue fever (mosquito) and leishmaniasis (sand-fly). However, it is supposed that schistosomiasis (water-snail) may decrease in response to climate change.

The stratospheric cooling due to depletion of ozone in the stratosphere by CFC and other aerosols may induce climate change. As a result of stratospheric ozone depletion the potential health risks would affect human health. The potential health risks are the increase in incidence of skin cancer in fair-skinned populations; eye lesions such as cataracts; and, perhaps, suppression of immune activity. So these health risks occur due to climate change. The climatic change over the past quarter-century certainly caused various incremental impacts on at least some health outcomes. Recent emergence of malaria and dengue in highland regions around the world is now attributed to climate factors or to the several other factors that are known to be significant determinants of transmission. Hot weather conditions cause noxious photochemical smog in urban areas. Warmer summers also record the incidence of food poisoning. Rising global temperature happens to be linked with the emergence of such health hazards.

Impacts of climate change on health happen to cause many maladies. These are –

1. Increased heat-related mortality and morbidity;
2. Decreased cold-related mortality in temperate countries;

3.  Greater frequency of infectious disease epidemics following floods and storms; and
4.  Substantial health effects following population displacement from sea level rise and increased storm activity.

We can identify the vulnerable groups to disease and injury for each of the potential impacts of climate change. For example, the socially excluded urban dwellers, the elderly and the poor are the most at risk of diseases and injury from thermal extremes. Populations living marginally and having no primary health care are increasingly prone to malaria and dengue with global warming. Temperate lands and high lands having low temperature conditions did not record almost any case of malaria and dengue earlier. The IPCC report shows instances of these diseases related to climate change has risen considerably in the last ten years. However, there remain still many uncertainties.

## Climate Change and the Vector-borne Diseases

A vector-borne disease is caused when the pathogenic microorganism is transmitted from an infected individual to another individual by an arthropod or other agent, sometimes with other animals serving as intermediary hosts. The transmission depends on many attributes and requires at least three different living organisms. These living organisms happen to be the pathologic agent, either a virus, protozoa, bacteria, or worm. The vectors are commonly arthropods, such as ticks or mosquitoes. Humans act as the host. There are also intermediary hosts such as domesticated and/or wild animals which usually act as a reservoir for the pathogen until susceptible human populations are exposed.

Vector-borne diseases are common worldwide and cause high morbidity and mortality amongst human populations. Vector-borne diseases affect nearly 50 percent of the world population. The vector-borne diseases are not distributed in the world evenly. Vector-borne diseases happen to be area specific. The distribution of the incidence of vector-borne diseases is concentrated in developing countries located in tropical and subtropical areas. Hot and humid weather conditions favour the growth of the vectors. Weather controls vector population dynamics and disease transmission, with temperature and humidity as key variables. So climate has direct relationship with the growth of vectors. Any form of climate change may induce the impact of vector-borne diseases.

The complexities of interactions between environment and host are best shown by the example of vector-borne diseases. The success of pathogens and vectors is determined partly by their reproductive rate. Malaria-carrying mosquito populations can increase tremendously within a very short time. Equally the *Plasmodium* parasite species proliferates rapidly in both mosquito and human hosts. In contrast, *tsetse* flies have a low reproductive rate and their populations take much longer time to increase under favourable conditions. Hence, infectious diseases transmitted by the *tsetse* fly (including human sleeping sickness) respond less rapidly to variations in climate than do many mosquito-borne infections. Vectors' ability to transmit disease is also affected by feeding frequency. Hard ticks (such as the vectors of Lyme disease) feed more frequently and for shorter periods than soft ticks. Hard ticks therefore tend to be much more efficient vectors of human diseases. Overall, high vector and pathogen reproductive capacity; preference for humans as a source of blood meals; low life cycle complexity; and high sensitivity to temperature changes result in an infectious disease that has high sensitivity to climate variability.

For better revelation of the potential impact of changes in the incidence of vector-borne diseases attributable to climate change requires examination of the following topics:

- Current status of vector-borne diseases around the world;
- Environmental parameters that affect the incidence of vector-borne diseases;
- Models that predict changes in the incidence of vector-borne diseases attributable to climate change;
- Possible direct and indirect effects of climate change on specific vector-borne diseases; and
- Programmes for surveillance, treatment, and control of vector-borne diseases.

### *Climate Sensitivities of Infectious Diseases*

Both the infectious agent (protozoa, bacteria, viruses, etc) and the associated vector organism (mosquitoes, ticks, sandflies, etc.) are tiny and do not possess thermostatic mechanisms. Their temperature and fluid levels are controlled directly by the local climate. So each infective or vector species can survive and reproduce within the climate envelope. Usually, the incubation time of a vector-borne infective agent within its vector organism is typically very sensitive to changes in temperature. Other climatic sensitivities are the level of precipitation, sea level elevation, wind and duration of sunlight.

Many infectious diseases are related to seasonal patterns of climate. It is not very certain to what extent changes in disease patterns will occur under the conditions of global climate change. Climate happens to be one of several important factors influencing the incidence of infectious diseases. Other important considerations include socio-demographic influences such as human migration and transportation; and drug resistance and nutrition; as well as environmental influences such as deforestation; agricultural development; water projects; and urbanisation. It is highly unlikely that climatic changes induce an isolated effect on disease; rather the effect is likely dependent on the extent to which humans cope with or counter the trends of other disease modifying influences.

Effect of climate variability on infectious diseases is specified largely by the unique transmission cycle of each pathogen. Transmission cycles dependent on a vector or non-human host happen to be more susceptible to external environmental influences than those diseases which include only the pathogen and human. Temperature, precipitation and humidity are important environmental factors for possible transmission.

## Effect of Temperature and Precipitation on Vector-borne Diseases

Temperature and precipitation are two key elements of weather and climate. Any modification of these elements is reflected in the climate change. Effects of temperature and precipitation on vectors and vector-borne pathogens are noted below to identify the relation between climate and diseases. Change of sea level has been the result of rising temperature worldwide causing melting of polar and highland ice and happens to be a part of the climate change process. Impact of sea level rise on selected vector-borne diseases has also been discussed.

### *Temperature Effects on Selected Vectors and Vector-borne Pathogens*

*Vector*

- Survival can decrease or increase depending on species;
- Some vectors have higher survival at higher latitudes and altitudes with higher temperatures;
- Changes in the susceptibility of vectors to some pathogens e.g. higher temperatures reduce size of some vectors but reduce activity of others;
- Changes in the rate of vector population growth;
- Changes in feeding rate and host contact (may alter survival rate);
- Changes in seasonality of populations.

*Pathogen*

- Decreased extrinsic incubation period of pathogen in vector at higher temperatures
- Changes in transmission season
- Changes in distribution
- Decreased viral replication.

### *Effects of Changes in Precipitation on Selected Vector-borne Pathogens*

*Vector*

- Increased rain may increase larval habitat and vector population size by creating new habitat
- Excess rain or snowmelt can eliminate habitat by flooding, decreasing vector population
- Low rainfall can create habitat by causing rivers to dry into pools (dry season malaria)
- Decreased rain can increase container-breeding mosquitoes by forcing increased water storage
- Epic rainfall events can synchronize vector host-seeking and virus transmission

- Increased humidity increases vector survival; decreased humidity decreases vector survival.

## Pathogen

Few direct effects are noted including humidity effects on malarial parasite development in the anopheline mosquito host.

## Vertebrate Host

- Increased rain can increase vegetation, food availability, and population size
- Increased rain can cause flooding: decreases population size but increases human contact.

### Increased Sea Level Effects on Selected Vector-borne Pathogens

Sea level change displays the following effects:

1. Alteration of estuary flow and changes in existing salt marshes and associated mosquito species,
2. Results in decreasing or eliminating selected mosquito breeding-sites (e.g. reduced habitat for *Culiseta melanura*).

## Properties of Vector-borne Diseases

Important properties in the transmission of vector-borne diseases are as follows:
1. Survival and reproduction rate of the vector;
2. Time of year and level of vector activity, specifically the biting rate;
3. Rate of development and reproduction of the pathogen within the vector;
4. Vectors, pathogens, and hosts each survive and reproduce within certain optimal climatic conditions and changes in these conditions can modify greatly these properties of disease transmission.

The most influential climatic factors for vector-borne diseases include temperature and precipitation. Sea level elevation, wind, and daylight duration are additional important considerations. The following paragraphs illustrate the effects of climatic sensitivity.

### Temperature Sensitivity

Extreme temperatures often results in fatal conditions for the survival of disease-causing pathogens but incremental changes in temperature may cause varying effects. A vector living in an environment where the mean temperature approaches the limit of physiological tolerance for the pathogen, a small increase in temperature may be dangerous to the pathogen. Alternatively, a vector living in an environment of low mean temperature, a small increase in temperature may result in increased development, incubation and replication of the pathogen.

Temperature may modify the growth of disease carrying vectors by changing their biting rates, as well as control vector population dynamics and change the rate of contact with humans. Finally, a shift in temperature regime can change the length of the transmission season. Disease carrying vectors may adapt to changes in temperature by altering geographical distribution. An emergence of malaria in the cooler climates of the African highlands may be the result of the mosquito vector adapting in a new habitat to cope with increased ambient air temperatures. It may also be assumed that vectors could have undergone an evolutionary change to adapt in increasing temperatures. There is recent instance to suggest that the pitcher-plant mosquito (*Wyeomia smithii*) can adapt genetically to survive the longer growing seasons associated with climate change. The change in response was linked to a marked genetic shift within the mosquito species. It is supposed that the mosquito population have now greater ability to evolve genetically with increased selection pressure. So, disease carrying vectors may have an analogous microevolution which helps adaptation to altered seasonal patterns associated with global climate change.

### Precipitation Sensitivity

Variability in precipitation may be directly related to the cause of infectious disease outbreaks. Increased precipitation favours the extension of area for the larval habitat and creates new breeding grounds. So, increased precipitation helps a growth in food supplies which in turn support a greater population of vertebrate reservoirs. On the other hand heavy rainfalls in odd seasons may cause flooding and decrease vector populations by eliminating larval habitats and creating unsuitable environments for vertebrate reservoirs.

However, flooding may drive out insect or rodent vectors from their habitat to seek shelter in houses. With their intrusion into human shelters vector-human contact happens to rise. Epidemics of leptospirosis, a rodent-borne disease, have been reported from Brazil due to the occurrence of floods. In the wet tropics unusual drought turns the rivers sluggish creating more stagnant pools for breeding the vectors.

### Humidity Sensitivity

Transmission of vector-borne diseases, particularly for insect vectors, is also largely humidity sensitive. Under dry conditions mosquitoes and ticks desiccate easily and can not survive. Saturation deficit appears to be one of the most critical determinants for climatic control on diseases. Important illustrations are dengue fever and Lyme disease.

**Table 11.1:** Potential Health Impact of Above-Average Rainfall

| Event | Type | Description | Potential Health Impact |
|-------|------|-------------|-------------------------|
| Heavy Precipitation | Meteorological | Extreme event | Increase in mosquito abundance or decrease (if breeding site washed) |
| Flood | Hydrological | Rivers overflowing the banks | Changes in mosquito abundance/contamination of surface water |
| Flood | Social | Damage of property/crops | Changes in mosquito abundance/contamination of surface water with faecal matter/pollutants |
| Flood | Catastrophic flood | Greater flood damages | Changes in mosquito abundance/contamination of surface water with faecal matter/pollutants, increased risk of respiratory/diarrhoeal diseases, loss of food and population displacement |

**Table 11.2:** Potential Health Impact of Below-Average Rainfall

| Event | Type | Description | Potential Health Impact |
|-------|------|-------------|-------------------------|
| Drought | Meteorological | Evaporation> water absorption, and moisture decreases | Changes in vector abundance if vector breeds in dry river beds |
| Drought | Agricultural | Drier than normal causing less crop production | Depends on socio-economic factors |
| Drought | Social | Reduction in food supply/income, reduction in water supply/quality | Food shortage/illness/malnutrition/increased risks of infection and diseases |
| Drought | Food shortage/famine/drought disaster | Food shortages leading to death | Deaths from malnutrition and infectious diseases, health impacts associated with displacement of population |

## *Sea Level Sensitivity*

Sea level rise associated with climate change may result in the decrease or even elimination of breeding habitats for salt-marsh mosquitoes. Bird and mammalian hosts of this ecological niche depending on feeding of the vectors may be threatened by extinction. This may also help the elimination of viruses endemic to this habitat. On other hand inland intrusion of salt water may convert former fresh water habitats into salt-marsh areas which could support vector and host species displaced from former salt-marsh habitats.

**Table 11.3:** Water and Food-borne Agents and Climate

| Pathogen Groups | Pathogenic Agent | Food-borne Agents | Indirect Water-borne Agents | Direct Weather Effect | Weather Effect |
|---|---|---|---|---|---|
| Viruses | Enteric Viruses (e.g. Hepatitis A Virus & Coxsackie B virus | shell fish | Groundwater | Storms with rainfall increase transport of faecal and waste water sources | Survival increases with reduced temperature and sunlight (UV rays*) |
| Bacteria Cyanobacteria Dinoflagillates | Vivrio (e.g. V. Vulnificus, V. Parahaemo-lyticus, V. Cholerae), Anabaena spp, Gymnodinium Pseydibutzschia spp. | Shell fish | Recreational, Wound infections | Enhanced Zoo Plankton blooms | Salinity and temperature related growth in marine environment |
| Protozoa | Enteric Protozoa (e.g. Cyclospora, Crytosporidium) | Fruits and vegetables | Recreational and drinking water | Storms with rainfall increase transport of faecal and waste water sources | Temperature associated with malnutrition and infectivity of Cyclospora |

* Also applies to bacteria and protozoa

## Vector-borne Diseases

Potential changes in humidity and temperature can induce changes in the geographic ranges and life cycles of plants, animals, insects, bacteria, and viruses. The range of many plant pests tend to shift northward by several hundred kilometres. Such changes are also noticed in the insects that cause diseases to both humans and animals.

### *Global Distribution of Vector-borne Diseases*

In 1992 the World Health Organization (WHO) Commission on Health and Environment in its publication, '*Our Planet, Our Health*', reported that "The conditions in which environmental factors are most prominent are tropical diseases, which to a large extent are caused by infection by parasites requiring one or more intermediate hosts and vectors for their development."

Vector-borne diseases are mostly dominant in the tropics and subtropics and relatively rare in temperate zones. These diseases appear to be the serious health problems in developing countries. WHO estimated the number of people infected by the most serious vector-borne diseases—malaria, 270 million; schistosomiasis, 200 million; lymphatic filariases, over 90 million; onchocerciasis, nearly 18 million; leishmaniasis, 12 million; dracunculiasis, 1 million; and African trypanosomiasis, 25,000 new cases per year (1990). However, the available data appear to be the underestimate as many of the cases are neither reported nor registered in the developing countries.

Moreover, most of the deaths had been due to infectious and parasitic diseases for the population under 5 years old, 5 years and older, and all ages. Although vector-borne diseases are presently considered a major health hazard only in tropical and subtropical regions, in the coming decades the climate change would induce conditions suitable for outbreaks of infectious diseases in temperate regions as well. EPA's 1989 Report to Congress on '*The Potential Effects of Global Climate Change on the United States*' already outlined five vector-borne diseases, namely, Lyme disease, Rocky Mountain spotted fever, Malaria, Dengue fever, and viral Encephalitis as of great concern in USA due to global warming.

**Table 11.4:** List of Vector-Transmitted Diseases in Terms of Global Importance

| Disease | Prevalence |
|---------|-----------|
| Malaria | Global (80-90% of cases in Africa, 100-200 million people infected; 1-2 million deaths annually) |
| Schistosomiasis (bilharzia) | Global (largest occurrence in Africa; a debilitating disease; an estimated 200 million people infected) |
| Japanese encephalitis (brain fever) | South, South-East and East Asia (closely related to irrigated rice production; epidemic outbreaks with high mortality rates among children) |
| Lymphatic filariasis (elephantiasis) | Global ( mostly urban, with the exception of Central Africa where it is linked to irrigation and South/South-East Asia where it is linked to weed-infested reservoirs and to latrines either in the field or in nearby communities) |
| River blindness (onchocerciasis) | West and Central Africa and foci in Central America (the Onchocerciasis Control Programme has eliminated the disease as a public health problem in a large part of West Africa) |

*Source:* Adapted from Feacham et al., 1977

### *Malaria*

In the early part of the twentieth century the extent of human malaria was checked through a global campaign, particularly in the developed countries in temperate regions. It was possible due to social and economic uplift and through successful eradication programmes. But still the disease remains highly endemic in most countries lying in the tropics, even extending northwards in Turkey, Syria, Iraq, Iran, Afghanistan, Pakistan and China.

The extent of malaria was largely controlled by taking action against the parasite by widespread use of antimalarial drugs for chemotherapy and chemosuppression and those against the anopheline vector. The control of malaria transmission was achieved through the destruction of anopheline vector by the use of residual insecticides on the interior surfaces of human dwellings. In addition larvicidal measures including biological control were also initiated including the environmental management to reduce breeding places of vectors. The success of spraying programmes for the interruption of malaria transmission depended upon the vulnerability of local vector species of *Anopheles* in each endemic area. The more endophilic (indoor-resting) vectors such as *A. labranchiae* in Italy or *A. minimus* in northern India are usually not difficult to control as their survival-rates are low to maintain malaria transmission in areas with adequate insecticidal coverage.

Even after a prolonged campaign with spraying of larvicidal chemicals malaria has not been eradicated as expected. In many parts of the world where malaria was known to be eradicated new cases of vector-borne diseases are emerging. It is now known that as a result of indiscriminate use of pesticides for public health or agriculture purposes, many vector populations have become resistant to the insecticides used. 50 species of anopheline are known to be resistant to one or more pesticides. Of these 11 species happen to be the important vectors of malaria. A number of the vectors are not affected by both organochlorine and organophosphorus insecticides, some of these are also resistant to carbamates and pyrethroids. Among these multiresistant vectors are *A. albimanus* in Mexico and Central America, the *A. culicifacies* complex in India and Sri Lanka, *A. pharoensis* in Egypt, *A. sacharovi* in Turkey and *A. stephensi* in India and Iran. Even, some vector populations exhibit exophilousy to such an extent that makes it difficult to control malaria transmission by means of vector control based on house spraying alone. The environmental conditions and the genetic background of the vector population control the success.

*P. falciparum* was first known to be resistant to chloroquine in the early sixties. There had been reports of growing resistance in subsequent years with areas extending and developing in Southeastern Asia and South America. In recent years it has become increasingly serious menace in Africa. Resistance of *P. falciparum* to other antimalarial drugs is also rising in these areas. As a result treatment for malaria has become more difficult. It is now urged to prevent transmission through vector control and by means of personal protection.

Many vectors of malaria and other diseases happen to be morphologically identical species and remain in the groups of sibling species. Such species are reproductively separate from each other as they do not normally interbreed where mixed populations occur. Despite the morphological similarity of sibling species, they show ecological and behavioural differences and have varying capacity as vectors. *Anopheles* complexes of sibling species are:

- *A. balabacensis complex:* Within the *leucosphyrus* group the *balabacensis* complex forms a subset, and within this complex in a subset of *dirus* forms at least four species are found.
- *A. culicifacies complex:* At least four species (A, B, C, D) are found in the Indian subcontinent alone.
- *A. gambiae complex:* Six species exist in tropical Africa (*arabiensis, bwambae, gambiae, melas, merus* and *quadriannulatus*).
- *A. maculatus complex:* At least four and possibly seven siblings are found.
- *A. maculipennis complex:* Eight species in Eurasia (*atroparvus, beklemishevi, labranchiae, maculipennis, martinus, melanoon, messeae,* and *sacharovi*) and four species in North America (*aztecus, earlei, freebomi* and *occidentalis*) are found.
- *A. punctulatus complex:* At least five species in the Australasian zone exist.

*Ecological Conditions*

A low-flow regime forms pool of stagnant or nearly stagnant water in the riverbed creating suitable breeding sites for malaria vectors. Periodic flushing with storm flow can eliminate the risk. In the estuaries low flow may result in salt intrusion where anophelines breeding in brackish water may flourish, such as *Anopheles sundaicus* (South-East Asia), *Anopheles melas* (west coast of Africa) and *Anopheles merus* (east coast of Africa). In cases with the formation of temporary sandbars creating coastal lagoons the breeding will flourish. Hydrologic conditions having standing water in them may fascilitate schistosomiasis transmission.

On the other hand dams and impoundments may cause a variety of health risks, in part because of ecological change (mosquito and snail propagation along shallow shorelines, associated with aquatic weeds, and blackfly breeding on spillways), and in part because of demographic changes. The risk could be minimised with the measures like periodic reservoir fluctuation, steepening of the shorelines, controlling aquatic weeds, locating settlements away from the reservoir. The blackfly problem may be solved by constructing the dams with two spillways that can be used alternately.

Due to rise of water table water-logging conditions emerge in many areas where many mosquito vector species thrive. Area with good drainage condition checks the growth of vector compared with the areas having poor drainage conditions. Standing water in agricultural fields by various types of irrigation (surface, contour and furrow irrigation) cause greater health risks than others (sprinkler, central pivot or drip irrigation). In the case of surface irrigation, canal lining benefits environmental and health concerns alike with less loss of precious water. Irrigation with alternate wetting and drying of paddy fields and synchronised cropping of rice appears to be effective against vector-borne diseases. However, a fall in the water table, in some parts of the world may help the growth of *Phlebotomine* sandflies which live in semi-arid conditions and transmit leishmaniasis, in its visceral form a fatal illness.

### Lymphatic Filariasis

Human lymphatic filariasis is widely found in humid tropical Africa, the Americas, Asia, and numerous islands in the Pacific Ocean. Local epidemiology of the disease depends largely on the behaviour and ecology of the various species of mosquito vectors.

Occurrence of Lymphatic Filariasis differs from Malaria or Arboviruses. Malaria or Arboviruses may be infected by the bite of a single infective vector, whereas filarial infections require repeated inoculation of infective larvae, perhaps hundreds per year so that the worms can reproduce successfully and produce microfilaraemia. Often the disease happens to be asymptomatic initially with subsequent episodes of acute inflammation of the lymphatic system and fever. Ultimately it may reach the chronic stage of "elephantiasis" developing only after many years. By that time the microfilariae usually disappear from the bloodstream of the patient.

Two types of Filariasis are identified. These are *Bancroftian Filariasis* and *Brugerian Filariasis*.

### Bancroftian Filariasis

*W. bancrofti* is becoming increasingly prevalent in the cities of tropical countries due to transmission by *Culex quinquefasciatus* and *C. pipiens*. These species breed prolifically in polluted water. So these species are concentrated in urban situation where environmental status is poor. These species now exhibit widespread resistance to insecticides. So, appropriate environmental management and source reduction can control them.

In rural environment *W. bancrofti* is mainly transmitted by some of the same *Anopheles* spp. which happens to be the principal vector of malaria. Such strains of the parasite are nocturnally periodic. In some locations (e.g. Africa and Malaysia) urban *Culex quinquefasciatus spp.* fail to transmit as it seems to be apparently refractory to anopheline-adapted strains of *W. bancrofti*. The reverse situation is observed in other areas (e.g. India, Malaysia) where *Culex*-adapted strains of *W. bancrofti* can not be transmitted by anophelines. In Polynesia transmission is done by the *Aedes (Stegomyia) scutellaris* group which bites during daytime and evening. In some forests of Southeast Asia there are also subperiodic strains of W. bancrofti transmitted by the *Aedes (Finlaya) niveus* group.

### Brugian Filariasis

*Brugia Malayi* is found only in South Asia. Now its distribution and prevalence have been minimised by control of vector *Mansonia* spp. It is the simple method of removal of the host-plants from their breeding-places. Subperiodic strains occur in swamp-forest habitats, transmitted mainly by the *M. bonneaedives* group as a zoonotic cycle. Periodic strains of *B. Malayi* are transmitted also by *Anopheles* spp. and are usually not zoonotic.

### Brugia Timori

*Brugia Timori* is usually found in Indonesian islands of Flores and Timor and nearby locations. It is apparently not zoonotic. The only known vector is *Anopheles barbirostris*. *B. Timori* appears to be not transmitted by various other mosquito species such as *A. subpictus*.

### Yellow Fever

Yellow fever is endemic in tropical Africa from 15° N to 15° S latitudes and in the north and eastern parts of south america. Yellow fever emerged as epidemic in parts of Middle America. The disease had its origin in Africa and was transported to America along with its principal vector A. aegypti during the period of slave trade in the 16th century. It caused epidemic amongst the population in the coastal towns of South America and the USA as far north as New York. African slaves appeared to be immune not to suffer severely from yellow fever but mortality among the Europeans was recorded high. More recently, yellow fever emerged as epidemic in Ethiopia in 1960-1962 with an estimated 15,000 deaths and also in 1986/87 in West Africa. The epidemics were transmitted by A. *aegypti* in America. Probably some time, the virus moved into the forests in the interior of Brazil. Now it is recognised that jungle yellow fever is prevalent amongst the monkeys transmitted by the mosquitoes which breed in the forest environment. At present yellow fever in urban location has been controlled by immunisation. However, little success has been achieved in lowering the risk of rural yellow fever and in the enzootic areas of Africa and America where human cases occur every year. The urban vector, *A. aegypti* is re-emerging in areas in South America from where it had earlier been eradicated. Interestingly yellow fever has never been reported from Asia despite the presence of the vector *A. aegypti* and considerable populations of susceptible monkeys.

### Dengue and Dengue Haemorrhagic Fever

Dengue is the most important arboviral disease of humans. Arboviruses are characterised as being arthropod-borne viruses. Dengue occurs worldwide, particularly in tropical and subtropical regions. In recent decades, emergence of dengue has created increasing urban health problem in tropical countries. The disease is thought to have occurred due to ineffective vector and disease surveillance; inadequate public health infrastructure; population growth; unplanned and uncontrolled urbanisation; and increased travel. The main vector of dengue is the domesticated mosquito known as *Aedes aegypti*. It requires clear water for breeding. So it breeds in urban environments in any open containers filled with water. *Aedes aegypti* breed in small containers, such as plant pots, which contains water. *Aedes albopictus*, which can tolerate colder temperatures, also can transmit Dengue. Dengue is seasonal and usually associated with warmer and more humid weather. Increased rainfall in many locations can affect the vector density and transmission potential. Rainfall may affect the breeding of mosquitoes.

Dengue appears to be endemic in many of the tropical countries of Asia, the Pacific and Caribbean regions. Sometimes it emerges as periodic epidemics. Dengue haemorrhagic fever is a severe, often fatal disease characterised by fever, shock, acute haemorrhage with high mortality rate. It attacks usually the younger children. Dengue and associated fever are now known to occur in many of the cities and towns of Southeast Asia. It appears to be one of the leading causes for hospitalisation and death in children in tropical Asia. Dengue Haemorrhagic fever has also been reported in the Caribbean region with sporadic occurrence from time to time.

Dengue was probably transported to the Mediterranean region from East Africa in the late 19th century, through the slave trade out of Zanzibar and via the Red Sea ports. There is evidence to suggest that dengue originated from foci in tropical Asia and from there spread to Africa.

The peri-domestic, highly anthropophilic *Aedes aegypti* happens to be the principal vector of all the serotypes and the sole vector in the America and Australia. Areas infested by *A. aegypti* mosquitoes appear to be potentially dengue-risk zones. Other species like *Aedes albopictus* cause dengue as the sole vector in some rural areas of Southeast Asia. In Southwest Pacific region *A. aegypti*, as well as members of the *A. scutellaris* complex remain to be the principal or sole vectors. Dengue haemorrhagic fever usually occurs in urban centres infested by *A. aegypti*. It is now believed that monkeys also act as host for spread of dengue involving canopy feeding monkeys and forest mosquitos of the Aedes niveus group which feed on man and monkeys. There might be transovarial transmission of the virus in *A. aegypti* and *A. albopictus* in nature.

## Encephalitis

Encephalitis means an inflammation of the brain, but it usually refers to brain inflammation caused by a virus. Arbovirus-related encephalitides include a group of acute inflammatory diseases affecting parts of the brain, spinal cord, and meninges. In mild cases, these infections cause feverish headaches or aseptic meningitis. In severe cases those symptoms are associated with stupor, coma, convulsions (in infants), and occasionally spastic paralysis. Encephalitis may be of any types like Japanese Encephalitis, California Encephalitis, Eastern Equine Encephalitis, St. Louis Encephalitis and Western Equine Encephalitis.

### Japanese Encephalitis

Encephalitis means an inflammation of the brain, but it usually refers to brain inflammation caused by a virus. Japanese encephalitis is a type of encephalitis which caused widespread epidemics in Japan and the Republic of Korea. The disease is known to occur in many Asian countries from Far East to South Asia. The disease is found in Siberia, Japan, Republic of Korea, China, Indonesia, Singapore, Malaysia, Thailand, Viet Nam, Burma, Nepal, India and Sri Lanka. India experienced the epidemics of Japanese Encephalitis in 1973 and 1977-78, characterised by high mortality.

Mosquito is the vector of virus for encephalitis. The virus is borne in nature by mosquitoes and non-human vertebrates. Human becomes prey of the virus accidentally. Encephalitis occurs under warm climate. In temperate regions like Japan, the disease spreads in under warm weather conditions. However, in the warm tropics it could happen during any season, although the risk is higher during and immediately after the rainy season when the mosquito population rises. So, this disease is seasonal and weather specific which helps breeding of virus carrying mosquitoes. *Culex tritaeniorhynchus*, a rice field breeding mosquito appears to be the principal vector which feeds mainly on large animals and birds. *C. gelidus* (predominantly a pig biter) and *C. vishnui* group mosquitoes are also vector carriers. The virus of encephalitis has also been found in other Culex spp. as well as Aedes and Anopheles mosquitoes.

Japanese encephalitis dominantly occurs in rural areas. Encephalitis is mostly concentrated in many countries of southeastern Asia, associated with rice cultivation where virus carrying mosquitoes can breed in the rice fields. Transmission occurs mostly by bite. Transovarial transmission of the virus may also occur by some Aedes species.

## California Encephalitis

Encephalitis has also been reported from other parts of the world. Everywhere it is caused by Arboviruses. It is also widely found among the animals. One variety is known to be the California Encephalitis found in America. Apart from the Americas, cases of the disease have been reported from Africa and Europe. Generally they indicate sub-clinical symptoms in man, but occasional acute cases occur. Rodents and lagomorphs are the main wild hosts and reservoirs in North America, where ten types of virus in this group are found.

## Eastern Equine Encephalitis

Eastern Equine Encephalitis widely occurs in both, North America and South America. Such outbreaks are common in the eastern coastal areas of Canada, USA, extending towards the Caribbean and South America as far south as Argentina. Horses remain the principal victims and human cases are rare and sporadic. Species of Aedes (Ochlerotatus) are the principal vectors to man and horses. The virus happens to be enzootic in birds and is caused by ornithophilic Culex spp. and Culiseta spp. through transovarial transmission. Pheasants are particularly susceptible.

*St. Louis Encephalitis*

St. Louis Encephalitis occurs widely in North America and South America. Human exposures are rare. But occasional outbreaks may cause disease in appreciable number of humans. *Culex tarsalis* in Western USA, *C. quinquefasciatus* in Mid. USA and *C. nigripalpus* in USA appear to be the principal transmitter.

*Western Equine Encephalitis*

Western Equine Encephalitis bears the characteristics of Eastern Equine Encephalitis, except that rodents are the principal carriers of this type of encephalitis. Nearly 26 species of mosquitoes probably cause the epidemics by transmitting the virus. In North America, important vectors are found breeding in salt marshes. These are: *Aedes taeniorhynchus, A. sollicitans, Culex portesi* and *Psorophora ferox*. Epidemics emerge extending from Texas to Peru affecting both horses and people. The resultant effects of Western Equine Encephalitis are severe and often fatal.

### Rodent-borne Diseases

Rodents cause a number of diseases whether as intermediate infected hosts or as hosts for arthropod vectors such as ticks. Certain rodent-borne diseases like leptospirosis, tularaemia and viral haemorrhagic diseases are caused by flooding. So above-average precipitation may lead to occurrence of these diseases. Plague, Lyme disease, tick borne encephalitis (TBE) and hanta-virus pulmonary syndrome (HPS) are other diseases associated with rodents and ticks.

   Rodent populations are found to grow more in temperate regions following mild wet winters. It occurs more frequently following winter-spring periods with above-average precipitation. Climatic conditions help the growth of rodents and promote breeding of flea populations with more food sources. Ticks also happen to be climate sensitive. Infection by hanta-viruses is usually caused by inhalation of airborne particles from rodent excreta. Alternate drought and heavy rainfall conditions may lead to growing rodent population, on one hand reducing populations of the rodents' natural predators during dry period and subsequent high rainfall increasing food availability in the form of insects and nuts. Rise in cases of hanta-virus was associated with the increase of rodent populations due to prevalence of two wet, relatively warm winters in the southern USA in 1998.

### Tick-Borne Diseases

Rocky Mountain spotted fever and Lyme disease are known to be tick-borne diseases and cause public health problems. These two diseases are borne by different species of ticks.

The tick-borne diseases appear to be particularly sensitive to any change that may affect the geographic range of their hosts and, consequently, the range of the vector, or carrier. Seasonality of environmental factors such as temperature, humidity, and vegetation also control tick populations. Climate must be warm enough to promote progression through the life cycles, humid enough to prevent the drying out of eggs, and cold enough in winter to initiate the resting stage. Higher temperatures may help the growth of the organism that is transmitted by the carrier, such as a virus. The rate of infection of the carrier would rise by these mechanisms. So climate has specific relationship with the tick-borne diseases. Tick populations also vary with the natural vegetation of an area. The occurrence of Rocky Mountain spotted fever is related to natural vegetation and changes in climate.

Though higher temperatures favour the presence of the agent, but they may also disturb the life cycle of some tick species. Accordingly warmer temperatures may reduce both tick survival and the spread of diseases they carry. As for many tick-borne diseases, the opportunity for a tick to acquire the infective agent from an infected animal is limited to the short period when the level of the agent in the blood of the host is high enough for the tick to receive an infective dose.

### *Diarrhoeal Illness*

Many enteric diseases have a seasonal pattern, indicating its sensitivity to climate. In the tropical countries the frequency of diarrhoeal diseases ascends the peak during the rainy season. Floods and droughts are also linked to increased risk of diarrhoeal diseases. On one hand heavy rainfall can pollute water supplies with the high run-off, on the other hand drought conditions causes the scarcity of fresh water leading to an increase of diseases related to public sanitation.

The contaminated water are infested with cholera, cryptosporidium, *E.coli*, giardia, shigella, typhoid, and viruses such as hepatitis A. Outbreaks of cryptosporidiosis, giardia, leptospirosis and other infections happen to be associated with heavy rainfall events in countries where public water supply becomes polluted due to contamination. An association between drinking water turbidity and gastrointestinal illness is evidenced in most countries of the world. A study of waterborne disease outbreaks in USA has shown that nearly 50 percent of these diseases are significantly associated with extreme rainfall.

Transmission of enteric diseases may also be enhanced by high temperatures. High temperature exerts a direct effect on the growth of disease organisms in the environment. In many countries the rise of monthly temperature changes shows a direct impact on the incidences of enteric diseases. It is estimated that 1° C rise in temperature would cause 3 percent increase in diarrhoeal diseases.

In fine, there is strong effective relationship between several important communicable diseases and climate on variable temporal and geographical scales. This is universal for all vector-borne diseases, many types of enteric illnesses and certain water-related diseases.

Relationships between year-to-year variations in climate and communicable diseases are most evident where these climate variations are marked, and in vulnerable populations in poor countries.

## Impacts of Extreme Climate on Health

Extreme climate events are known to be more frequent with climate change. Climate extremes can have devastating effects on human societies. We have the records of widespread disasters, famines and disease outbreaks triggered by droughts and floods in the past. These disruptive events cause adverse impact on health, particularly in the developing countries where public sanitation remains unsatisfactory. We can recognise two categories of climatic extremes. These are:

- Simple extremes of climatic statistical ranges, such as very low or very high temperatures
- *Complex events:* droughts, floods, or hurricanes.

Climate change results in the frequency of extreme events. Climatic anomalies occur inducing a sudden climatic shift in many cases. Extreme events are also associated with the intensity of storms and resultant heavy deluge. Climate variability is evidenced at various temporal scales (by day, season and year) and remains an inherent characteristic of climate, whether or not the climate system is subject to change. El Nino Southern Oscillation (ENSO) now appears to be one such event causing serious consequences of adverse climate in the time scale. In sensitive regions, ENSO events may cause significant inter-annual perturbations in temperature and/or rainfall in a cycle of 2–7 year, though uncertain. A range of physical, ecological and social mechanisms indicate the relationship between extremes of climate and disease.

## El Niño and Infectious Diseases

We have already noted the relationship between rainfall and diseases spread by insect vectors which breed in water, and are therefore dependent on surface water availability. Mosquitoes are the principal vector which transmits malaria and viral diseases such as dengue and yellow fever. The fact of mosquito abundance to rainfall events is now well established. Mosquitoes need access to breed in the stagnant water both under wet and dry conditions.

Large year-to-year fluctuations of natural disasters, some of which can be explained by El Niño, are described as the El Niño disaster cycle.

- The risk of a natural disaster is highest in the years during and after the appearance of El Niño and lowest in the years before.
- El Niño events in 1982-83 and 1997-98, were the largest in the twentieth century.
- El Niño is associated with death and disease, most of which result from weather-related disasters such as floods and droughts.

## ENSO

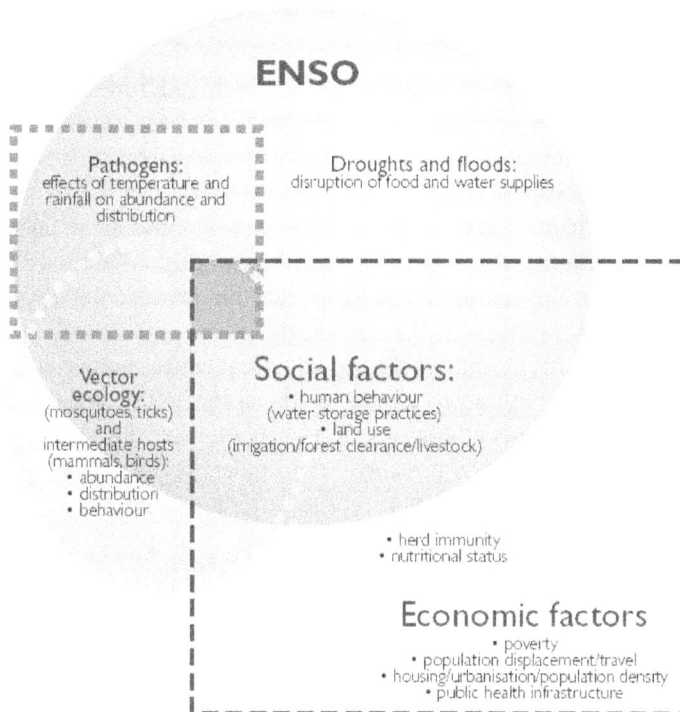

**Figure 11.1:** Impact of ENSO on Health

- In 1997 Central Ecuador and Peru receied rainfall more than 10 times normal, which caused flooding, extensive erosion and mudslides with loss of lives, destruction of homes and food supplies.
- In the same year nearly 10% of all health facilities in Peru were damaged.
- The 1991-92 El Niño brought the worst drought in southern Africa, which affected nearly 100 million people.
- Ecuador, Peru and Bolivia suffered serious malaria epidemics after heavy rainfall in 1983 El Niño. The epidemic in Ecuador was aggravated by displacement of population owing to the flooding.
- During the 1997 El Niño droughts hit Malaysia, Indonesia and Brazil, exacerbating the huge forest fires. Smoke inhalation from these fires was a major public health problem in these countries, with countless people visiting health centres with respiratory problems.

Recently, there has been growing recognition of the links between El Niño and disease. For example, doctors for many years were intrigued by the cycles of malaria in some countries roughly every 5 years. Such cycles in India, Venezuela and Columbia have now been linked to El Niño. Pronounced changes in the incidence of *epidemic diseases* can occur in parallel with extreme weather conditions associated with the El Niño cycle.

*Malaria*

Malaria shows sensitivity to climate in desert and highland fringe areas where rainfall and temperature, respectively, are critical parameters for disease transmission. In these areas higher temperatures and/or rainfall associated with El Niño enhance the effective transmission of malaria. It is now evidenced that ENSO events have causal effect on weather conditions, particularly rainfall in many parts of the world. Even the monsoon rainfall in India and South Eastern Asia is related to the ENSO events. We have discussed the ENSO events in the Chapter 8 of this book in detail. Such areas record unprecedented drought or excessive rainfall during ENSO events.

It has been found that drought in the previous year causes increased malaria mortality. There are several possible reasons for this relationship. Drought-related malnutrition may enhance an individual's susceptibility to infection. Famine conditions may accelerate the rate of mortality. India recorded high mortality due to epidemics of malaria following the El Niño event in 1877. Many deaths had been recorded after the end of the drought. The reason is that drought-breaking rains increased vector abundance to spread the disease vigorously.

Many parts of South America were affected by ENSO-related climate anomalies. Serious epidemics in the northern countries of South America emerge mainly in the year following the El Niño event. In 1983 following a strong El Niño event, malaria epidemics spread over Ecuador, Peru and Bolivia. In Venezuela and Colombia, the incidence of malaria also increased in the post-El Niño year. El Niño is associated with a reduction of the normal high rainfall regime in much of Colombia, as well as an increase in mean temperature, increase in dew point, and decrease in river discharges. Southern Africa and east of the Sahara demonstrate ENSO-related rainfall anomalies. The 1997/98 El Niño caused heavy rainfall and flooding in Kenya, after two years of drought giving rise to incidence of malaria. From January to May 1998, a major epidemic of falciparum malaria broke out in Kenya. The malaria epidemic was compounded by widespread food shortages due to drought for two consecutive previous years.

*Dengue*

Dengue is a seasonal disease, related to warm and humid climatic conditions. ENSO may act indirectly by causing changes in water storage practices brought about by disruption of regular supplies.

Between 1970 and 1995, the frequency of epidemics of dengue in the South Pacific was positively correlated with the Southern Oscillation Index (SOI). High positive values of the SOI (denoting La Niña conditions) indicate much warmer and wetter conditions than the average which favours the breeding of mosquitoes. Such positive correlation has also been found between SOI and dengue in other countries. These include American Samoa, Nauru, Tokelau, Wallis and Western Samoa where positive correlations between SOI and local temperature and/or rainfall have been observed. During La Niña conditions, these five islands experience wetter and warmer climate than normal conditions. The effect of ENSO was also evidenced in South East Asia for the occurrence of dengue. In Viet Nam the number of dengue cases increased during the El Niño years. Many countries in Asia were affected with an unusually high level of dengue and dengue haemorrhagic fever in 1998, which may be caused by ENSO related weather. Positive linkages between ENSO and dengue epidemics have also been reported from French Guyana, Indonesia, Colombia and Suriname which experience warmer temperatures and less rainfall during El Niño years.

## Heatwaves and Cold Spells

In recent years many studies have indicated the relationship between temperature and mortality. The effect of a hot day remains only for a few days affecting the mortality. In contrast, a cold day induces an effect that continues for weeks. In many temperate countries mortality rates rise in winter than death rates in summer. However, the causes of this winter excess can not be well explained. Most probably, different mechanisms play the role in heat and cold related mortality. Cold related mortality in temperate countries may be caused partly by the occurrence of seasonal respiratory infections. On the other hand heatwaves cause some clinical syndromes such as heatstroke. But very few deaths occur directly by heat. Blood viscosity rises for exposure to high temperatures. The heat stress may cause heart attack or stroke. On the other hand the low temperatures are associated with increased blood pressure and fibrinogen levels.

Individual sensitivity to extreme weather events varies. If the frequency and intensity of heatwaves rise, the risk of death and serious illness would be enhanced particularly for the older age groups, those with pre-existing cardio-respiratory diseases, and the urban poor. The effects of an increase in heatwaves often would be aggravated by rising humidity and urban air pollution. Thermal stress has been rising in middle to high latitude cities, especially in populations living in houses of unsuitable architecture and limited air conditioning.

Impacts of heatwave in urban populations will be intensified in near future in the cities of temperate countries causing several hundred extra deaths every summer. Climate change may also have adverse effects on thermal stress-related mortality in cities in developing countries. Warmer winters and fewer cold spells, however, will decrease cold-related mortality in many temperate countries.

Rise in the frequency of extreme events like storms, floods, droughts and cyclones would also aggravate human health in different ways. These natural hazards may result in direct loss of life and injury and aggravate health indirectly due to loss of shelter; population displacement; contamination of water supplies; loss of food production (leading to hunger and malnutrition); increased risk of infectious disease epidemics (including diarrhoeal and respiratory diseases); and damage to infrastructure for provision of health services. Increased frequency and intensity of cyclones on a regional scale may cause disasters with epidemics and colossal loss of human life. In recent decades climate-related disasters have resulted in hundreds of thousands of deaths in countries such as China, Bangladesh, Venezuela and Mozambique.

## Impact of Heatwave on Mortality

It is really difficult to define a heatwave as responses to very high temperatures vary between populations and within the same population over time. There is no standard international definition of a heatwave. Nature of heatwave also varies from hot tropical countries to cool temperate countries. Day maximum temperature around 40° C may not be fatal in warm tropical regions, but may be fatal and even cause deaths in the cool temperate regions. Heatwave is usually more fatal in the cool temperate countries than in the hot tropical countries. Global warming with steady rise of temperature appears to be a menace in many countries of the temperate regions during the summer months. Chicago in USA registered 514 deaths (12 per 100000 populations) and 3300 excess emergency hospitalisation, caused by a heatwave in July, 1995. Athens recorded a heatwave in 1987 causing heat related death of 926 persons. India experienced heatwave in 1998 when nearly 2600 persons died. In Orissa alone 1300 people died when temperature rose to 49.5° C.

The elderly persons and those with pre-existing illness are the most vulnerable groups. During heatwaves, excess mortality is mostly recorded among these vulnerable groups. Much of this excess mortality is caused due to cardiovascular, cerebrovascular and respiratory disease. An increase in the frequency and intensity of heatwaves would increase the numbers of additional deaths due to hot weather.

### *Vulnerability to Temperature-related Mortality*

Indicators of vulnerability to heat and cold may be noted as follows:
- Age and disease profile
- Socioeconomic status
- Housing conditions
- Prevalence of air conditioning
- Behaviour (e.g. clothing).

Age appears to be the predominating factor. Both individual and population level studies provide strong and consistent evidence that age is a risk factor for heat-related mortality. Studies vary on the age at which the vulnerability is increased. There are physiological reasons why the elderly people are more vulnerable. Individual risk factors for dying in the heatwave are considered to be chronic illness; confinement in bed; unable to care for themselves; isolated and without air conditioning.

### *Winter Mortality*

Mortality rate also shows distinctly a seasonal variation. In many temperate countries death rates rise during winter. It happens to be 10-25 percent higher in winter than those in summer. The major causes of winter death are due to cardiovascular, cerebrovascular, circulatory and respiratory diseases. Influenza, occurring annually, also has large effect on winter mortality rates. However, such mortality is not related to winter cold spell. Social and behavioural adaptations to cold happen to be an important factor in preventing winter deaths in high latitude countries.

Sensitivity to cold weather (measured as the percentage increase in mortality per 1° C change in temperature) is recorded higher in warmer regions. Mortality increases to a greater extent with a sharp drop in temperature in regions with warmer winters, in populations with less home heating facilities and where people have inadequate winter clothing. The elderly (aged 75 and over) persons are largely vulnerable to winter death. In winter 30 percent more persons of this vulnerable group die. This vulnerability may be related to a combination of physiological susceptibility, behavioural factors and socioeconomic disadvantage. Excess winter mortality is an important issue in the United Kingdom where there has been much debate on the role of poor housing, poor fuel consumption and other socioeconomic issues for the elderly population. It has also been found that inadequate home heating and socioeconomic deprivations are other important reasons for excess death in winter.

### **Effects on Air Pollution**

Weather conditions may have impacts on the transportation of air-borne pollutants, pollen production and levels of fossil fuel pollutants caused by household heating and energy consumption. Climate change may give rise to the concentration of ground level ozone. The effects of climate change and/or weather on other pollutants though not very certain may be of great concern in the coming years. Climate change is supposed to increase the risks of forest and rangeland fires and associated smoke hazards. Major fires in 1997 in south-east Asia, North America and South America had intensified the respiratory and eye problems with the generated smoke. There had been two to three fold increases in respiratory disease and 14 percent decrease in lung function in school children in Malaysia.

Record shows that *Aeroallergens* are rising. It is estimated that two fold rise of carbon dioxide levels from about 300 to 600 ppm will induce a four-fold increase in the production of ragweed pollen. Pollen counts from birch trees (the main cause of allergies in Northern Europe) have increased with rising temperature associated with global warming.

**UV Radiation and Health**

We are all exposed to UV radiation from the sun. Many people are also exposed to artificial sources of UV radiation used in industry, commerce and recreation. Emissions from the sun include visible light, heat and UV radiation.

The UV region has the wavelength ranging from 100 nm to 400 nm. UV radiation is divided into three bands. These are:

- UVA (315-400 nm)
- UVB (280-315 nm)
- UVC (100-280 nm).

In the atmosphere, all UVC and approximately 90 percent of UVB radiation are absorbed by ozone, water vapour, oxygen and carbon dioxide. UVA radiation is less affected by the atmosphere. Therefore, the UV radiation reaching the earth's surface has UVA radiation with a small UVB component.

**Environmental Factors for UV Radiation**

Level of UV radiation depends on several factors. These are height of the sun, latitude, cloud cover, altitude and, role played by ozone and ground reflection.

**1. *Sun height***

Height of the sun in the sky determines the UV radiation level. Higher the sun in the sky, the higher will be the UV radiation level. Declination of the sun varies during day hours and also in different periods of the year. So UV radiation varies with time of day and time of year, with maximum levels found when the sun is at its maximum elevation, at around midday (solar noon) during the summer months.

**2. *Latitude***

Latitude is the other control of solar radiation. Areas close to the equator get maximum solar radiation throughout the year. Hence in the areas closer to the equator the higher will be the UV radiation levels.

**3. *Cloud Cover***

UV radiation levels remain to be the highest under cloudless skies. Even with cloud cover, UV radiation levels can be high due to the scattering of UV radiation by water molecules and fine particles in the atmosphere.

## 4. *Altitude*

Altitude is the other factor controlling the UV radiation level. At higher altitudes air is thinner than the lower altitudes. A thinner atmosphere filters less UV radiation. With a rise of every 1000 metres in altitude, UV levels increase by 10 percent to 12 percent.

## 5. *Ozone*

Ozone in the atmosphere can absorb UV radiation. Due to absorption by ozone less UV radiation can reach the earth's surface. Due to temporal variation of ozone in the atmosphere UV radiation levels vary over the year and during day hours.

## 6. *Ground Reflection*

Albedo is the ground reflection of solar rays. Different surfaces have different albedo. So UV radiation is reflected or scattered to varying extents by different surfaces, e.g. snow can reflect as much as 80 percent of UV radiation, dry beach sand about 15 percent, and sea foam about 25 percent.

## Health Effects of UV Radiation

UV radiation possesses both beneficial and adverse properties. UV radiation in sunlight produces vitamin D which is beneficial to human health. Limited exposure to UV radiation will help in maintaining good health. Even UV radiation is used to treat several diseases, including rickets, psoriasis, eczema and jaundice.

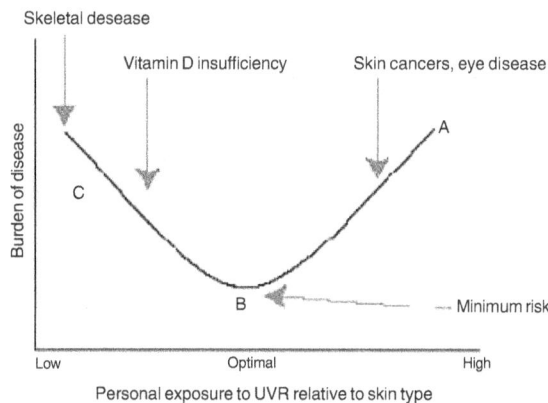

**Fig 11.2.:** Relationship of Exposure to UVR and Burden of Diseases

But prolonged exposure to UV radiation may cause acute and chronic health effects on the skin, eye and immune system. Sunburn (erythema) appears to be the common effect of excessive UV radiation exposure. Over the longer term, UV radiation induces degenerative changes in cells of the skin, fibrous tissue and blood vessels leading to premature skin aging, photodermatoses and actinic keratoses. Inflammatory reaction of the eye is the other long term effect of UV radiation. In the most serious cases, skin cancer and cataracts can occur. Non-melanoma skin cancers, e.g. basal cell carcinomas and squamous cell carcinomas, are not usually fatal and can be surgically removed. Every year nearly 2 to 3 million people are affected by non-melanoma skin cancer. On the otherhand malignant melanoma appears to be the deadly skin cancer. Approximately 130,000 malignant melanomas occur globally each year, substantially causing high mortality rates in fair-skinned populations. An estimated 66,000 deaths occur annually from melanoma and other skin cancers.

The other effect of UV radiation is on human eye causing cataract and other diseases. WHO estimated that nearly 12 to 15 million people become blind from cataracts every year in the world, of which up to 20 percent may be caused or enhanced by sun exposure. It is also now known that environmental levels of UV radiation may suppress cell-mediated immunity. As a result the risk of infectious diseases rises and the efficacy of vaccinations is reduced. The risk of UV radiation-related health effects on the eye and immune system is independent of skin type.

**Ozone Depletion and UV Radiation-related Health Effects**

Ozone plays important role in the stratosphere in the absorption of UV radiation, particularly UV B radiation. Due to such absorption lower level of UV radiation can reach the earth's surface and cause lesser health hazards. Stratospheric ozone layer declined for at least two decades, with losses of about 10 percent in winter and 5 percent in summer and autumn in various locations of the earth's surface, particularly over the Antarctic, and also in such diverse locations as Europe, North America and Australia. Researchers also observe depletion of ozone over the North Pole. The problem is becoming more and more critical each year. From the recent satellite imageries the depletion of ozone over the poles has been confirmed.

The major culprit for the depletion of ozone layer happens to be chloroflurocarbon (CFC11 and CFC12). F. Sherwood Rowland and Mario Molina, who shared the Nobel Prize in chemistry in 1995 for their discovery of the role of CFCs in the destruction of ozone, suggested that most of the chlorine atoms would combine with ozone and initiate a chain reaction (see the formation of ozone in Chapter 1). As a result of chain reaction, a single chlorine atom can remove as many as 100,000 molecules of ozone. Rowland and Molina predicted that if industry continued to release a million tons of CFCs into the atmosphere each year, atmospheric ozone would eventually drop by 7 to 13 percent. With the depletion of ozone in the atmosphere, more ultraviolet radiation would reach the earth's surface. It is estimated that increased exposure would lead to higher incidence of skin cancer, cataracts, and damage to the immune system and to slower plant growth. As some CFCs would persist in the atmosphere for more than 100 years, these would last through the twenty-first century. Rowland and Molina called for the ban on further release of CFCs in the air. Alerted by the ozone depletion in the atmosphere, the United States of America, Canada, Norway and Sweden banned the use of CFCs in aerosol sprays in late nineties. Many other countries also followed their pathways.

The ozone layer in the atmosphere is now affected by CFC through the break down of ozone. So the ozone layer is being depleted enhancing the penetration of UV rays to the earth's surface. As the ozone layer becomes thinner, the protective filter provided by the atmosphere is progressively reduced. Consequently, human beings and the environment are exposed to higher UV radiation levels, and especially higher UVB levels that have the greatest impact on human health, animals, marine organisms and plant life.

It is estimated that nearly 1.5 million DALYs (Disability Adjusted Life Years) are lost every year due to excessive UV radiation exposure. Human exposure to solar ultraviolet radiation poses an important public health concern. Skin cancer and malignant melanoma are among the most adverse health effects, but a series of other health effects are also recognised. On the other hand, a moderate degree of UV exposure is beneficial for the production of Vitamin D which is essential for bone health. Low Vitamin D levels are likely to be associated with other chronic diseases. So the public health policy on ultraviolet radiation needs to aim at preventing the disease burden associated both with excessive and with insufficient UV ray exposure.

## Types of Health Impacts of Stratospheric Ozone Depletion

Stratospheric ozone depletion can induce many possible impacts on health. These are noted below:

### *Effects on Skin*

- Malignant melanoma
- Non-melanocytic skin cancer—basal cell carcinoma, squamous cell carcinoma
- Sunburn
- Chronic sun damage
- Photodermatoses.

### *Effects on the Eye*

- Acute photokeratitis and photoconjunctivitis
- Climatic droplet keratopathy
- Pterygium
- Cancer of the cornea and conjunctiva

- Lens opacity (cataract)—cortical, posterior subcapsular
- Uveal melanoma
- Acute solar retinopathy
- Macular degeneration.

### Effects on Immunity and Infection

- Suppression of cell mediated immunity
- Increased susceptibility to infection
- Impairment of prophylactic immunisation
- Activation of latent virus infection.

### Other Effects

- Cutaneous Vitamin D production
  - Prevention of rickets, osteomalacia and osteoporosis
  - Possible benefit for hypertension, ischaemic heart disease and tuberculosis
  - Possible decreased risk for schizophrenia, breast cancer, prostate cancer
  - Possible prevention of Type 1 (usually insulin dependent) diabetes
- Non-Hodgkin's lymphoma
- Altered general well-being
  - Sleep/wake cycles
  - Seasonal effective disorder
  - Mood.

### Indirect Effects

- Effects on climate, food supply, infectious disease vectors, air pollution, etc.

### Skin Cancer

Solar radiation appears to be a cause of skin cancer (melanoma and other types), particularly in fair-skinned humans. The most recent assessment by the United Nations Environment Programme (1998) projected appreciable rise in skin cancer incidence due to stratospheric ozone depletion. The incidence of skin cancer, i.e. cutaneous malignant melanoma, has been rising steadily in white populations over the past few decades. This is particularly evident in areas of high UVR exposure such as South Africa, Australia and New Zealand. Human skin pigmentation had been the evolutionary process for many millennia, probably to meet the competing demands of protection from the deleterious effects of UVR and maximisation of the beneficial effects of UVR. Skin pigmentation shows a clear latitudinal distribution in indigenous populations. A strong relationship exists between the incidence (and mortality) of all types of skin cancer and latitude, at least amongst the homogeneous populations. Latitude expresses the amount of UVR reaching the earth's surface. This is due partly to the differing thickness of the ozone layer at different latitudes, and partly to the angle at which solar radiation passes through the atmosphere. The spectacular rise in skin cancers in these populations over recent decades indicates the complexities of post-migration geographical vulnerability and modern behavioural patterns.

Future impacts of ozone depletion on skin cancer incidence in European and North American populations have been studied in recent years. It is estimated for the expected rise of skin cancer incidence in the US white population, based on the situation of ozone depletion. The first situation is that there is no restriction on CFC emissions. The second situation outlines a 50 percent reduction in the production of the five most important ozone-destroying chemicals by the end of 1999 in terms of Montreal Protocol of 1987. In the third situation the production of 21 ozone-depleting chemicals is reduced to zero by the end of 1995 under the Copenhagen Amendments to that Protocol. Under the first situation, it is presumed that there would be a relative increase in total skin cancer incidence by 5-10 percent in "European" populations living between 40° N and 52° N by 2050 (based on a 1996 baseline of 2,000 cases of skin cancer per million per year in the United States and 1,000 cases per million per year in northwest Europe).

Different actions have been taken to control the emission of CFC and the Ozone Hole is now supposed to be stabilised. However it is anticipated that by 2050 there will be more penetration of ultraviolet radiation intensifying the sunburn and incidence of skin cancer.

### *Effects on Eye*

High intensity UVR also damages the eye's outer tissues causing "snow blindness". Acute exposure of the eye to high levels of UVR, particularly in settings of high light reflectance such as snow-covered surroundings, can cause painful inflammation of the cornea or conjunctiva. This is known as snow-blindness. Common eye diseases caused by UV rays happen to be photokeratitis and photoconjunctivitis. These are the ocular equivalent of acute sunburn. Pterygium is a common condition that usually causes nasal conjunctiva, sometimes with extension to the cornea. It is particularly common in populations in areas of high UVR or high exposure to particulate matter.

Other eye disorders associated with UVR are usually rare, but may occur. Such disorders cause significant morbidity to the affected persons. Acute solar retinopathy, or eclipse retinopathy, happens to be one such morbidity which needs immediate medical attention. It is actually the sun burn of the retina as one looks at the sun during eclipse through naked eye. Different types of cataract formation are associated with UV rays.

**Table 11.5:** Effects of Stratospheric Ozone Depletion

| Latitude (Month) | Decrease in Ozone (%) | Ozone (Dobson Unit) | Biologically Effective Irradiance (W/m²) | Increase in BE | Increase in RAF | Calculated Time (min) for 50% Immuno-suppres-sion |
|---|---|---|---|---|---|---|
| 40° N January | 0 | 335.6 | 0.073 | 0.0 | ..... | 350 |
| | 5 | 318.8 | 0.075 | 3.0 | 0.60 | 340 |
| | 10 | 302.0 | 0.078 | 6.3 | 0.63 | 327 |
| | 20 | 268.5 | 0.083 | 13.5 | 0.68 | 307 |
| 40° N July | 0 | 307.9 | 0.278 | 0.0 | ...... | 92 |
| | 5 | 292.5 | 0.285 | 2.5 | 0.50 | 90 |
| | 10 | 277.1 | 0.292 | 5.3 | 0.53 | 87 |
| | 20 | 246.3 | 0.310 | 11.5 | 0.58 | 82 |

*Abbreviation:* BE = Biologically Effective Irradiance & RAF = Radiation Amplification Factor.

### *Effects on Human Immunity*

UVR exposure may also induce both local and whole-body immuno-suppression. Cellular immunity is also damaged by the UV rays. UVR-induced immuno-suppression therefore may affect the patterns of infectious disease and influence the occurrence and progression of various autoimmune diseases.

It now evidenced from the laboratory experiments that ultraviolet radiation suppresses components of both local and systemic immune functioning. An increase in ultraviolet radiation exposure therefore may increase the occurrence and severity of infectious diseases and, in contrast, reduce the incidence and severity of various autoimmune disorders. There is now growing interest among the scientists to study the influence of ultraviolet radiation upon immune system function, Vitamin D metabolism and the consequences for human disease risks.

### *Effect on Infectious Disease Patterns*

It is conjectured that higher UV rays exposure may suppress the immune responses to infection of the human host. The total UVR dose required for immune suppression is likely to be less than that required for skin cancer induction. In animals, high UVR exposure has been found to decrease host resistance to viruses (such as influenza and cytomegalovirus), parasites (such as malaria) and other infections (such as *Listeria monocytogenes* and *Trichinella spiralis*).

Increased exposure to UV rays also affects the efficacy of vaccines. The scientists are concerned that increased exposure to UVR due to stratospheric ozone depletion can lower the effectiveness of vaccines, particularly BCG, measles and hepatitis. BCG vaccine efficacy demonstrates a latitudinal gradient with reduced efficacy at lower latitudes. Seasonal differences in vaccine efficacy have also been observed for hepatitis B. The potential health effect of increased UV exposure would be the reduced vaccine efficacy particularly for vaccines that require host immune responses to intra-dermally administered antigens.

## Non-Hodgkin's Lymphoma

There has been significant rise in the occurrence of Non-Hodgkin's Lymphoma (NHL) throughout the world in recent years. Increased exposure to UV rays may be the probable reason for the rise of incidence. NHL symptoms are similar to non-melanoma skin cancer. Chronic immunosuppression appears to be a risk factor for NHL and UVR has immunosuppressive effects on humans. In future NHL will be a disease of concern with increased UV rays exposure.

### *Indirect Effect*

Moreover UV rays can damage the molecular chemistry of photosynthesis both on land (terrestrial plants) and at sea (phytoplankton). This would have worldwide effect on food production causing nutritional and health problems.

## Beneficial Effect for Autoimmune Diseases

Recent advancements in the study of photoimmunology and epidemiology suggest that UVR may play a beneficial role in autoimmune diseases such as multiple sclerosis (MS), type 1 diabetes mellitus (IDDM) and rheumatoid arthritis (RA).

Each of these autoimmune diseases indicates a breakdown in immunological self-tolerance caused by an inducing agent such as an infectious micro-organism or a foreign antigen. A cross-reactive auto-immune response occurs and a "self-molecule" is no longer self-tolerated by the immune system. Latitudinal gradient is also reflected in multiple scelerosis. Differential UV-induced immune suppression of autoimmune activity has been noticed. At lower latitudes where MS prevalence is lower high levels of UVR exposure may slow down the immune over-activity that occurs in MS. For type 1 diabetes the epidemiological studies indicate a possible beneficial role for UVR. Disease becomes dominant with increasing latitude. For rheumatoid arthritis, dietary supplementation with Vitamin D has been related to lower levels of disease activity. Overall, the epidemiological features of these three autoimmune diseases are consistent with a protective effect for high personal UVR exposure.

## Impact of Natural Disaster on Health

Natural disasters related to climate are floods, droughts, thunderstorms and cyclones. Different climatic anomalies cause many disasters. There has been an increasing trend in natural disaster impacts worldwide. El Nino is one of them. Floods, droughts, storms and cyclone also cause impacts on human health.

El Niño displays appreciable effect on the total number of persons affected by natural disasters. Worldwide, disasters triggered by droughts are twice as frequent during the year after the onset of El Niño than other years. On an average in an El Niño year, around 35 per 1000 persons are affected by a natural disaster. This happens to be four times greater than the rate in non-El Niño years. The risk is more manifested for famine disasters. Global disaster caused by El Nino happens to be the consequences of drought.

In 1997/98 Kenya was nearly devastated by flooding and excess rainfall. Ecuador and northern Peru experienced severe flooding and mudslides along the coastal regions which severely damaged the local infrastructure. On the other side, Guyana, Indonesia and Papua New Guinea had experienced drought conditions. Although not all natural disasters in 1997/98 should be attributed to the El Niño event, the impacts of disasters range from 21,000 to 24,000 deaths.

Extreme weather events directly cause death and injury and exert substantial indirect health impacts. These indirect impacts occur as a result of damage to the local infrastructure, population displacement and ecological change. The health impacts of natural disasters may be noted as follows:

1.  Physical injury
2.  Decreases in nutritional status, especially in children
3.  Increases in respiratory and diarrhoeal diseases due to crowding of survivors, often with limited shelter and access to potable water
4.  Impacts on mental health which may be long lasting in some cases
5.  Increased risk of water-related and infectious diseases due to disruption of water supply or sewage systems, population displacement and overcrowding
6.  Release and dissemination of dangerous chemicals from storage sites and waste disposal sites into flood waters.\

### *Floods*

Flood poses danger to human populations. Drowning and injuries are common effects of flood. Local infrastructure can be affected severely during a natural disaster. Flood usually causes damage to buildings and equipment, including materials and supplies, roads and transport. Drainage and sewerage systems break down pollute the water supply systems. During and following both catastrophic and non-catastrophic flooding considerable risk to health occurs if the floodwaters become contaminated with human or animal waste. A study in populations displaced by catastrophic floods in Bangladesh in 1988 found that diarrhoea was the most dominant morbidity followed by respiratory infection. Watery diarrhoea caused most of the death for all age groups under 45. In both rural Bangladesh and Khartoum, Sudan, the proportion of severely malnourished children had risen immediately after flooding. However, in the developed countries, both physical and disease risks from flooding happen to be greatly reduced by a well maintained flood control and sanitation infrastructure and public health measures, such as monitoring and surveillance activities to detect and control outbreaks of infectious disease. However, the recent experience of flooding in Central Europe, in which over 100 people died, demonstrated that floods even pose a major threat on health and welfare in industrialised countries. Floods also cause psychological morbidity. An increase in psychological symptoms and post-traumatic stress disorder, including 50 flood-linked suicides, occurred in the two months following the major floods in Poland in 1997. Variety of respiratory symptoms have been reported due to dampness in the home caused by occasional flooding.

**Table 11.6:** Impact of Extreme Climatic Events—1990

| Regions | Events | Killed (thousands) | Affected (millions) |
|---|---|---|---|
| Africa | 247 | 10 | 104.3 |
| Eastern Europe | 150 | 5 | 12.4 |
| Eastern Mediterranean | 139 | 14 | 36.1 |
| Latin America & Caribbean | 298 | 59 | 30.7 |
| Southeast Asia | 286 | 458 | 427.4 |
| Western Pacific | 381 | 48 | 1199.8 |
| Developed Countries | 577 | 6 | 40.8 |
| Total | 2078 | 601 | 1851 |

### *Droughts*

A drought is a period of abnormally dry weather which persists long enough to cause hydrologic imbalance. It appears to be a period of deficient moisture in the soil such that there is inadequate water required for plants, animals and human beings. Four general types of drought may be identified, all which have impact on humans. These are:
1. *Meteorological Drought:* Precipitation is unusually low for a particular region;
2. *Agricultural Drought:* Amount of moisture in the soil is no longer sufficient for crops under cultivation;
3. *Hydrological Drought:* Surface water and groundwater supplies remain below normal;

4. *Socioeconomic Drought:* Lack of water lowers the economic capacity of people to survive, affecting the non-agricultural production.

The health impacts on populations are manifested by declining food production. Famine often occurs when a pre-existing situation of malnutrition aggravates. The health consequences of drought appear to be the emergence of diseases resulting from malnutrition. Outbreaks of malaria emerge due to changes in vector breeding sites and malnutrition enhances susceptibility to infection.

**Table 11.7:** Disaster Types and Risk of Communicable Diseases
(As related to climatic events)

| Type | Person to Person (Perception) | Water-borne | Food-borne | Vector-borne |
|---|---|---|---|---|
| Hurricane/Cyclone | M | H | M | H |
| Tornado | L | L | L | L |
| Heatwave | L | L | L | L |
| Coldwave | L | L | L | L |
| Flood | M | H | M | H |
| Famine | H | H | M | M |

H = High
M = Medium
L = Low

### *Storms and Tropical Cyclones*

Storms and tropical cyclones appear to be devastating in low-lying and environmentally degraded areas with poor infrastructure having high density of population. Storm surge happens to be the principal cause of instant deaths by drowning. Storms and cyclones are associated with torrential widespread rainfall which causes extensive floods. Floods add misery with the consequence of crop damage, livestock deaths and diseases associated with contamination of water.

Bangladesh has experienced some of the most serious impacts of tropical cyclones in the last century as a result of combination of meteorological and topographical conditions and the inherent vulnerability of a low-income, poorly resourced population. Improved early warning systems can reduce the impacts of cyclones and storms.

### *Forest Fires*

Burns and smoke inhalation are the direct effects of fires on human health. Loss of vegetation on slopes accelerates soil erosion and enhances the risk of landslides causing more misery to the people. The conditions become aggravated when an urban sprawl extends into surrounding hilly and forest areas. Smokes and dusts of forest fire may also lead to air pollution. Air pollution causes increased mortality and morbidity in susceptible persons, and increased risk of hospital and emergency admissions. In recent years it has been evidenced that forest fires in many parts of the world are related to global warming.

## Seasonality of Infectious Disease

Mostly all infectious diseases show a seasonal pattern. Winter mortality is high in temperate and high latitudes indicating the critical morbidity in winter months.

Cyclic influenza occurs usually in the late fall, winter and early spring in North America. This disease pattern may be related to increased likelihood of transmission due to indirect social or behavioural adaptations to the cold weather such as staying indoors. Another reason may be related to pathogen sensitivities to climatic factors such as humidity. Many other infectious diseases show cyclic seasonal patterns, which may be controlled by climate to some extent.

In most regions around the world, enteric diseases indicate significant seasonal fluctuations. In Scotland, campylobacter infections are characterised by short peaks in the spring. In Bangladesh and India, cholera outbreaks are common during the monsoon season when drinking water becomes polluted. In Peru, cyclospora infections become dominant in the summer and subside in the winter. Some vector-borne diseases (e.g. malaria and dengue fever) also show their dominance in the months of heavy rainfall and humidity. Epidemics of infectious diseases emerge during the hot and dry season and subside soon after the beginning of the rainy season in sub-Saharan Africa. Seasonal fluctuations in the occurrence of infectious disease suggest its relationship with climatic factors.

## Disease Burden due to Climate Change

Climate change is likely to affect the pattern of deaths from exposure to high or low temperatures. However, the effect on actual disease burden is difficult to quantify. We have little information on how deaths, during thermal extremes, were caused to sick/frail persons who would have died soon anyway.

In 2030, the estimated risk of diarrhoea will rise up to 10 percent higher in some regions even when there would be no climate change. It expresses the particular exposure-response relationship. Estimated effects on malnutrition differ markedly among regions. By 2030, the relative risks for unmitigated emissions, relative to no climate change, vary from a significant increase in the Southeast Asia region to a small decrease in the Western Pacific. Overall, although the estimates of changes in risk appear to be somewhat flexible due to regional variation in rainfall, they refer to a major existing disease burden covering large numbers of people.

The estimated proportional changes in the numbers of people killed or injured in coastal floods would be large, although they refer to low absolute burdens. Impacts of inland floods are supposed to rise by a similar proportion, and would generally cause a greater acute hike in disease burden. While these proportional increases are similar in developed and developing regions, the baseline rates could be much higher in developing countries.

Changes in various vector-borne infectious diseases can be predicted. Malaria in regions bordering current endemic zones would be subject to changes. Smaller changes would occur in currently endemic areas. Most temperate regions would remain unsuitable for transmission, because either they remain climatically unsuitable (e.g., most of Europe) or socioeconomic conditions are likely to remain unsuitable for reinvasion (e.g., southern United States).

Presently the total current estimated burden appears to be small relative to other major risk factors. However, in contrast to many other risk factors, climate change and its associated risks would be increasing rather than decreasing over time.

A changing climate is likely to affect all of such conditions like drinking water, sufficient food, secured shelter, and good social conditions, on which depend the quality of life and public health. IPCC suggest that a warming climate is likely to cause a few localised benefits, such as decreased winter deaths in temperate climates, and increases in food production in some, particularly high latitude regions. Public health services and high living standards are improving to protect people from specific impacts in some countries, particularly the developed countries. It is not probable that climate change would cause malaria to re-emerge in northern Europe or North America. But the situations appear different in other countries and regions of the world. It has now been ascertained that the health effects of a rapidly changing climate are likely to cause negative impacts, particularly in the poorest communities and countries, which have contributed least to greenhouse gas emissions for probable climate change. Some of the health effects could be:

- There would be higher frequencies of heatwaves. Recent analyses indicate that human-induced climate change significantly increased the likelihood of the European summer heatwave of 2003.
- More variable precipitation patterns are likely to affect the supply of freshwater, multiplying the risks of water-borne disease.
- Rising temperatures and variable precipitation are likely to decrease the production of staple foods in many of the poorest regions, turning them food deficit areas and increasing risks of malnutrition.
- Rising sea levels would cause the risk of coastal flooding, and affect the shelter of many thousands people displacing them. More than fifty percent of the world's population now lives within 60 km of the sea. The Nile delta in Egypt, the Ganges-Brahmaputra delta in Bangladesh, and many small islands, such as the Maldives, the Marshall Islands and Tuvalu appear to be the most vulnerable regions, which are likely to be submerged.

- Changes in climate would probably extend the transmission seasons of important vector-borne diseases, and alter their geographic range, potentially bringing them to regions that lack population immunity and/or a strong public health infrastructure.

Nevertheless, WHO concluded that the effects of the climate change that has occurred since the mid-1970s may have resulted in deaths of over 150,000 persons in 2000. It also affirmed that these impacts are likely to be severe in the future.

## Impact of Climate Change on Public Health in the Cities of Developing Countries

Climate change appears as an emerging threat to global public health, particularly in the crowded cities of the developing countries with poor infrastructure for protection of public health. It is also highly inequitable, as the greatest risks are to the poorest people in the cities, who have contributed least to Greenhouse Gas (GHG) emissions. The rapid economic development and the concurrent urbanisation of poorer countries have turned the developing-country cities vulnerable to health hazards from climate change. Common vulnerability factors appear to be their coastal location, exposure to the urban heat-island effect, high levels of outdoor and indoor air pollution, high population density, and poor sanitation. There are opportunities but lack of actions improving health and cutting GHG emissions most obviously through policies related to transport systems, urban planning, building regulations and household energy supply. As a result these cities have become the largest current global health burdens, including approximately 800,000 annual deaths from ambient urban air pollution, 1.2 million from road-traffic accidents, 1.9 million from physical inactivity, and 1.5 million per year from indoor air pollution.

The global mean temperature is likely to rise by 1.4-5.8° C between 1990 and 2100 with associated changes in the hydrological cycle. These will generate a range of health impacts. World Health Organization (WHO) concluded that the effects of the climate change that have occurred since the mid-1970s may have resulted in a net increase of over 150,000 deaths in 2000. It also concluded that these impacts are likely to be higher in the coming decades. The largest health risks would be for children in the poorest communities. Nearly 50 percent of the world population now lives in urban areas compared with only 15 percent in 1900. City living is expected to reduce vulnerability in several important respects, including lower rates of diseases such as malaria, and often higher individual incomes and better access to health services. Unfortunately the urban poor share most of the disease burdens and remain to be most vulnerable. It is estimated that over 900 million people, one-third of the global urban population, and more than 70 percent of urban dwellers in developing countries, now live in slum-like conditions. These are associated with low incomes, poor housing and provision of basic services, and no effective regulation of pollution or ecosystem degradation. Number of these people is likely to reach about 2 billion by 2020. Health care programmes for this group will therefore become increasingly important and global concern associated with the climate change.

## Further Readings

Bouma, M.J. and Dye, C. (1997) "Cycles of malaria associated with El Niño in Venezuela", *Journal of the American Medical Association* **278**.

Bouma, M.J. *et al.* (1997) "Predicting high-risk years for malaria in Colombia using parameters of El Niño Southern Oscillation", *Tropical Medicine and International Health* **2**.

Brown, V. *et al.* (1998) "Epidemic of malaria in north-eastern Kenya", *Lancet* **352**.

Checkley, W. *et al.* (2000) "Effects of the El Niño and ambient temperature on hospital admissions for diarrhoeal diseases in Peruvian children", *Lancet* **355**.

Committee on Climate, Ecosystems, Infectious Diseases, and Human Health. (2001) *Under the Weather: Climate, Ecosystems, and Infectious Disease.* National Research Council Board on Atmospheric Sciences and Climate, Division of Earth and Life Sciences. Washington, DC, USA: National Academy Press.

Easterling, D.R. *et al.* (2000) "Climate extremes: observations, modelling and impacts", *Science* **289**.

Epstein, P.R. (2000) "Is global warming harmful to health"? *Scientific American* **283(2)**.

Epstein, P.R. (1999) "Climate and health", *Science* **285**.

Fagan, B. (1999) *Floods, Famines and Emperors. El Niño and the Fate of Civilisations.* Basic Books, New York, USA.

Gagnon, A. *et al.* (2002) "The El Niño Southern Oscillation and malaria epidemics in South America", *International Journal of Biometeorology* **46**.

Gill, C.A. (1920) "The relationship of malaria and rainfall", *Indian Journal of Medical Research* **7(3)**.

Gubler, D.J. (1998) "Dengue and dengue haemorrhagic fever", *Clinical Microbiology Review* **11**.

Intergovernmental Panel on Climate Change (2001) *Climate Change 2001: Third Assessment Report (Volume I).* Cambridge, UK Cambridge University Press.

Kovats R. *El Niño and health.* (1999), World Health Organization, Geneva, Switzerland.

Langford, I.H. and Bentham, G. (1995) "The potential effects of climate change on winter mortality in England and Wales", *International Journal of Biometeorology* **38**.

Lindsay, S. and Martens, W.J.M. (1998) "Malaria in the African Highlands: past, present and future", *Bulletin of the World Health Organization* **78**.

Lindgren, E. and Gustafson, R. (2001) "Tick-borne encephalitis in Sweden and climate change", *Lancet* **358**.

MacDonald, G. (1957) *The Epidemiology and Control of Malaria.* Oxford University Press, Oxford, UK.

Malakooti, M.A. *et al.* (1998) "Re-emergence of epidemic malaria in the highlands of western Kenya", *Emerging Infectious Diseases* **4(4)**.

Martens, W.J.M. *et al.* (1995) "Potential impact of global climate change on malaria risk", *Environmental Health Perspectives* **103**.

Martens, W.J.M. *et al.* (1995) "Climate change and vector-borne diseases: a global modeling perspective", *Global Environmental Change* **5**.

McCarthy, J.J. *et al.* (2001) "Climate change 2001: impacts, adaptation, and vulnerability", *Contribution of Working Group II to the Third Assessment Report of the Intergovernmental Panel on Climate Change*, Cambridge University Press, Cambridge, UK.

McMichael, A.J. *Human Frontiers, Environments and Disease*, Cambridge University Press, 2001, Cambridge, UK.

McMichael, A.J. *et al.* eds. (1996) *Climate Change and Human Health,* World Health Organization (WHO/EHG/96.7), Geneva, Switzerland.

Mouchet, J. *et al.* (1998) "Evolution of malaria in Africa for the past 40 years: Impact of climatic and human factors", *Journal of the American Mosquito Control Association* **14**.

Murray, C.J.L. and Lopez, A.D. eds. (1996) "Global burden of disease", *Global Burden of Diseases and Injury Series: Vol. 1.* Boston, USA, Harvard School of Public Health, Harvard University.

Nicholls, N. (1993) ENSO, "Drought and flooding rain in south-east Asia", In: *South-east Asia's Environment Future: the Search for Sustainability,* Brookfield, H. and Byron, Y. Eds. United Nations University Press/Oxford University Press, Tokyo, Japan.

Parry, M. *et al.* (2001) "Millions at risk. Defining critical climate change targets and threats", *Global Environmental Change* **11**.

Patz, J. *et al.* (2000) "The potential health impacts of climate variability and change for the United States: executive summary of the report of the health sector of the US National Assessment", *Environmental Health Perspectives*, **108**.

Patz, J.A. and McGeehin, M.A. (2000) "The potential health impacts of climate variability and change for the United States: Executive summary of the report of the health sector of the U.S. national assessment", *Environmental Health Perspectives* **108(4)**.

Patz, J.A. *et al.* (1996) "Global climate change and emerging infectious diseases", *Journal of the American Medical Association* **275**.

Poveda, G. *et al.* (2000) "Climate and ENSO variability associated to malaria and dengue fever in Colombia", In: *El Niño and the Southern Oscillation, multiscale variability and global and regional impacts*, Eds. Diaz H.F. and Markgraf, F., Cambridge University Press, Cambridge, UK.

Reiter, P. (2001) "Climate change and mosquito-borne disease", *Environmental Health Perspectives,* **109**, supplement 1.

Reiter, P. (2000) "Malaria and global warming in perspective?" *Emerging Infectious Diseases,* **6(4)**.

Watson, R.T. *et al.* (1998) Eds. "The regional impacts of climate change, An assessment of vulnerability", *A special report of IPCC Working Group II*, Cambridge University Press, Cambridge, UK.

Woodward, A.J. *et al.* (2000) "Protecting human health in a changing world: the role of social and economic development", *Bulletin of the World Health Organization,* **78**.

Woodward, A. *et al.* (1998) "Climate change and human health in the Asia Pacific: who will be most vulnerable?" *Climate Research* **11**.

World Bank (1992) *World Development Report, Development and the Environment*, Oxford University Press, Oxford, UK.

World Health Organization (2002) *The World Health Report 2002,* World Health Organization, Geneva, Switzerland.

World Health Organization (1998) *Malaria* -WHO Fact Sheet No. 94, World Health Organization, Geneva, Switzerland.

World Health Organization (1988b) "El Niño and its health impacts", *Weekly Epidemiological Record* **20**.

# Chapter 12

## CLASSIFICATIONS OF CLIMATE

**Approach to Classification of Climates**

Variations in climate are observed in different places of the earth. We know the factors for variations in climatic processes on a global and regional scale. Even in this mosaic of variations a nearly uniform pattern may be observed in places. A region having uniform characteristics of weather and climate sustainable for a long period is termed a *climatic region.*

Classification is important in scientific studies. Certainly lots of information could be available. For convenience and better understanding of the subject we attempt to link the data and process it into groups and categories. The processed and classified data may be archived and further used as and when required. Climatic classification is also attempted for this purpose. Just as we can identify regions of broadly similar climate, we can also find distant regions of the world where the same set of climate conditions occurs due to similar geographical reasons. One of the common classifications is based on the work of Koppen, a German climatologist who was the first to classify climate on the basis of precise measurements. His classification involves both annual and monthly levels of temperature and precipitation.

The purpose of classification is to organise a set of data or information about something to effectively communicate it in an informative way. Classification helps synthesise information into smaller units that are more easily understood. When considering the earth's climate, there is such an enormous amount of information that one has to break it down into areas of commonality to understand it easily. Climatologists have therefore created several ways to organise the wealth of information about earth's climate to bring order and understanding to it.

*Climate Classification Systems*

Three fundamental types of classifications of climate have been attempted. One important type is the *empirical systems* of classification, based on observable features. The Koppen's classification belongs to empirical systems based on observations of temperature and precipitation. These are two climate characteristics that can be measured conveniently. It is fairly easy to collect air temperature readings with a thermometer and precipitation with some sort of collecting device that can measure the amount of precipitation. Climates are grouped based on annual averages and seasonal extremes.

*Genetic classification systems* are those based on the cause of the climate. A genetic system relies on information about the climatic elements like solar radiation, air masses, pressure systems, etc. The causes of climatic processes are identified and the reasons are specified to determine the type of climate. The atmospheric science is advancing everyday and new explanations are derived. We still have a long way to go before we have a complete understanding of the causes of climate. These are inherently the most difficult type of classifications that depend on the multitude of variables needed. *Applied classification systems* are those approached for a purpose, or as an outgrowth of a particular climate-associated problem. The Thornthwaite's classification system has been attempted based on potential evapotranspiration, which will specify the water requirements for all practical purposes. C.W. Thornthwaite and his associates attempted to formulate a water budget technique to assess the water demand under different environmental conditions. His classification system was intended to predict the supply and demand for water in different climate regions.

Climate classification systems are the ways of classifying the world's climates. A climate classification may correlate closely with a biome category, as climate exerts a major influence on biological life in a region. In this regard the Köppen's classification of climates appears to be the most popular classification scheme.

### History of Classification of Climates

The Greek origin of the word 'climate' relates to the Sun's declination, *clima* meaning the slope of the earth (i.e. latitude). The seasonal variation of this declination depends only on latitude. The Greeks divided the world climate into groups limited by the latitudinal belts— 1.Tropic of Cancer, 2.Tropic of Capricorn, 3. Arctic Belt and 4. Antarctic Belt. In between theses belts we find five climatic types. Egyptian and Helenistic writers knew the causative effect of latitude on climate.

In modern climatology the German scientists worked painstakingly to distinguish and delineate climatic regions based on instrumental observations for the last two centuries. In 1817 Alexander von Humboldt, doyen of modern geography, represented annual-mean temperatures on a world map. Wladimir Koppen (1846-1940) revised this map and plotted seasonal temperature range in 1884, devising his climate classification. Koppen, a biologist by profession, based his classification on plant characteristics (after Carolus Linnaeus in 1735) related to climatic elements and that of clouds (by Luke Howard in 1802). His classification was hierarchical, with major categories subdivided, and then subcategories divided again, and so on. In fact Koppen was a student of botany at St. Petersburg, later completing a Ph.D. at Heidelberg on the effect of temperature on plant growth. At the highest level his system is based on five sets of temperature limits. These were developed from his early division of thermal zones suited to various kinds of vegetation in 1884. He continued to revise his system several times, notably in 1918 and in 1936.

Interestingly Alfred Wegner, the postulator of continental drift theory, became associated with Koppen. In 1924 Koppen accompanied by his son-in-law Alfred Wegener (1880-1930) left Germany to Graz in Austria. Incidentally the universities in Germany banished Wegener for his theory of continental drift. In Graz, Koppen deduced geological climates in support of the continental drift theory. Also, he became associated with Rudolf Geiger (1894-1981) and collaborated with him for his 1936 system of climate classification. Geiger later revised the classification many times after the death of Koppen and took initiative popularising the Koppen's classification. Geiger also formed the discipline of microclimatology based on his wealth of observations understanding the climate near the ground and its variations due to topography and land use.

Tor Bergeron (1891-1971) developed a different method of climate classification in 1928, based on the causes of climate. In his *genetic system* he defined a location according to the frequencies of various kinds of airmass prevailing there. Each airmass was marked in terms of the latitude at which its temperature had been determined previously, and the kind of surface there, either marine or continental. Helmut Landsberg (1906-1985) applied this principle for the classification of climate in Central Europe and Pennsylvania in the late 1930's. However, this classification was not popular, as the number of classes was too small and the system had the dearth of considerable information.

In Australia, Griffith Taylor (1880-1963) challenged the validity of the large fraction of the Australian continent categorised as *desert* (BW) or *semi-desert* (BS) by Koppen. He considered the basis of such division and felt such division would mislead in assessing the potential for growth in Australia.

In USA, Warren Thornthwaite (1892-1963) devised a classification of climate differently. He proposed a hierarchical classification in 1931, essentially in terms of the annual pattern of soil-moisture conditions. Soil-moisture conditions were determined in a complicated manner based on the monthly input as rain, and implicitly on the output as evaporation, indicated by temperature. Many studies indicated that the Thornthwaite's classification was more sensible than Koppen's, except at low latitudes. Later, Thornthwaite related the soil moisture more explicitly in his landmark article, 'Approach toward a rational classification of climate', in 1948. He applied an empirical formula estimating evaporation, which was revised the same year by Howard Penman (1909-1984) in England.

## Classification of Climates

All these classification systems have been attempted in solving practical problems. Now we use complex statistical procedures to group the climates of places and to define areas of 'similar' climates.

## Koppen's Classification of Climates

The Koppen's classification is one of the most widely used system of climate classification. Koppen first proposed a classification around 1900 based on the concept that native vegetation is the best expression of climate. The climate zone boundaries were selected with vegetation distribution in mind. He revised his classification in 1918 and 1936 imputing the parameters like temperatures, rainfall and seasonality. It combined average annual and monthly temperatures and precipitation, and the seasonality of precipitation.

Koppen considered that the growth of vegetation in a region is largely controlled by the effectiveness of precipitation. It simply does not depend on the amount of precipitation, but also depends on the amount of evaporation. So precipitation and temperature are closely related in characterising the natural vegetation of a region. He considered that the same amount of precipitation in winter or in the cold climate will be more effective than it is in summer or warm climate.

Koppen basically used 5 capital letters to signify the macro-climate. These are namely:

A – Tropical forest climate, B – Dry climate, C – Warm Temperate Rainy climate, Mild Winter, D. Cold Forest climate, Severe Winter and E. Polar climate.

In addition Koppen used many small letters as subscripts to specify the seasonal characteristics.

- *Subdivisions based on temperature*
  - Subscript 'a' :   hot summer
  - Subscript 'b' :   warm summer
  - Subscript 'c' :   cool summer
  - Subscript 'd' :   cold winter (D climates only)
  - Tundra (ET) :   summer temperatures above freezing
  - Ice cap (EF) :   temperatures below freezing year-round
- *Subdivisions based on precipitation*
  - Subscript 'f' :   no dry season
  - Subscript 'w':   dry winter
  - Subscript 's' :   dry summer
  - Subscript 'm':   seasonal monsoon
  
  *For B Climates*
  - Steppe (BS) :   semi-arid (steppe has about twice the precipitation as desert)
  - Desert (BW):   arid
  - Subscript 'h' :   hot and dry (mean annual temperature at or above 64° F)
  - Subscript 'k' :   cool and dry (mean annual temperature below 64° F)

In the table the characteristics of Koppen's Classification have been summarised:

Table 12.1 shows that 4 types (A, C, D and E) are classified based on temperature. Koppen followed Alfonse de Candolle in determining the temperature limit of the climatic types. Candolle showed that the growth of tropical trees would not grow properly below

**Table 12.1:** Characteristics of Koppen's Classification

| Climate Type | Temperature Index | Symbol | Dry Period | Limit of Dryness or Cold | Remarks |
|---|---|---|---|---|---|
| Tropical Forest Climate | >18° C | A | f, s, w<br><br>(Rainfall>6 cm.) | – | f means no rain, s means rain in summer & w means rain in winter |
| Dry Climate | h (hot)>18° C<br>k (cold)<18° C | B | – | S (steppe)<br>W (desert) | |
| Warm Temperate Rainy Climate | Warmest month Temp.>10° C<br><br>Coldest month Temp.>–3° C to 18° C | C | f, s, w<br><br>(Rainfall atleast 1.2 cm in a month) | – | s<br><br>(Driest in summer) |
| Cool Forest Climate | Warmest month Temp.>10° C<br><br>Coldest month Temp. <–3° C | D | f, s, w<br><br>(Similar to C) | – | (Similar to C) |
| Polar Climate | Warmest month Temp.0-10° C (Tundra)<br><br>Warmest month Temp.<0° C (Permafrost) | E | – | T (Tundra) warmest month temp.0-10° C<br><br>F (Permafrost) warmest month temp.<0° C | |

18° C. In C type of climate the temperature of the warmest month was shown to be > 10° C and of the coldest month ranging –3° C to 18° C. On the other hand in D type of climate the temperature of the warmest month could be > 10° C, but the temperature of the coldest month would not exceed –3° C. In summer the temperature was shown to be > 10° C, because the temperature below 10° C could affect the tree growth. The temperature of the coldest month was noted as <–3° C, as the ice could cover the surface for few weeks at this temperature. In E type of climate temperature of the warmest month was shown ranging 0° C to 10° C, at least for a day it would be frost free in the Tundra climate. From the following table we can establish the relationship between the climate classification by Koppen and Candolle. In his approach to climate classification Koppen studied the works done by other experts in the field and judiciously applied or incorporated their findings.

**Table 12.2:** Climatic Zones of the World

| Climates | Symbol | Koppen's nomenclature | Equivalent to De-Candolle zone | Climatic limits zone |
|---|---|---|---|---|
| Tree 1 | A | Tropical Rainy | Megathermal | Coldest month temperature above 18° C (64° F) |
| Tree 1 | C | Warm Temperate | Mesothermal | Coldest month between –3° C (27° F) and 18° C (64° F) |
| Tree 1 | D | Boreal | Microthermal | Coldest month below –3° C (27° F) warmest above 10° C (50° F) |
| Snow 2 | E | Snow | Hekistothermal | Warmest month below 10° C (50° F) |
| Dry 3 | B | Dry | Xerophilous | Annual rainfall less than r (index of dryness) |

*Note:* 1 - Abundant rainfall for forest vegetation, 2 - Too cool for tree growth, 3 - Too dry for tree growth.

Koppen classified B type climate based on aridity or dryness. For this he calculated the aridity index relating temperature and rainfall. Evaporation data could not be available worldwide. Relationship between temperature and rainfall could indicate the effectiveness of rainfall subject to temperature conditions. His calculation of aridity index may be understood from the following table.

**Table 12.3:** Calculation of Aridity Index

| Seasonal distribution of rainfall (relating temperature) | Steppe Grassland (BS) / Forest (f) & Desert (BW) limit | Steppe Grassland (BS) limit |
|---|---|---|
| Maximum rainfall in winter | $r/t = 1$ | $r/t = 2$ |
| Rainfall uniformly distributed throughout the year | $r/(t+7) = 1$ | $r/(t+7) = 2$ |
| Maximum rainfall in summer | $r/(t+14) = 1$ | $r/(t+14) = 2$ |

Here, r = Total annual rainfall/snowfall (in cm.) and t = Annual average temperature (in ° C)

In winter the rate of evaporation is low. So, in winter dryness index is assumed lower (e.g. $r/t = 1$ for the boundary of desert and $r/t = 1{\sim}2$ for steppe).

High rainfall is required in summer to maintain the effective rainfall due to higher rate of evaporation. So, the limit of index is considered maximum 'rainfall in summer' higher (e.g. r/(t + 14) = 1 for the boundary of desert and r/(t + 14) = 1~2 for steppe).

### *Characteristics of Koppen's Classifications*

In later revisions of Koppen's classifications of climate no change was suggested in the macro-divisions, but many symbols were incorporated to signify the micro-variations. Incorporation of these symbols made the classifications more acceptable and useful. The characteristic features of Koppen's classification are noted below.

## A. Tropical Climate

*Symbols Used Stating the Climatic Characteristics*

A = Temperature exceeds 18° C.

f = No dry period. Rainfall in every month exceeds 60 cm.

s = Summer dry. In summer the rainfall is below 6 cm. in a month. In winter rainfall exceeds 6 cm. in a month.

m = Monsoon climate—brief dry period. However, the soils remain moist due to heavy rainfall in the wet months. Tropical rainforest grows. Am, the monsoon climate is intermediate between Af and Aw climate.

w = Maximum rainfall in the autumn.

w′ = Two maximum rainy periods separated by two dry periods.

S = Dry period during the summer solstice (Rare).

I = Temperature difference between the warmest and coldest months is less than 5° C.

g = Annual temperature distribution in the Gangetic type of climate. Maximum temperature is recorded in the preceding month of Monsoon.

### *Subdivisions of Tropical Climate*

Three principal subdivisions are noted in the Tropical climate. These are – 1. Tropical Rainforest Climate (Af), 2. Tropical Monsoon Climate (Am) and 3. Tropical Wet and Dry or Savana Climate (Aw).

*Tropical Rainforest Climate*

In this climate type all twelve months record average precipitation of at least 60 mm (2.36 inches). This climate type is usually located within 5°-10° latitude, North and South of the equator. In some eastern-coast areas, they may extend even upto 25° away from the equator. In this climate type ITCZ is usually located where the trade winds converge. This climate is dominated by the Doldrums, where Low Pressure System exists almost throughout the year. No distinct seasons could be identified. This climate happens to be uniformly and monotonously wet throughout the year. In many places the period of high sun and longer days is distinctly wettest. A few places with this climate are found at the outer edge of the tropics, almost exclusively in the southern hemisphere (e.g. Santos, Brazil). Due to heavy rainfall and warm conditions tropical vegetation grows abundantly characterising the rainforest vegetation.

*Regions*

1. Amazon River Basin 2. Congo River Basin 3. Eastern Coast of Central America 4. Eastern Coast of Brazil 5. Eastern Coast of Madagscar 6. Eastern and North Eastern India 7. Southern Bangladesh 8. Malaysia 9. Indonesia 10. Phillipines.

*Stations Showing Temperature and Precipitation for 12 Months*

### Af - Andagoya, Columbia 5° N, Elevation: 65 m

|  | Jan. | Feb. | Mar. | Apr. | May | June | July | Aug. | Sept. | Oct. | Nov. | Dec. | Year |
|---|---|---|---|---|---|---|---|---|---|---|---|---|---|
| Temp. ° C | 27 | 27 | 28 | 28 | 27 | 27 | 27 | 27 | 27 | 27 | 27 | 27 | 27 |
| Precip. mm | 554 | 519 | 557 | 620 | 655 | 655 | 572 | 574 | 561 | 563 | 563 | 512 | 6905 |

### Af - Iquitos, Peru 4° S, Elevation: 104 m

|  | Jan. | Feb. | Mar. | Apr. | May | June | July | Aug. | Sept. | Oct. | Nov. | Dec. | Year |
|---|---|---|---|---|---|---|---|---|---|---|---|---|---|
| Temp. ° C | 27 | 27 | 27 | 27 | 26 | 26 | 25 | 27 | 27 | 27 | 27 | 27 | 26 |
| Precip. mm | 256 | 276 | 349 | 306 | 271 | 199 | 165 | 157 | 191 | 214 | 244 | 217 | 2845 |

### Tropical Monsoon Climate

Tropical Monsoon climate is most commonly found in South Asia and West Africa. The monsoon circulation imparts the climatic characteristics. The wind directions change according to the seasons. This climate is characterised by the wet period with high rainfall coinciding the summer solstice. The driest month recording less than 60 mm rainfall coincides with the winter solstice.

In some littoral areas the easterly *trade-winds* bring enough precipitation during the winter months to prevent the climate from becoming a tropical wet-and-dry climate. Jakarta, Indonesia and Nassau, Bahamas are included among these locations.

*Regions*

1. Generally pole-ward of Af type climate areas 2. Northern, Eastern and North Eastern India 3. Interior Mayanmar 4. Indo-China Peninsula 5. Northern Australia 6. Borderlands of Congo River Basins 7. South Central America 8. Western Central America 9. Extreme Southern Florida 10. Caribbean Islands.

*Stations Showing Temperature and Precipitation for 12 Months*

### **Am - Kolkata, India** 22.5° N, Elevation: 6 m

|  | Jan. | Feb. | Mar. | Apr. | May | June | July | Aug. | Sept. | Oct. | Nov. | Dec. | Year |
|---|---|---|---|---|---|---|---|---|---|---|---|---|---|
| Temp. °C | 20 | 23 | 28 | 30 | 31 | 30 | 29 | 29 | 30 | 28 | 24 | 21 | 27 |
| Precip. mm. | 13 | 24 | 27 | 43 | 121 | 259 | 301 | 306 | 290 | 160 | 35 | 3 | 1582 |

### **Am - Mangalore, India** 13° N, Elevation: 22 m

|  | Jan. | Feb. | Mar. | Apr. | May | June | July | Aug. | Sept. | Oct. | Nov. | Dec. | Year |
|---|---|---|---|---|---|---|---|---|---|---|---|---|---|
| Temp. °C | 27 | 27 | 28 | 29 | 29 | 27 | 26 | 26 | 26 | 27 | 27 | 27 | 27 |
| Precip. mm. | 5 | 2 | 9 | 40 | 233 | 982 | 1059 | 577 | 277 | 206 | 71 | 18 | 3467 |

### *Tropical Wet and Dry or Savanna Climate*

This tropical climate type shows distinct dry season, with the driest month having precipitation less than 60 mm. This climate type is usually located at the outer margins of the tropical zone, but occasionally found in an inner-tropical location (e.g., San Marcos, Antioquia, Colombia). This climate type lies intermediate between the rainforest climate (Af) and semiarid (Bsh) climate. Sometimes *As* is used in place of *Aw* if the dry season occurs during the time of high sun and longer days. This is the case in parts of East Africa (Mombasa, Kenya) and Sri Lanka (Trincomalee), for instance. The dry season occurs during the period of lower sun and shorter days.

*Regions*

I. The marginal areas north and south of Af climate type in Africa 2. Extreme eastern part of Brazil 3. Similar locations north and south of Af climate in other parts of the world.

*Stations Showing Temperature and Precipitation for 12 Months*

### Aw - Darwin, Australia 12.5° S, Elevation: 27 m

|              | Jan. | Feb. | Mar. | Apr. | May | June | July | Aug. | Sept. | Oct. | Nov. | Dec. | Year |
|--------------|------|------|------|------|-----|------|------|------|-------|------|------|------|------|
| Temp. ° C    | 28   | 28   | 28   | 28   | 27  | 25   | 25   | 26   | 28    | 29   | 29   | 29   | 28   |
| Precip. mm.  | 341  | 338  | 274  | 121  | 9   | 1    | 2    | 5    | 17    | 66   | 156  | 233  | 1563 |

## B. Dry Climate

Dry Climate prevails usually in the arid and semi-arid regions of the lower latitudes. Cold deserts are, however, found even in the mid-latitude region.

*Symbols used Stating the Climatic Characteristics*

B = Dryness is measured according to the table.

h (heiss = in German hot) = Hot climate. Annual average temperature exceeds 18° C. k = Temperature of the hottest month drops below 18° C.

s = Dry summer. Rainfall in the winter wet months is 3 times more than dry summer.

w = Dry winter. Rainfall in the summer wet months is 10 times more than the dry winter.

n (nebel) = Continous fog, found in the coastal areas due to the impact of cold ocean currents.

### Subdivisions of Dry Climate

We can distinctly identify four subdivisions of Dry Climate. These are–1. Hot Desert (Bwh) 2. Cold Desert (Bwk) 3. Hot Steppe (Bsh) and 4. Cold Steppe (Bwk). These climates are identified in places where precipitation is less than potential evapotranspiration. The threshold is determined in terms of the calculation as stated in the Table 12.3.

If the annual precipitation is less than half the threshold for Group B, it is classified as *BW* (desert climate); if it is less than the threshold but more than half the threshold, it is classified as *BS* (steppe climate).

A third letter indicates the temperature. Usually, *h* signifies low latitude climate (average annual temperature above 18° C) and *k* signifies middle latitude climate (average annual temperature below 18° C). Commonly the practice is to use *h* to indicate that the coldest month has an average temperature > 0° C (32° F), and *k* denoting that at least one month averages below 0° C.

In some desert areas, along the west coasts of continents at tropical or near-tropical locations, cooler conditions prevail than encountered elsewhere at comparable latitudes. The temperatures remain low due to the impact of cold ocean currents and frequent fog and low clouds occur even under dry climate. This climate is sometimes marked as *BWn* and examples can be found at Lima, Peru and Walvis Bay, Namibia.

*Regions (Bwh)*

1. Saudi Arabia 2. Iran 3. Northwestern India and adjoining Pakistan 4. Central and Western Australia 5. Southern and Western Africa 6. Northern Africa 7. Northern Mexico and 8. Southwestern USA.

*Hot Deserts (Bwh)*

1. Sahara Desert 2. Sonorian Desert 3. Thar Desert 4. Austrlian Desert 5. Arabian and Middle Eastern Deserts and 6. Kalahari Desert.

*Stations Showing Temperature and Precipitation for 12 Months*

### BWh - Berbera, Somalia 10.5° N, Elevation: 8 m

|  | Jan. | Feb. | Mar. | Apr. | May | June | July | Aug. | Sept. | Oct. | Nov. | Dec. | Year |
|---|---|---|---|---|---|---|---|---|---|---|---|---|---|
| Temp. ° C | 25 | 26 | 27 | 29 | 32 | 37 | 37 | 37 | 34 | 29 | 26 | 26 | 30 |
| Precip. mm. | 8 | 2 | 5 | 12 | 8 | 1 | 1 | 2 | 1 | 2 | 5 | 5 | 52 |

### BWh - Alice Springs, Australia 23.5° S, Elevation: 579 m

|  | Jan. | Feb. | Mar. | Apr. | May | June | July | Aug. | Sept. | Oct. | Nov. | Dec. | Year |
|---|---|---|---|---|---|---|---|---|---|---|---|---|---|
| Temp. ° C | 28 | 28 | 25 | 20 | 15 | 12 | 12 | 14 | 18 | 23 | 26 | 27 | 21 |
| Precip. mm. | 44 | 34 | 28 | 10 | 15 | 13 | 7 | 8 | 7 | 18 | 29 | 29 | 252 |

*Regions (Bwk)*

1. Caspian Sea region extending towards Northern China and Mongolia 2. Southern portions of South America and 3. Ladakh in India.

*Cold Deserts (Bwk)*

1. Russian Turkestan Desert 2. Takla Makan Desert 3. Gobi Desert 4. Arizona and Nevada Deserts and 5. Mohave Desert.

*Station Showing Temperature and Precipitation for 12 Months*

### BWk - Lovelock, Nevada, USA 40° N, Elevation: 1211 m

|  | Jan. | Feb. | Mar. | Apr. | May | June | July | Aug. | Sept. | Oct. | Nov. | Dec. | Year |
|---|---|---|---|---|---|---|---|---|---|---|---|---|---|
| Temp. °C | 0 | 4 | 6 | 10 | 15 | 20 | 24 | 22 | 18 | 12 | 5 | 0 | 11 |
| Precip. mm. | 16 | 16 | 11 | 14 | 13 | 14 | 4 | 6 | 7 | 13 | 13 | 15 | 143 |

*Regions (Bsh)*

1. South Africa (beyond 20° S) 2. Borderlands of Australian Desert 3. Portions of Southern South America 4. Portions of India (bordering Thar and rainshadow areas of Western Ghat) 5. Bordering the deserts of Northwestern Africa and 6. Borderlands of Saudi Arabian Deserts.

*Station Showing Temperature and Precipitation for 12 Months*

### BSh - Monterrey, Mexico 26° N, Elevation: 512 m

|  | Jan. | Feb. | Mar. | Apr. | May | June | July | Aug. | Sept. | Oct. | Nov. | Dec. | Year |
|---|---|---|---|---|---|---|---|---|---|---|---|---|---|
| Temp. °C | 14 | 17 | 20 | 23 | 26 | 27 | 28 | 28 | 26 | 22 | 18 | 15 | 22 |
| Precip. mm. | 18 | 23 | 16 | 29 | 40 | 68 | 62 | 151 | 78 | 26 | 29 | 20 | 606 |

*Regions (Bsk)*

1. Western Great Plains of USA 2. South Central Canada and 3. Bordering Cold Desert areas extending from Caspian Sea eastward to China and Mongolia.

*Station Showing Temperature and Precipitation for 12 Months*

### BSk - Denver, Colorado, USA 40° N, Elevation: 1611 m

| | Jan. | Feb. | Mar. | Apr. | May | June | July | Aug. | Sept. | Oct. | Nov. | Dec. | Year |
|---|---|---|---|---|---|---|---|---|---|---|---|---|---|
| Temp. ° C | −1 | 1 | 4 | 9 | 14 | 19 | 23 | 22 | 17 | 11 | 4 | −1 | 10 |
| Precip. mm. | 14 | 16 | 34 | 45 | 63 | 43 | 47 | 38 | 28 | 26 | 23 | 15 | 391 |

## C. Subtropical Climate

Subtropical climate dominates in the subtropics and even extends to the mid-latitude region beyond the Tropics of Cancer and Capricorn.

*Symbols Used Stating the Climatic Characteristics*

C = Temperature exceeds 10° C in the warmest month. In the coldest month temperature ranges between −3° to 18° C.

f = No dry period. Rainfall in every month exceeds 1.2 cm.

s = Dry summer. In the driest month of summer rainfall is less than 1.2 cm. In the wettest month in winter the rainfall shall be atleast 3 times higher.

W = Dry winter. In the driest winter month the rainfall is below 1.2 cm. But in the wettest month in summer the rainfall shall be 10 times higher.

a = Hot summer. In the hottest month average temperature exceeds 22° C.

b = Mild summer. In the hottest month average temperature is less than 22° C.

c = Mild summer for a brief period – less than 4 month's duration. Temperature exceeds 10° C.

i = Similar to A.

g = Similar to A.

x = Maximum rainfall in late spring or early summer. Late summer is dry.

n = Similar to B.

### Subdivisions of Subtropical Climate

The subtropical climate is characterised by an average temperature above 10° C in their warmest months, and a coldest month average between −3° C and 18° C.

Subtropical climates are subdivided into the following types:
1. Mediterranean Climate (Csa/Csb) 2. Humid Subtropical Climate (Cfa/Cwa) 3. Maritime Temperate or Oceanic Climate (Cfb/Cwb) and 4. Maritime Subarctic or Subpolar Oceanic Climate (Cfc).

## *1. Mediterranean Climates*

*Mediterranean climates* (*Csa*, *Csb*) usually occur on the western sides of the continents between the latitudes of 30° and 45°. These climatic areas lie in the polar front region in winter, and thus record moderate temperatures and variable rainy weather. Summers are hot and dry, due to the domination of the subtropical high pressure systems, except in the immediate coastal areas, where summers are milder due to the influence of nearby cold ocean currents that may create fog condition but restrict rainfall.

## *Regions*

1. Mediterranean coastal areas in Southern Europe and Northern Africa 2. Cape Town area of South Africa 3. Californea Coast 4. Chile Coast 5. Iran Highlands and 6. Southern and Southwestern Australia.

## *Stations Showing Temperature and Precipitation for 12 Months*

### Cs - Santiago, Chile 33.5°S, Elevation: 512 m

|  | Jan. | Feb. | Mar. | Apr. | May | June | July | Aug. | Sept. | Oct. | Nov. | Dec. | Year |
|---|---|---|---|---|---|---|---|---|---|---|---|---|---|
| Temp. °C | 19 | 19 | 17 | 13 | 11 | 8 | 8 | 9 | 11 | 13 | 16 | 19 | 14 |
| Precip. mm. | 3 | 3 | 5 | 13 | 64 | 84 | 76 | 56 | 30 | 13 | 8 | 5 | 360 |

### Cs - Rome, Italy 42°N, Elevation: 131 m

|  | Jan. | Feb. | Mar. | Apr. | May | June | July | Aug. | Sept. | Oct. | Nov. | Dec. | Year |
|---|---|---|---|---|---|---|---|---|---|---|---|---|---|
| Temp. °C | 8 | 8 | 10 | 13 | 17 | 22 | 24 | 24 | 21 | 16 | 12 | 9 | 15 |
| Precip. mm. | 76 | 88 | 77 | 72 | 63 | 48 | 14 | 22 | 70 | 128 | 116 | 106 | 881 |

## 2. Humid Subtropical Climates

*Humid Subtropical climate (Cfa, Cwa)* is usually found in the interiors of continents, or along the east coast of the continents between the latitudes of 25° and 40° (46° N in Europe). Contrary to the summer conditions of the Mediterranean climate, here the summers are humid. Humid condition is caused by the unstable tropical air masses, or onshore Trade Winds. In Eastern Asia winters remain dry due to the location of the Siberian high pressure system over the area. For this reason the region also record very severe cold than other places at similar latitude. On the other hand due to the influence of East Asian Monsoon summers are very wet.

*Regions*

1. Southeastern USA 2. Southeastern South America 3. Coastal Southeastern South Africa 4. Eastern Asia and 5. Eastern Australia.

*Station Showing Temperature and Precipitation for 12 Months*

### Cf - New Orleans, USA 30° N, Elevation: 1 m

|  | Jan. | Feb. | Mar. | Apr. | May | June | July | Aug. | Sept. | Oct. | Nov. | Dec. | Year |
|---|---|---|---|---|---|---|---|---|---|---|---|---|---|
| Temp. ° C | 12 | 13 | 16 | 20 | 24 | 27 | 28 | 28 | 26 | 21 | 16 | 13 | 20 |
| Precip. mm. | 98 | 101 | 136 | 116 | 111 | 113 | 171 | 136 | 128 | 72 | 85 | 104 | 1371 |

## 3. Maritime Temperate or Oceanic Climates

*Maritime Temperate climate* or *Oceanic climate (Cfb, Cwb)* climate usually occurs on the western sides of continents between the latitudes of 45° and 55°. This climatic region is located immediately pole-ward of the Mediterranean climates. However, in Australia this climate is found immediately pole-ward of the humid subtropical climate and even in somewhat lower latitude. In Western Europe, this climate occurs in coastal areas up to 63° N latitude. The climate is dominated by the polar front throughout the year, leading to variable and often overcast weather. Summers are cool due to cloud cover, but mild winters are experienced than in other comparable latitudes.

### Maritime Subarctic or Subpolar Oceanic Climate

*Maritime Subarctic Climate* or *Subpolar Oceanic Climate (Cfc)* occurs in the region poleward of the Maritime Temperate climates, and is confined either to narrow coastal strips on the western poleward margins of the continents. In the northern hemisphere this climate also extends to the islands off the coastal belts.

*Regions (both Cfb and Cfc)*

1. Coastal Oregon, Washington, British Columbia and Southern Alaska 2. Northwestern Europe 3. Interior South Africa 4. Southeast Australia and New Zealand and 5. Southern Chile.

*Station Showing Temperature and Precipitation for 12 Months*

## Cf - Buenos Aires, Argentina 34.5° S, Elevation: 27 m

|  | Jan. | Feb. | Mar. | Apr. | May | June | July | Aug. | Sept. | Oct. | Nov. | Dec. | Year |
|---|---|---|---|---|---|---|---|---|---|---|---|---|---|
| Temp. ° C | 23 | 23 | 21 | 17 | 13 | 9 | 10 | 11 | 13 | 15 | 19 | 22 | 16 |
| Precip. mm. | 103 | 82 | 122 | 90 | 79 | 68 | 61 | 68 | 80 | 100 | 90 | 83 | 1026 |

## D. Temperate Climate

Temperate Climate is experienced on the poleward margins of mid-latitude regions.

*Symbols used Stating the Climatic Characteristics*

D = Temperature above 10° C in the warmest month and below –3° C in the coldest month.

f = Similar to C.

s = Similar to C.

w = Similar to C.

a = Similar to C.

b = Similar to C.

c = Similar to C.

d = Average temperature in the coldest month below –38° C.

### *Subdivisions of Temperate Climate*

The *Temperate* climate is characterised by an average temperature above 10° C in their warmest months, and a coldest month average below –3° C. This climatic type is usually found in the interiors of continents, or along the east coasts of the continents, north of 40° N latitude. In the Southern Hemisphere, Group D climates are extremely rare due to the smaller land masses in the middle latitudes and the almost nearly absent south of 40° S latitude, except in some highland locations in New Zealand that have heavy winter snows.

Group D climates are subdivided into–1. Hot Summer Continental Climate (Dfa/Dwa Dsa) 2. Warm Summer Continental Climate (Dfb/Dwb/Dsb) 3. Continental Sub-arctic or Taiga Climate (Dfc/Dwc/Dsc) 4. Continental Sub-arctic Climate with severe winter (Dfd/Dwd).

*Hot Summer Continental Climate*

*Hot Summer Continental Climate* (*Dfa*, *Dwa*, *Dsa*) displays three sub-types. *Dfa* climate usually occurs in the middle-high latitude region lying between < 30° and > 40°. In eastern Asia *Dwa* climate is found further south and influenced by the Siberian high pressure system. The divergent air here causes winters to be dry, and summers are usually very wet due to the effect of East Asian monsoon circulation.

*Warm Summer Continental Climate*

*Warm Summer Continental* (*Dfb*, *Dwb*, *Dsb*) includes three subtypes. Of these *Dfb* and *Dwb* climates are observed immediately north of Hot Summer Continental climate, generally in the middle-high latitude region extending from > 40° to < 50° in North America and Asia, and also in central and eastern Europe and Russia. This climate type is located between the Maritime Temperate and Continental Sub-arctic climates, where it even extends up to the latitude > 50° and lower 60° latitude.

*Dsb* resembles the characteristics of *Dsa*, but is found at even higher altitudes or higher latitudes. In North America *Dsb* climate extends further poleward than in Eurasia since the Mediterranean climate exists further pole-ward.

*Continental Sub-arctic* or *Taiga Climate*

*Continental Subarctic* or *Taiga climate* (*Dfc*, *Dwc*, *Dsc*) varies into three sub-types. Of these *Dfc* and *Dwc* climates occur poleward than other Group D climates, mostly located between > 50° N and < 60° N latitude. It might even occur as far north as 70° latitude. *Dsc*, like *Dsa* and *Dsb*, is confined exclusively to highland locations near areas where Mediterranean climate occurs. This type is extremely rare.

*Continental Sub-arctic Climates with Extremely Severe Winters*

*Continental Sub-arctic climates with extremely severe winters* (*Dfd*, *Dwd*) shows variation in two types. These types of climates are found only in eastern Siberia. The locations of this climate, most notably Verkhoyansk and Oymyakon are veritable synonyms for extreme, severe winter cold.

*Regions*

Dfa and Dwa types

> 1. Eastern and Midwestern USA (Atlantic coast to 100th meridian poleward of C climate type) 2. East Central Europe 3. Northern China 4. Manchuria 5. Northern Korea and 6. Honshu.

Dfb and Dwb types

> 1. New England 2. The Great Lakes Region, 3. South Central Canada 4. Southeastern Scandinavia 5. Eastern Europe 6. West Central Asia 7. Eastern Manchuria and 8. Hokkaido.

Dfc, Dwc, Dfd and Dwd types

> 1. Northern North America (Newfoundland to Alaska) and 2. Northern Eurasia (from Scandinavia through Siberia to the Bering Sea and Okhotsk Sea).

*Station Showing Temperature and Precipitation for 12 Months*

**Dw - Calgary, Canada** 51° N, Elevation: 1140 m

|  | Jan. | Feb. | Mar. | Apr. | May | June | July | Aug. | Sept. | Oct. | Nov. | Dec. | Year |
|---|---|---|---|---|---|---|---|---|---|---|---|---|---|
| Temp. °C | −10 | −9 | −4 | 4 | 10 | 13 | 17 | 15 | 11 | 5 | −2 | −7 | 4 |
| Precip. mm. | 17 | 20 | 26 | 35 | 52 | 88 | 58 | 59 | 35 | 23 | 16 | 15 | 444 |

## E. Polar Climate

Polar Climate is confined in the polar region extending from the polar circles to poles.

*Symbols Used Stating the Climatic Characteristics:*

T = Tundra – temperature ranges in the warmest month from 0° C to 10° C.

F = Permafrost – temperature in the warmest month below 0° C.

### Subdivisions of Polar Climate

These climates are characterised by average temperatures below 10° C in all the months of the year. Two principal types are distinguished–1. Tundra Climate and 2. Permafrost or Ice Cap Climate.

*1. Tundra Climate:*

*Tundra Climate (ET)* is distinctly identified having the warmest month with an average temperature ranging between 0° C and 10° C. Such climate occurs on the northern edges of the North American and Eurasian landmasses, and on nearby islands. This climate is also found along the outer fringes of Antarctica (particularly the Palmer Peninsula) and nearby islands.

Tundra Climate is also found at high elevations outside the polar region above the tree line.
*Regions:*

1. Arctic Ocean borderlands of North America, Greenland and Eurasia, 2. Antarctic Peninsula and 3. Some polar islands.

*Station Showing Temperature and Precipitation for 12 months:*

**ET - Barrow, Alaska, USA** 72° N, Elevation: 9 m.

|  | Jan. | Feb. | Mar. | Apr. | May | June | July | Aug. | Sept. | Oct. | Nov. | Dec. | Year |
|---|---|---|---|---|---|---|---|---|---|---|---|---|---|
| Temp. °C | −25 | −28 | −26 | −19 | −7 | 1 | 4 | 3 | −1 | −10 | −19 | −24 | −13 |
| Precip. mm. | 5 | 4 | 4 | 4 | 4 | 8 | 22 | 25 | 16 | 12 | 6 | 4 | 113 |

*2. Permafrost* or *Ice Cap Climate*

*Permafrost* or *Ice Cap Climate (EF)* region is permanently covered with ice sheets. Average temperatures below 0° C are recorded in all the twelve months. This climate is dominant in Antarctica (e.g., Scott Base) and in inner Greenland (e.g., Eismitte or North Ice).

*Regions*

1. Interior Greenland, 2. Most of the Antarctica (except Antarctic Peninsula) and 3. Permanently frozen areas of Arctic Ocean and associated islands.

*Station Showing Temperature and Precipitation for 12 Months*

## EF - Eismitte, Greenland 71° N, Elevation: 2953 m

|  | Jan. | Feb. | Mar. | Apr. | May | June | July | Aug. | Sept. | Oct. | Nov. | Dec. | Year |
|---|---|---|---|---|---|---|---|---|---|---|---|---|---|
| Temp. °C | −42 | −47 | −40 | −32 | −24 | −17 | −12 | −11 | −11 | −36 | −43 | −38 | −29 |
| Precip. mm. | 15 | 5 | 8 | 5 | 3 | 3 | 3 | 10 | 8 | 13 | 13 | 25 | 111 |

## H. Highland Climate

Highland climate is found in high altitude region. The climatic characteristics differ from other climate types in lower region of the same latitude. Trewartha included this type later in the Koppen's classification.

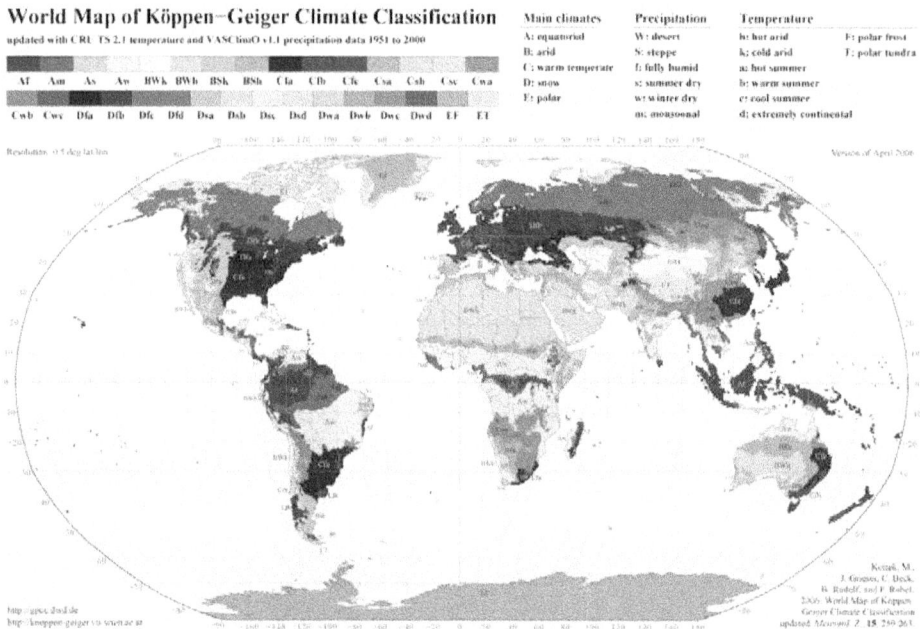

**Figure 12.1:** Koppen-Geiger Classification of World Climates

Source: Kottek M. et al, Meteorol Z **15**

*Number of Climate Types in Koppen's Classification*

**Table 12.4:** Twenty-five Climate Types in Koppen's Classification

| A Type | B Type | C Type | D Type | E Type | H Type |
|--------|--------|--------|--------|--------|--------|
| Af | Bsh | Cfa | Dfa | ET | H |
| Am | Bsk | Cfb | Dfb | EF | |
| Aw | Bwh | Cfc | Dfc | | |
| | Bwk | Csa | Dfd | | |
| | | Csb | Dwa | | |
| | | Cwa | Dwb | | |
| | | Cwb | Dwc | | |
| | | | Dwd | | |
| 3 | 4 | 7 | 8 | 2 | 1 |

*Climate Chart of the Revised Koppen's Classification*

The following table shows the *climate chart* of Koppen's classification (revised).

**Table 12.5:** Climate Chart of Koppen's Classification (Revised)

| A | Tropical Humid | Af – Tropical wet | No dry season |
|---|---|---|---|
| A | Tropical Humid | Am – Tropical Monsoonal | Short dry season; heavy monsoon rain in other months |
| A | Tropical Humid | Aw – Tropical Savanna | Winter dry season |
| B | Dry | BWh – Subtropical desert | Low-latitude desert |
| B | Dry | BSh – Subtropical steppe | Low-latitude dry |
| B | Dry | BWk – Mid-latitude desert | Mid-latitude desert |
| B | Dry | BSk – Mid-latitude steppe | Mid-latitude dry |
| C | Mild Mid-Latitude | Csa – Mediterranean | Mild with dry, hot summer |
| C | Mild Mid-Latitude | Csb – Mediterranean | Mild with dry, warm summer |
| C | Mild Mid-Latitude | Cfa – Humid subtropical | Mild with no dry season, hot summer |
| C | Mild Mid-Latitude | Cwa – Humid subtropical | Mild with dry winter, hot summer |
| C | Mild Mid-Latitude | Cfb – Marine west coast | Mild with no dry season, warm summer |
| C | Mild Mid-Latitude | Cfc – Marine west coast | Mild with no dry season, cool summer |

| D | Severe Mid-Latitude | Dfa – Humid continental | Humid with severe winter, no dry season, hot summer |
| D | Severe Mid-Latitude | Dfb – Humid continental | Humid with severe winter, no dry season, warm summer |
| D | Severe Mid-Latitude | Dwa – Humid continental | Humid with severe, dry winter, hot summer |
| D | Severe Mid-Latitude | Dwb – Humid continental | Humid with severe, dry winter, warm summer |
| D | Severe Mid-Latitude | Dfc – Sub-Arctic | Severe winter, no dry season, cool summer |
| D | Severe Mid-Latitude | Dfd – Sub-Arctic | Severe, very cold winter, no dry season, cool summer |
| D | Severe Mid-Latitude | Dwc – Sub-Arctic | Severe, dry winter, cool summer |
| D | Severe Mid-Latitude | Dwd – Sub-Arctic | Severe, very cold and dry winter, cool summer |
| E | Polar | ET – Tundra | Polar tundra, no true summer |
| E | Polar | EF – Ice Cap | Perennial ice |
| H | Highland | | |

*Merits of Koppen's Classification*

1. In Koppen's classification climatic regions are indicated by specific symbols. The symbols are also easy to remember. By remembering the symbols one can easily narrate the climate type.
2. All the climate types are categorised empirically from the available climatic data worldwide. Due to reliable and nearly constant database this classification will be acceptable and useful in future also.
3. Later it was found that the Koppen's classification could be related to the General Circulation of the Atmosphere. Trewartha incorporated this principle in later revision of Koppen's classification.

*Demerits of Koppen's Classification*

1. Koppen's classification of world climates is basically based on climatic database available throughout the world. In many parts of the world statistical records are poor and not even available due to non-existence of observatory at micro-level. In many cases for delimitation of climatic boundaries assumptions had been made on the basis of natural vegetation types. Micro-climate differs due to physical landscape, but it was never considered as a factor for climate variation. For example the Pudget

Sound region in the west coast of North America had been shown as Mediterranean climate. But the natural vegetation, with tall trees like Douglas Firs, completely differs from the Mediterranean vegetation. Here stunted trees and shrubs of Mediterranean climate are rarely found.

2. One of the most frequently-raised objections concerns the Temperate climate (C type), considered by many as broad and general, as it includes both Tampa, Florida and Cape May, New Jersey, for example. In his book, *'Applied Climatology'*, John Griffiths proposed a new subtropical zone, encompassing those areas with a coldest month of between 6° C and 18° C, effectively subdividing Group C into two nearly equal parts.

3. Of the five principal climate types four types (A, C, D and E) were specified on the basis of temperature. But in the determination of Dry climate (B type) the effectiveness of rainfall with the evaporation was considered for the delimitation of the dry *B* climates. The critics argued that their separation by Köppen into only two thermal subsets was inadequate. Griffiths and others suggested that the dry climates be placed on the same temperature continuum as other climates, with the thermal letter being followed by an additional capital letter '*S*' for steppe or '*W*' (or *D*) for desert. Griffiths also proposed an alternate formula for use as an aridity threshold: $R = 160 + 9T$, with *R* equalling the threshold, in millimetres of mean annual precipitation, and *T* denoting the mean annual temperature in degrees Celsius.

4. The accuracy of the 10° C warmest-month isotherm as the limiting line of the polar climates was also questioned. Otto Nordenskiöld, for example, devised an alternate formula: $W = 9–0.1\ C$, with *W* representing the average temperature of the warmest month and *C* that of the coldest month, both in degrees Celsius. This boundary appeared to be more coincident with the tree line, or the latitude pole-ward of which trees can not grow, than the 10° C warmest-month isotherm. The former tends to run poleward of the latter near the western margins of the continents, but at lower latitude in the interior landmass, the two lines cut across at or near the east coasts of both Asia and North America.

5. In some cases Koppen delimited the boundaries in consideration of physical landscape in mind. In other cases he did not consider that.

6. Koppen did not include the Highland climate in his classification as a separate type. Trewartha later included the Highland type in the revised Koppen's classification.

7. The boundaries of Koppen's classifications are rigid and fixed. In reality transitional climate exists between the two boundaries.

## Moderation of Koppen's Classification by Trewartha

G.T. Trewartha made an attempt to eliminate the errors in Koppen's classification of world climates. He included a new type, 'Highland Climate' in the revised classification. He also related the climate types to their genesis and climatic processes. In this sense the classification by Trewartha appears to be both empirical and genetical.

Trewartha divided the world climate into six major types. Of the six types five are classified based on temperature conditions and one based on precipitation. The climate types are:

*Based on Temperature*

A – Tropical: In the coldest month temperature exceeds 18.3° C.

C – Subtropical: For eight months temperature shall be 10° C or more.

D – Temperate: For four months temperature shall be 10° C or more.

E – Boreal: For one month (warmest month) temperature shall be 10° C or more.

F – Polar: In every month temperature shall be less than 10° C.

*Based on Precipitation*

B – Dry: Limit (where the potential evaporation equals the precipitation).

*Symbols Used by Trewartha to Specify the Climate Types*

A　=　Temperature of the coldest month in the coastal part exceeds 18.3° C. Frost never occurs.

r　=　Rainy, 10 to 12 months wet, 0 to 2 months dry.

w　=　Winter dry, more than 2 months dry.

s　=　Summer dry (normally not found).

B　=　Evaporation exceeds precipitation.

Boundary between Steppe and Wet = $R = \frac{1}{2} T - \frac{1}{4} Pw$, where R = Rainfall in inches, T = Temperature in ° F, Pw = Precipitation in the winter half.

Boundary between Desert and Steppe = $R = \frac{1}{2} T - \frac{1}{4} Pw / 2$, where R = Rainfall in inches, T = Temperature in ° F, Pw = Precipitation in the winter half.

W (in German language – wuste) = Desert or Dry, S = Semiarid or Steppe, h = Hot, average monthly temperature exceeding 10° C for 8 months or more.

K (in German language kalt) = Cold, average monthly temperature below 10° C for 8 months or more, s = Dry summer, w = Dry winter, n = continuous fog.

C　=　Temperature exceeds 10° C.

a　=　Warm summer, temperature exceeds 22.2° C in the warmest month.

b = Mild summer, temperature less than 22.2° C in the warmest month.

f = No dry period. Difference between the driest and wettest month shall be less than s and w.

s = Dry summer. Rainfall in the winter half shall be three times more than the summer half. Driest month in the summer records less than 3 cm. rainfall. Annual total rainfall exceeds 88.9 cm.

w = Dry winter. Rainfall in the summer half shall be ten times more than the winter half.

D = Temperature exceeds 10° C at least for 4 to 7 months.

o = Oceanic. Temperature of the coldest month exceeds 0° C (>2° C in the interior).

c = Continental. Temperature in the coldest months records less than 0° C.

a = Similar to C.

b = Similar to C.

f = Similar to C.

s = Similar to C.

w = Similar to C.

E = Temperature exceeds 10° C for 1 to 3 months in a year.

F = Temperature is below 10° C in all the months in a year.

t = Tundra. Temperature in the warmest months ranges 0° to 10° C.

i = Permafrost. All months record temperature below 0° C.

## Boundary Line between Different Climate Types

A/C boundary line = Equatorial coldest temperature limit—in marine location the temperature of the coldest month shall exceed 18° C.

C/D boundary line = Average monthly temperature shall be 10° C or more for at least 8 months.

D/E boundary line = Average monthly temperature shall be 10° C or more for at least 4 months.

t/i boundary line in F type climate = Temperature shall be at least 10° C in the warmest month.

B/A, B/C, B/D, B/E boundary lines = Evaporation equals precipitation.

BS/BW boundary line = Half of the index for B/A, B/C, B/D, B/E boundary lines.

h/k boundary line in Dry climate = Similar to C/D.

Do/Dc boundary line = Temperature in the coldest month ranges from 0° C to 2° C.

No boundary line for the division of H climate was attempted due to paucity of climatic data.

**Table 12.6:** Trewartha's Classification (Revised Koppen's Classification)

| Climate Divisions | Climate Types | Air Pressure and Pressure Belts | | Precipitation Characteristics |
|---|---|---|---|---|
| | | *Summer* | *Winter* | |
| A – Tropical Wet | Ar – Tropical rainy | ITC, Doldrums Equatorial Wly. | ITC, Doldrums Equatorial Wly. | Dry period not exceeding two months |
| | Aw – Tropical dry | ITC, Doldrums Equatorial Wly. | Dry Trade winds | Wet summer, dry winter |
| C – Subtropical | Cs – Subtropical dry, summer dry | Subtropical High (East) | Westerlies | Dry summer, winter rainfall |
| | Cf – Subtropical wet | Subtropical High (West) | Westerlies | Rainfall in all seasons |
| D – Temperate | Do – Oceanic | Westerlies | Westerlies | Rainfall in all seasons |
| | Dc – Continental | Westerlies | Westerlies/ Anticyclone | Rainfall in all seasons, maximum in summer, snowfall in winter |
| E – Boreal | E – Boreal | Westerlies | Polar air/Anticyclone | Meagre precipitation throughout the year |
| F – Polar | Ft – Tundra | Polar air | Polar air | Meagre precipitation throughout the year |
| | Fi – Permafrost | Polar air | Polar air | |
| B – Dry | Bs – Semiarid (Steppe) | Subtropical High | Subtropical High | Short wet season, Meagre rainfall, mostly in summer |
| | Bsh – Warm tropical, Subtropical semiarid | Dry Trade | Dry Trade, Anticyclone | |
| | Bsk – Cool temperate, Boreal | | | |

| Climate Divisions | Climate Types | Air Pressure & Pressure Belts | | Precipitation Characteristics |
| --- | --- | --- | --- | --- |
| | | Summer | Winter | |
| | Bw – Dry (Desert) | Subtropical High | Subtropical High | Dry conditions prevail |
| | Bwh – Tropical Hot, Subtropical dry | Dry Trade | Dry Trade | |
| | Bwk – Cool temperate, boreal desert | – | Continental anticyclone | Dry conditions prevail |
| H – Undifferentia-ted high land | – | – | – | – |

## *Description of Climate Types after Trewartha*

On both sides of the equator in low latitudes the torrid-zone extends with no winter, high temperature throughout the year and abundant rainfall. This is known as Tropical Wet Zone. Snowfall never occurs. Temperature in the coldest month exceeds 18.3° C. Two subtypes are recognised: 1. Tropical Moist Climate (Ar)—Rainy period continues for 10 to 12 months and 2. Tropical Moist-Dry Climate (Aw)—The winter season remains dry for at least 2 months.

In the Tropical Moist zone the Inter Tropical Convergence Zone (ITCZ) and Doldrums are located. Here the winds converge at the centre of low. As a result the rainfall is high. Due to high rainfall the rainforests have grown over the area.

In the Tropical Moist-Dry zone ITCZ exists and Equatorial Westerlies blow in the summer months. In the winter months anticyclone develops and dry Trade winds blow over the region. The Savanna Grasslands and Deciduous forest stand as natural vegetation.

In the wet mid-latitude region 3 principal climate types are recognised. These are: 1. Subtropical Climate Type (C), 2. Temperate Climate Type (D) and 3. Boreal Climate Type (E).

In the Subtropical Climate Zone (C) the temperature drops below 0° C in the interior of the continents. But along the coastal belt on the margin of equator the temperature rarely drops below 18° C in the coldest month. Here the temperature remains 10° C and above for 8 months. Two sub-types are identified – 1. Subtropical Wet (Cfw) and 2. Subtropical Summer (Cs). Subtropical Wet zone is found on the eastern side of the continents. Here in summer the subtropical anticyclone dominates having westerly component of unstable wind systems. But in winter this zone comes under the influence of cyclone and westerly winds and gets adequate rainfall. Subtropical Summer climate is popularly known as Mediterranean climate. This zone is found on the western side of the continents ranging 35°-45° latitudes. Summer months are usually dry, as the dominating anticyclone restricts the entry of the westerlies. In winter months this anticyclone shifts to lower latitude allowing the entry of the westerlies as cyclones. So rainfall occurs over this zone in winter.

In the Temperate Climate Zone monthly average temperature remains 10° C and above for 4 to 7 months. Two sub-types are recognised: 1. Temperate Oceanic Climate (Do) and 2. Temperate Continental Climate (Dc). The boundary between the two sub-types is delineated by 0° C isotherm in the coldest month. Winter is mild over the Temperate Oceanic Climate Zone, but severe in the Temperate Continental Climate Zone.

Boreal Climate (E) is found in the middle-high latitudes. Here cool summer is short, but severe long winter is experienced. Frost-free season is also very brief. Average temperature remains 10° C and above only for 1 to 3 months. Coniferous forest is the dominant natural vegetation. Due to severity of climate human population is low and settlements are rare. In the high latitudes Polar climate (F) is found. In this climate type there is no summer. Average temperature in all the months remains below 10° C. Two principal sub-types are recognised. These are: 1. Tundra Climate (Ft) – where the temperature of the warmest month is above 0° C, but lower than 10° C and 2. Permafrost Climate (Fi) – where the temperature remains below 0° C in all the months.

The classification of B type climate is not simply based on temperature. Inadequacy of precipitation (evaporation exceeds precipitation) has been considered to delineate the boundary of B type climate. So the B type climate is not only seen in the tropical region, but also found in subtropical, temperate and even in boreal climate. Two sub-types of B climate are found: 1. Desert Region (Bw) and Steppe Region (Bs). Even these are further subdivided on the basis of temperature. These are: 1. Tropical or Subtropical Desert (Bwh), 2. Tropical or Subtropical Semi-arid or Steppe (Bsh), 3. Temperate or Boreal Desert (Bwk) and 4. Temperate or Boreal Steppe (Bsk). If the average (monthly) temperature exceeds 10° C for 8 months then it is termed warm or *wurst* (w). Where the average (monthly) temperature exceeds 10° C for less than 8 months then it is called cold or *kalt* (k). In the dry belt of Tropical and Subtropical region subtropical high pressure and trade winds dominate. Here we find subsidence of upper air causing divergence of air on the surface. This particular type of climate is evidenced in the interior of the continents and also in the rain-shadow areas of high mountains.

*Deviation of Trewartha's Classification from the Koppen's Classification*

We can identify the differences of Trewartha's classification from the Koppen's classification as follows:

1.  In B type climate Koppen considered the average temperature of 18° C delineating the boundary of Bs/Bk. In the revised classification by Trewartha it was specified as 0° C in the coldest month.

2.  Koppen considered –3° C in the coldest month to determine the boundary between C/D climates. Trewartha considered it to be 0° C.

3.  In C type of climate sub-types were identified on the basis of precipitation in general. But in the revised classification the importance of seasonal precipitation had been stressed. In D type climate also similar procedures had been adopted.
4.  Koppen defined the E type climate as the polar climate. But in the revised classification by Trewartha E type climate had been recognised as the Boreal climate. This division had been closely correlated with the coniferous vegetation. The E type climate of Koppen's classification had been marked as F type climate by Trewartha.
5.  Koppen identified 4 (A, C, D and E) thermal provinces and 1 (B) dry province. Trewartha revised it to be 5 (A, C, D, E and F) thermal provinces and 1 (B) dry province.
6.  Koppen extended the characteristics of low land climate to the high land region. Trewartha considered the Highland (H) as a separate entity. Due to paucity of data he marked this climate as undifferentiated.

**Thornthwaite's Classification of World Climates**

In USA, C.W. Thornthwaite attempted to devise a classification of world climates with a different approach. Like Koppen he also considered that the climatic elements like temperature, precipitation, their seasonal distribution determine the climate types. But he was not satisfied with the divisions made by Koppen. He correlated the effective temperature with the vegetation growth to identify the climate boundaries. Thornthwaite also categorised his classifications on empirical basis.

Thornthwaite first proposed his classification in 1931 to prepare a climatic map of North America, later revised in 1933. Thornthwaite's first world classification of climate was presented in 1931-1933. It differed from earlier attempts by others in that it assigned the importance to the moisture factor. In 1948 he proposed an entirely new system based fundamentally on the ideas presented above. He assigned the central place in this new classification to the *potential evapotranspiration* (PE), which he defined as the evapotranspiration that would occur from a vegetation-covered surface if the soil moisture conditions were adequate for unrestricted transpiration.

Different types of vegetation differ in their potential evapotranspiration because they absorb different amounts of solar radiation. The solar radiation and albedo factor appear to have control over the potential evapotranspiration. In other words the thermal efficiency is a measure of potential evapotranspiration. Ångströms calculated the albedos of a few different surfaces as follows: grassland—0.26; oak woodland—0.175; and pine forest—0.148. So the pine forest might absorb 16 percent more energy than the grassland but the proportion available for evapotranspiration varies through a narrower percentage range. Most common garden vegetables and field crops absorb about the same amount of solar energy as grassland. Some types of forest may have a higher albedo than grass although we have no records of them; if such exist they would transpire less than grassland.

In practice, it is very difficult to measure potential evapotranspiration by an instrument directly. The most reliable measurements of evaporation and transpiration could be related to climatic elements in an effort to derive at valid and practical empirical relationship, based on the monthly or seasonal data from irrigation and drainage projects and on daily observations from carefully operated evapotranspirometer tanks. Thornthwaite observed that when adjustments were made for variations in day length there was a close relation between mean temperature and potential evapotranspiration. Through study of the available data he devised a formula for the computation of potential evapotranspiration for any place at known latitude, where temperature and precipitation records could be available.

Thornthwaite continued to work on classification of climates. He finally revised his classification again in 1955.

### Thornthwaite's 1931 and 1933 Classifications

Following Livingstone, Thornthwaite devised such a *Thermal Index* which can relate the growth of vegetation. He called this thermal index as the *Coefficient of Temperature Efficiency*. The monthly coefficient was defined as T/E Ratio and the summation of 12 months coefficients was called the T/E Index. In the polar areas favourable conditions for the growth of vegetation are absent. So T/E Index in the polar areas appears to be very low and over the pole zero (0). On the other hand tropical region possesses the most favourable conditions for the growth of vegetation. So, T/E Index is the highest in the tropical areas. T/E index was calculated as follows:

$i' = T - 32/4$, where T = Monthly (° F) and i′ = Monthly Temperature Index (T/E Ratio).

T/E Annual Index was calculated by the summation of 12 month's T/E Ratios.

$I' = \sum_{12\ months} T-32/4$ (I′ = Annual T-E index).

Based on this T/E index (Annual) Thornthwaite divided the world climate into 6 Temperature Provinces.

**Table 12.7:** Temperature Provinces after Thornthwaite

| Temperature Province | T/E Index (Annual) |
|---|---|
| A | >128 |
| B | 64-127 |
| C | 32-63 |
| D | 16-31 |
| E | 1-15 |
| F | 0 |

In Table 12.7 Annual T/E index for 'A' province was assumed to be 128. Thornthwaite determined this annual index (128) assuming the annual average temperature to be 75° F. The calculation is shown below:

$$T - 32/4 = 128/12 \text{ or } 12T - 32 \times 12 = 4 \times 128 \text{ or } 12T = 512 + 384 \text{ or } T = 896/12 = 74.66 \ (75° \text{F})$$

Similarly Thornthwaite calculated the Precipitation Efficiency Index (P-E) for the climatic classification. He devised the following formula to determine the monthly P/E on the basis of temperature and precipitation data.

$P/E = 11.5 \ (P/T - 10)^{10/9}$, where P = Precipitation (inches) and T = Temperature (° F).

From the summation of 12 months P/E ratios P/E index was calculated.

$$P/E \text{ Index (Annual)} = \sum_{12 \text{ months}} 11.5 \ (P/T - 10)^{10/9}$$

Thornthwaite divided the world climate into 5 Humidity Provinces and related the characteristic natural vegetation with each Humidity Province. Humidity Provinces are shown in the following table.

**Table 12.8:** Humidity Provinces after Thornthwaite

| Humidity Province | Characteristic Natural Vegetation | P/E Index (Annual) |
|---|---|---|
| A′ (Wet) | Rainforest | >128 |
| B′ (Moist) | Forest | 64-127 |
| C′ (Semi-moist) | Grassland | 32-63 |
| D′ (Semi-arid) | Steppe | 16-31 |
| E′ (Arid) | Desert | <16 |

These Humidity Provinces were subdivided into 4 sub-types based on characteristic seasonal precipitation. These were:

r = Adequate rainfall in all seasons

s = Inadequate rainfall in summer

w = Inadequate rainfall in winter

d = Inadequate rainfall in all seasons.

After Thornthwaite 120 $(6 \times 5 \times 4)$ types of world climate could be recognised. Thornthwaite himself identified 32 types of world climates. These are shown in the following table.

**Table 12.9:** World's Climate Types after Thornthwaite

| AA′r | BA′r | CA′r | DA′w | EA′d | D′ | E′ | F′ |
|------|------|------|------|------|-----|-----|-----|
| AB′r | BA′w | CA′w | DA′d | EB′d | | | |
| AC′r | BB′r | CA′d | DB′w | EC′d | | | |
| | BB′w | CB′r | DB′s | | | | |
| | BB′s | CB′w | DB′d | | | | |
| | BC′r | CB′s | DC′d | | | | |
| | BC′s | CB′d | | | | | |
| | CC′r | | | | | | |
| | CC′s | | | | | | |
| | CC′d | | | | | | |
| 3 | 10 | 7 | 6 | 3 | 1 | 1 | 1 |

*Derivation of Thornthwaite's P/E Index*

Precipitation efficiency depends on the rate of evaporation in a particular place. Unfortunately data on evaporation can not be found from wide areas on the earth's surface. So Thornthwaite decided to calculate precipitation efficiency based on temperature records. For this he analysed the statistical data of 21 centres in the south-western part of USA.

Firstly he fixed the standard values for temperature, precipitation and evaporation from the climographs. Next the standard P/T-10 values were calculated for P/E ratios. In the following table the standard values are enumerated.

**Table 12.10:** Standard Values for the Calculation of P/E Ratio

| $x = P/E$ | $y = P/T\text{-}10$ | $x = P/E$ | $Y = P/T\text{-}10$ |
|-----------|---------------------|-----------|---------------------|
| –0 | –0 | 1.0 | –1111 |
| –1 | –0139 | 1.5 | –1608 |
| –2 | –0262 | 2.0 | –2080 |
| –3 | –0385 | 2.5 | –2515 |
| –4 | –0484 | 3.0 | –296 |
| –5 | –0597 | 4.0 | –833 |

Plotting the values on log graph we get the following equation:

$y = kx_n$

So, $\log y = \log k + n.\log x$,

Say, $u = \log x$, $v = \log y$ and $w = \log k$

Then, $v = w + n.u$

For convenience two values of P/E and P/T-10 have been taken and employed in the equation, then:

| x | u (log x) | y | v (log y) |
|---|---|---|---|
| ·2 | −·69897 | ·0262 | −1.58170 ........(i) |
| 2.0 | ·30103 | ·208 | −·68194 ........(ii) |

From the first equation we get,     $-1.58170 = w - ·69897.n$
$-·68194 = w + ·30103$

Subtracting we have     $·89976 = n$   or   $n = ·9$ (approx)

Putting the value of n in the first equation we get:

$$w = ·69897 \times ·9 \text{ or } w = - ·953$$

So,    $\log k = - ·953$ or $9.047 - 10$ or $k = ·1114$

Putting the value of k in the principal equation (i.e. $y = kxn$) we get, $y = ·1114 \times ·9$

Or, $P/T - 10 = ·1114 (P/E)9/10$   or   $(P/E)9/10 = P/T - 10 \times 1/·1114$

Multiplying both sides by 10/9 we have, $P/E = (1/·1114) 10/9 \times (P/T - 10)10/9$

$$\text{So,} \quad P/E = 11.5(P/T - 10)^{10/9}$$

So, the value of P/E can be calculated if the Precipitation (P) and Temperature (T) records are available.

### Thornthwaite's 1948 Classification of World Climates

For his classification of world climates Thornthwaite applied a new concept. He called this *Evapotranspiration.* Thornthwaite found that when adjustments were made for variations in day length there was a close relation between mean temperature and potential evapotranspiration. Through study of the available data he produced a formula that permits the computation of Potential Evapotranspiration (PE) for any place whose latitude is known and for which temperature records are available. Thornthwaite calculated Potential Evapotranspiration (PE) from the available records of average monthly temperature and length of the day. He considered the number of days as 30 for a month and 12 hours for a day-length. His formula for determination of PE is stated below:

PE (in centimetre) = $1.6(10T/I)_a$, Here I = Summation of 12 month's value of $(t/5)_{1.514}$ and 'a' is the multiplying factor. From the study of accumulated soil moisture (monthly moisture budget) *monthly surplus* and *monthly deficit* could be determined. Monthly Moisture Index may be calculated by the given formula as noted below:

Im = (100S – 60D)/PE

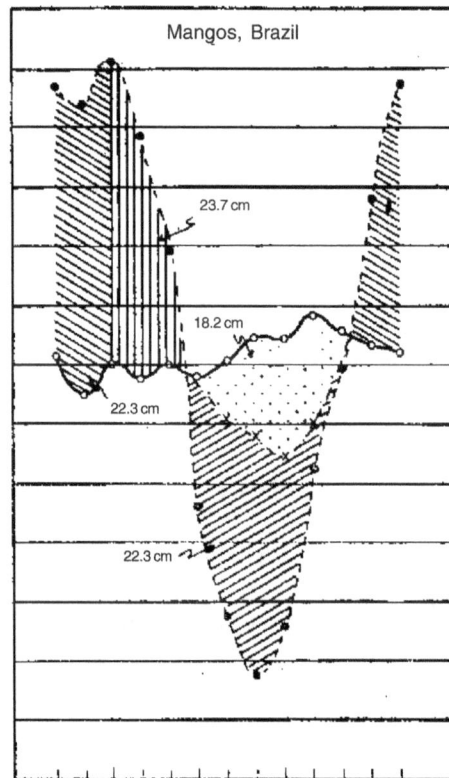

**Figure 12.2:** Precipitation and Potential Evapotranspiration for Mangos, Brazil

Here the moisture has been multiplied by 0.6, as the surplus moisture is taken by the deep rooted perennial vegetation in the next season. In this classification temperature efficiency is calculated from PE, because the evaporation process not only depends on precipitation but also on temperature. The Thermal Provinces have been identified by the ratio of PE/Im. The divisions of Thermal Provinces are given in Table 12.11.

**Table 12.11:** Moisture Provinces and Thermal Provinces

| Moisture Index (Im) | Moisture Provinces | Potential Evapotranspiration Index – PE (cm) | Thermal Provinces |
|---|---|---|---|
| >100 | Perhumid (A) | >114 | Megathermal (A′) |
| 20 to100 | Humid ($B_1 - B_4$) | 57 – 114 | Mesothermal ($B'_1$–$B'_4$) |
| 0 to 20 | Moist Subhumid ($C_2$) | 28.5 – 57 | Microthermal ($C'_1$–$C'_2$) |
| –33 to 0 | Dry Subhumid ($C_1$) | 14.2 – 28.5 | Tundra (D′) |
| –67 to –33 | Semiarid (D) | <14.2 | Frost |
| –100 to –67 | Arid (E) | – | – |

In 1955, Thornthwaite revised his classification marginally. In this revision Potential Evapotranspiration (PE) and precipitation are considered the bases of four basic climatic criteria: *moisture adequacy, thermal efficiency, seasonal distribution of moisture adequacy* and *summer concentration of thermal efficiency.* Each was represented by an index value and the boundaries were drawn quantitatively. From the adjustments of storage water in the soil the difference between mean monthly amounts of P and PE would indicate the monthly surplus (S) or deficit (D). Moisture adequacy could be expressed by the monthly *Moisture Index* (Im) as stated in the following formula:

Im = 100 (S – D) / PE,

If the soil moisture is assumed to be constant then Im = 100 (P/PE – 1)

Annual Moisture Index could be calculated by the summation of 12 monthly values of moisture index (Im). Thornthwaite identified 9 Moisture Types (Provinces) based on the calculations of Moisture Index. Table 12.12 shows the limits of annual moisture index for 9 climatic moisture types (provinces).

Thornthwaite considered the 'Thermal Efficiency' simply as a measure of *Potential Evapotranspiration,* expressed in centimetres. Annual index of thermal efficiency was calculated by the summation of 12 monthly values. Again the summer concentration of thermal efficiency was calculated as the percentage of cumulative summer values in relation to annual index.

**Table 12.12:** Moisture Index (Annual) Limit for Nine Climatic Moisture Types

| Climatic Moisture Types (Provinces) | Moisture Index |
|---|---|
| A – Perhumid | 100 and above |
| $B_4$ – Humid | 80 to 100 |
| $B_3$ – Humid | 60 to 80 |
| $B_2$ – Humid | 40 to 60 |
| $B_1$ – Humid | 20 to 40 |
| $C_2$ – Moist Subhumid | 0 to 20 |
| $C_1$ – Dry Subhumid | –33.3 to 0 |
| D – Semiarid | –66.7 to – 33.3 |
| E – Arid | –100 to – 66.7 |

**Table 12.13:** Thermal Efficiency and its Summer Concentration

| Thermal Efficiency | | Summer Concentration | |
|---|---|---|---|
| Type | Index (centimeter) | Type | Concentration (percent) |
| A′ – Megathermal | 114 and above | a′ | Below 48.0 |
| $B'_4$ – Mesothermal | 99.7 to 114.0 | $b'_4$ | 48.0 to 51.9 |
| $B'_3$ – Mesothermal | 85.5 to 99.7 | $b'_3$ | 51.9 to 56.3 |
| $B'_2$ – Mesothermal | 71.2 to 85.5 | $b'_2$ | 56.3 to 61.6 |
| $B'_1$ – Mesothermal | 57 to 71.2 | $b'_1$ | 61.6 to 68 |
| $C'_2$ – Microthermal | 42.7 to 57 | $c'_2$ | 68 to 76.3 |
| $C'_1$ – Microthermal | 28.5 to 42.7 | $c'_1$ | 76.3 to 88.0 |
| D′ – Tundra | 14.2 to 28.5 | d′ | above 88.0 |
| E′ – Frost | Below 14.2 | | |

The seasonal distribution of moisture adequacy was determined on the basis of aridity and humidity indices. In moist climates *aridity index* was indicated by the annual water-deficit expressed as percentage of annual PE. On the other hand, Thornthwaite calculated the *Humidity Index* in Dry Climates by annual water surplus as percentage of annual PE. The following table shows the Seasonal Moisture Adequacy in moist and dry climates.

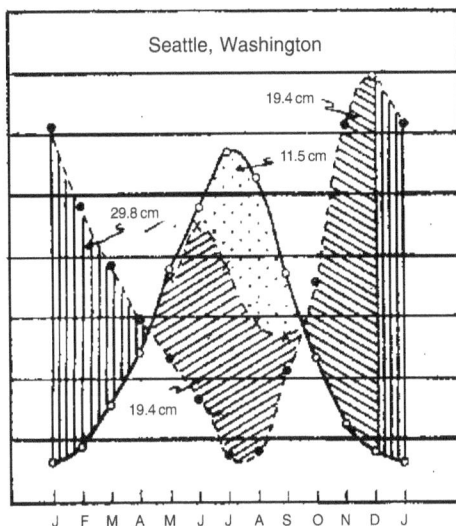

**Figure 12.3:** Precipitation and Potential Evapotranspiration for Seattle, Washington

**Table 12.14:** Seasonal Moisture Adequacy

| Moist Climates (A, B, C$_2$) | Aridity Index |
|---|---|
| r – Little or no water deficit | 0 to 10 |
| s – Moderate summer deficit | 10 to 20 |
| w – Moderate winter deficit | 10 to 20 |
| s$_2$ – Large summer deficit | above 20 |
| w$_2$ – Large winter deficit | above 20 |
| Dry Climates (C$_1$, D, E) | Humidity Index |
| d – Little or no water surplus | 0 to 16.7 |
| s – Moderate summer surplus | 16.7 to 33.3 |
| w – Moderate winter surplus | 16.7 to 33.3 |
| s$_2$ – Large summer surplus | above 33.3 |
| w$_2$ – Large winter surplus | above 33.3 |

By combining the above stated 4 elements of classification Thornthwaite attempted the classification of world climates symbolising with 4 letters. Due to complexity of classification he could not represent the classifications on a world map, though continental maps representing separate elements were prepared. The following table shows the Thornthwaite's classification for the selected stations.

**Table 12.15:** Thornthwaite's Classification for Selected Stations

| Annual Water Budget Criteria | Alice Springs, Australia (EA'da') | San Fransisco, U.S.A. (C₁B'₁da') | Moscow, Russia (C₁C'₂b'₁) |
|---|---|---|---|
| Precipitation (cm) | 24.5 | 55.1 | 63.1 |
| Thermal Efficiency (PE in cm) | 116.2 | 70.2 | 55.4 |
| Summer Concentration (percentage of PE) | 43.8 | 33.3 | 64.4 |
| Surplus (cm) | 0 | 3.6 | 10.4 |
| Deficit (cm) | 91.7 | 18.7 | 4.5 |
| Humidity Index (percent) | 0 | 5.1 | 18.8 |
| Aridity Index (percent) | 78.9 | 26.6 | 8.1 |
| Moisture Index | –78.5 | –21.5 | 10.7 |

### Merits and Demerits of Thornthwaite's Classification of Climates

*Merits*

1. Thornthwaite's 1931 classification resembled the Koppen's classification. The classification had been done empirically based on observed climatic data.
2. In Thornthwaite's classification climatic boundaries were not drawn simply based on temperature and precipitation like Koppen's classification.
3. In Thornthwaite's classification number of letters as symbols were few. It was advantageous to remember.
4. The empirical formula for determination of PE by Thornthwaite is being widely used in various water balance studies.

*Demerits*

1. Thornthwaite devised the formula for PE, based on the analysis of data in 21 centres of South-western USA and later applied this formula for classification of world climates. This application had not been tested for all cases.
2. In the Thornthwaite's classification number of climatic types had been three times more than the Koppen's classification.

### Genetic Classification of Climates

Climate of any region is determined by climatic processes. 'General Circulation of the Atmosphere' plays significant role in determining the characteristic climate over a place.

Other factors are also responsible for the genesis of a climate type. The climate of a particular place is the function of a number of factors. These factors include:

1) Influence of air mass
2) Latitude and its influence on solar radiation received
3) Location of global high and low pressure zones
4) Heat exchange from ocean currents
5) Distribution of mountain barriers
6) Pattern of prevailing winds
7) Distribution of land and sea
8) Altitude

At a macro-level, the first three factors play the most significant role influencing the climate of a region. These three factors indicate latitudinal effectiveness through the following climatic features:

- Relative annual latitudinal location of the overhead Sun at solar noon.
- Intertropical convergence zone and its area of uplift, cloud development and precipitation.
- Subtropical high pressure zone and its associated descending air currents and clear skies.
- Polar front and its area of uplift, cloud development and precipitation.
- Polar vortex and its associated descending air currents and clear skies.
- Relative location of tropical/subtropical and polar air masses.

**Table 12.16:** Classification of Climates after Flohn

| Wind belts | Precipitation Characteristics |
|---|---|
| 1. Equatorial Westerlies belt | Permanently wet |
| 2. Tropical zone – winter Trade wind | Summer rainfall |
| 3. Subtropical dry zone – Trade wind/ Subtropical High Pressure belt | Dominantly dry |
| 4. Subtropical winter rainfall zone – Mediterranean climate type | Winter rainfall |
| 5. Mid-latitude Westerlies belt | Precipitation throughout the year |
| 6. Sub-polar zone | Limited precipitation throughout the year |
| 7. Boreal Rainfall in summer<br>A. Continental sub-type | Limited snowfall in winter |
| 8. High Polar zone | Meagre precipitation, summer rainfall, snowfall in early winter |

In any classification where the climate types are related to the cause of climate, it is known as *Genetic Classification of Climate.* In 1931, Hettner attempted to classify the climate types on the basis of wind systems, continentality, precipitation amount and duration, relative distance from the sea, altitude etc. In 1936, Allisov classified the climate types based on the dominant influence of air masses. Flohn contributed appreciably for the development of genetic classification. In 1950, he classified the world climates on the basis of global wind belts and precipitation characteristics.

In Table 12.16 temperature has not been considered as the basis of climatic division. The global pressure and wind belts have been considered as key factors. However, this classification to a great extent resembles the Kopppen's classification.

### *Strahler's Classification of Climates*

In 1951, Strahler devised a climatic classification mostly based on controlling air masses. He divided the world climate into three macro-groups. These are:
1. Low latitude climates—controlled by the equatorial and tropical air masses.
2. Mid-latitude climates—controlled both by the tropical and polar air masses.
3. High latitude climates—controlled by the polar and arctic air masses.

In this classification 5 types in the first division, 5 types in the second division and 4 types in the third division had been identified. Later Highland climate was included. In Strahler's genetic classification, 15 climate types based on air masses were specified. In the following table the Strahler's classification and its relation with the Koppen's classification has been shown.

### **Table 12.17:** Strahler's Classification

**A. First Group: Low Latitude Climate**—Controlled by the Equatorial and Tropical Air Masses

| Climate Type | Source region and limiting zone of air masses – climatic characteristics | Koppen's symbol |
|---|---|---|
| 1. Wet Equatorial climate – 10° N to 10° S (in Asia 10° N to 20° N) | Equatorial trough (convergent low pressure region) – controlled by Equatorial air mass (mE), convective rainfall, nearly uniform temperature all the year round. | Af |
| 2. Trade wind – littoral climate – 10° to 25° N & S | Tropical Easterlies – mT winds from the western part of the continents penetrate and cause rainfall over the littoral zone. Rainfall varies. | Am |
| 3. Tropical Desert and Steppe climate – 15° to 35° N & S | Source region of cT air mass – lying at the upper level of high pressure axis near the Tropic of Cancer and Tropic of Capricorn. | Bwh/Bsh |
| 4. West coastal Desert climate – 15° to 30° N & S | Lying near the margin of Sub-tropical high pressure axis – foggy desert climate along the west coast. | Bwh/Bwk |

| Climate Type | Source region and limiting zone of air masses – climatic characteristics | Koppen's symbol |
|---|---|---|
| 5. Tropical Dry and Wet climate | Seasonal shifting of mT and mE air masses with cT air mass – summer rainfall and dry winter. | Aw/Cwa |

## B. Second Group: Mid-latitude Climates—Controlled by Both Tropical and Polar Air Masses

| Climate Type | Source region and limiting zone of air masses – climatic characteristics | Koppen's symbol |
|---|---|---|
| 6. Moist Sub-tropical climate – 20° to 35° N & S | On the eastern side of the continents mT air mass dominates – this air mass blows from the western side of the maritime high pressure axis – summer temperature and rainfall high. In winter influenced by cP air mass – many cyclones formed due to interaction of mT and cP air masses. | Cfa |
| 7. Maritime west littoral climate – 40° to 60° N & S | On the windward west littoral areas of mid-latitudes many cyclones enter – cool and moist mP air mass dominates. | Cfb/Cfc |
| 8. Mediterranean climate – 30° to 45° N & S. | With the shifting of the high pressure belts equator-ward in winter the region comes under the control of mT air mass – moist winter, dry summer. | Csa/Csb |
| 9. Mid-latitude Desert and Steppe climate – 35° to 50° N & S | In the interior of the continents orographic barriers restrict the entry of mT and mP air masses – in summer cT and in winter cP air masses dominate. | Bwk/Bwk´/Bsk /Bsk´ |
| 10. Moist continental climate – 35° to 60° N. | Located in the central and eastern parts of the continents in the mid-latitude region – convergence zone of polar and tropical air masses – seasonal variations and changing weather – mT air mass dominates in summer and cP air mass dominates in winter. | Dfa/Dfb/Dwa/ Dwb |

## C. Third Group: High Latitude Climates—Controlled by Polar and Arctic Air Masses

| Climate Type | Source region and limiting zone of air masses – climatic characteristics | Koppen's symbol |
|---|---|---|
| 11. Continental sub-polar climate – 50° to 70° N | Source region of cP air mass – winter long, severe cold and dry, summer short and cool, meagre precipitation caused by cyclones. | Dfc/Dfd/Dwa/ Dwd |
| 12. Maritime sub-polar climate – 50° to 60° N & 45° to 60° S | In winter mP air mass dominates in the windward coastal islands along the Arctic front – precipitation relatively higher and annual range of temperature low. | ET |

| Climate Type | Source region and limiting zone of air masses – climatic characteristics | Koppen's symbol |
|---|---|---|
| 13. Tundra climate – north of 55° N & south of 50° S | Along the Arctic coast mP and A (Arctic) air masses converge to form fronts and cyclones – moist climate, no summer, low temperature – marine influence observed on climate. | ET |
| 14. Permafrost masses | Source region of Arctic and Antarctic air | EF |

Later Highland or Mountain climates (H) were included in the Strahler's classification.

### *Bio-climatic Classification of Climates*

Since the beginning of studies on climate classification the natural vegetation has been considered as the primary index. In recent years the bio-scientists are increasingly inclined to correlate the climate with life-systems. Leslie Holridge first proposed a climate classification based on biotic factors in 1947, revised in 1967.

*Holridge's Life-zones System*

The *Holdridge life zones-system* is a global bioclimatic scheme for the classification of land areas. It was first published in 1947, and updated in 1967. It was a relatively simple system, derived empirically from the available data with an objective of cartographic presentation. A basic assumption of the system was that both soil and climax vegetation could be presented on a map once with the knowledge of climate.

It was first devised for tropical and subtropical area and then the system was applied globally. The system has been found effectively applicable for the tropical vegetation zones, Mediterranean zones, and Boreal zones, but less applicable to cold oceanic or cold arid climates, where moisture appears to be the determining factor. The system is now increasingly used in assessing the possible changes in natural vegetation patterns due to global warming.

The indicators of the system are recognised as follows:
• Mean annual bio-temperature (logarithmic)
• Annual precipitation (logarithmic)
• Ratio of annual Potential Evapotranspiration (PET) to mean total annual precipitation.

Bio-temperature is computed on the factors like growing season length and temperature. It is the mean of all temperatures referred above freezing, with all temperatures below freezing adjusted to 0° C, as plants are dormant at these temperatures. Holdridge's system does not primarily consider elevation. The system is considered more appropriate to the complexities of tropical vegetation than other systems.

In the Holridges classification scheme the following classes are identified and applied in bio-climatic studies.

1. Polar desert, 2. Sub-polar dry tundra, 3. Sub-polar moist tundra, 4. Sub-polar wet tundra, 5. Sub-polar rain tundra, 6. Boreal desert, 7. Boreal dry scrub, 8. Boreal moist forest, 9. Boreal wet forest, 10. Boreal rain forest, 11. Cool temperate desert, 12. Cool temperate desert scrub, 13. Cool temperate steppe, 14. Cool temperate moist forest, 15. Cool temperate wet forest, 16. Cool temperate rain forest, 17. Warm temperate desert, 18. Warm temperate desert scrub, 19. Warm temperate thorn scrub, 20. Warm temperate dry forest, 21. Warm temperate moist forest, 22. Warm temperate wet forest, 23. Warm temperate rain forest, 24. Subtropical desert, 25. Subtropical desert scrub, 26. Subtropical thorn woodland, 27. Subtropical dry forest, 28. Subtropical moist forest, 29. Subtropical wet forest, 30. Subtropical rain forest, 31. Tropical desert, 32. Tropical desert scrub, 33. Tropical thorn woodland, 34. Tropical very dry forest, 35. Tropical dry forest, 36. Tropical moist forest, 37. Tropical wet forest, 38. Tropical rain forest.

## Climatic Classification Based on Biomes

The study on biomes is considered of paramount importance for the ecosystem. So an approach has been made to classify the climate types and relate it to the existing biomes. Here three basic climate groups have been identified after Strahler's genetic classification based on the controlling air masses on which biomes are superimposed.

Three macro-climate groups show the climate types as controlled by the air masses. These are:
- Low latitude climates—controlled by the equatorial and tropical air masses.
- Mid-latitude climates—controlled both by the tropical and polar air masses.
- High latitude climates—controlled by the polar and arctic air masses.

### Group 1: Low Latitudes Climates—Controlled by the Equatorial and Tropical Air Masses

In this group the following biomes could be recognised and correlated with the prevailing climate types.
1. Rainforest Biome — Climate Type: Moist Tropical Climate (Af)
2. Savanna Biome — Climate Type: Wet and Dry Tropical Climate (Aw)
3. Desert Biome — Climate Type: Dry Tropical Climate (Bw).

### Group 2: Mid-latitude Climates—Controlled both by the Tropical and Polar Air Masses

In this group the following biomes could be recognised and correlated with the prevailing climate types.
1. Steppe Biome — Climate Type: Dry Mid-latitude Climate (Bs)
2. Chapparal Biome — Climate Type: Mediterranean Climate (Cs)

3.   Grassland Biome        — Climate Type: Dry Mid-latitude Climate (Bs)
4.   Deciduous Forest Biome — Climate Type: Moist Continental Climate (Cf).

*Group 3: High Latitude Climates—Controlled by the Polar and Arctic Air Masses*
1.   Taiga Biome   — Climate Type: Boreal Forest Climate (Dfc)
2.   Tundra Biome — Climate Type: Tundra Climate (ET)
3.   Alpine Biome — Climate Type: Highland Climate (H).
We can identify 10 biomes as related to different climate types of the world as noted above.

---

## Further Readings

Angstrom, A. (1925) 'The albedo of various surfaces of ground', *Geografisca Annaler,* **H 4**.

Barry, R.G. and R.J.Chorley (1968) *Atmosphere, weather and climate,* Methuen & Co., London.

Brooks, C.E.P. (1948) 'Classification of climates', *Meteorology Magazine,* **77**.

Critchfield, H.J. (1975) *General Climatology,* Prentice Hall India Ltd., New Delhi.

Garnier, B.J. (1946) 'Climates of New Zealand: According to Thornthwaite's Classification', *Annals of Association of American Geographers,* **36**, **3**.

Hare, F.K. (1951) 'Climatic classification', *The London Essays in Geography,* Ed. L.D. Stamp & S.W.Woolridge, Harvard University Press, Cambridge.

Oliver, J.E. and J.J.Hidore (2002) *Climatology,* Pearson-Education, Delhi.

Saha, P.K. and P.K. Bhattacharya (1994) *Adhunik Jalavayu Vidya,* West Bengal State Book Board, Government of West Bengal, Calcutta.

Strahler, A.N. (1969) *Physical Geography,* Wiley, New York.

Thornthwaite, C.W. (1933) 'The climates of the earth', *Geographical Review,* **23**.

Thornthwaite, C.W. (1941) *Atlas of climatic types in the United States 1900 – 1939,* U.S. Dept. Agri. Misc. Publication **421**.

Thornthwaite, C.W. (1948) 'An approach towards a rational classification of climate', *Geographical Review,* **38**.

Trewartha, G.T. (1968) *An introduction to climate,* McGraw Hill Kogakushu Ltd., Tokyo.

# APPENDICES

# Appendices

## I. Temperature Conversion: Degrees Celsius to Degrees Farenheit ( °C → °F)

In this book Unit of Degrees Celsius has been mostly used. In some countries the Unit of Degrees Farenheit is still to practice. So, for convenience of the readers the conversion of temperatures from Degrees Celsius to Degrees Farenheit has been incorporated. (°F = 9/5°C + 32)

| °C | .0 | .1 | .2 | .3 | .4 | .5 | .6 | .7 | .8 | .9 |
|----|-----|-----|-----|------|------|------|------|------|------|------|
| -24 | -11.2 | -11.4 | -11.6 | -11.7 | -11.9 | -12.1 | -12.3 | -12.5 | -12.6 | -12.8 |
| -23 | -9.4 | -9.6 | -9.8 | -9.9 | -10.1 | -10.3 | -10.5 | -10.7 | -10.8 | -11.0 |
| -22 | -7.6 | -7.8 | -8.0 | -8.1 | -8.3 | -8.5 | -8.7 | -8.9 | -9.0 | -9.2 |
| -21 | -5.8 | -6.0 | -6.2 | -6.3 | -6.5 | -6.7 | -6.9 | -7.1 | -7.2 | -7.4 |
| **-20** | **-4.0** | **-4.2** | **-4.4** | **-4.5** | **-4.7** | **-4.9** | **-5.1** | **-5.3** | **-5.4** | **-5.6** |
| -19 | -2.2 | -2.4 | -2.6 | -2.7 | -2.9 | -3.1 | -3.3 | -3.5 | -3.6 | -3.8 |
| -18 | -0.4 | -0.6 | -0.8 | -0.9 | -1.1 | -1.3 | -1.5 | -1.7 | -1.8 | -2.0 |
| -17 | 1.4 | 1.2 | 1.0 | 0.9 | 0.7 | 0.5 | 0.3 | 0.1 | 0.0 | -0.2 |
| -16 | 3.2 | 3.0 | 2.8 | 2.7 | 2.5 | 2.3 | 2.1 | 1.9 | 1.8 | 1.6 |
| **-15** | **5.0** | **4.8** | **4.6** | **4.5** | **4.3** | **4.1** | **3.9** | **3.7** | **3.6** | **3.4** |
| -14 | 6.8 | 6.6 | 6.4 | 6.3 | 6.1 | 5.9 | 5.7 | 5.5 | 5.4 | 5.2 |
| -13 | 8.6 | 8.4 | 8.2 | 8.1 | 7.9 | 7.7 | 7.5 | 7.3 | 7.2 | 7.0 |
| -12 | 10.4 | 10.2 | 10 | 9.9 | 9.7 | 9.5 | 9.3 | 9.1 | 9.0 | 8.8 |

| °C  | .0   | .1   | .2   | .3   | .4   | .5   | .6   | .7   | .8   | .9   |
|-----|------|------|------|------|------|------|------|------|------|------|
| -11 | 12.2 | 12.0 | 11.8 | 11.7 | 11.5 | 11.3 | 11.1 | 10.9 | 10.8 | 10.6 |
| **-10** | **14** | **13.8** | **13.6** | **13.5** | **13.3** | **13.1** | **12.9** | **12.7** | **12.6** | **12.4** |
| -9  | 15.8 | 15.6 | 15.4 | 15.3 | 15.1 | 14.9 | 14.7 | 14.5 | 14.4 | 14.2 |
| -8  | 17.6 | 17.4 | 17.2 | 17.1 | 16.9 | 16.7 | 16.5 | 16.3 | 16.2 | 16.0 |
| -7  | 19.4 | 19.2 | 19.0 | 18.9 | 18.7 | 18.5 | 18.3 | 18.1 | 18.0 | 17.8 |
| -6  | 21.2 | 21.0 | 20.8 | 20.7 | 20.5 | 20.3 | 20.1 | 19.9 | 19.8 | 19.6 |
| **-5** | **23.0** | **22.8** | **22.6** | **22.5** | **22.3** | **22.1** | **21.9** | **21.7** | **21.6** | **21.4** |
| -4  | 24.8 | 24.6 | 24.4 | 24.3 | 24.1 | 23.9 | 23.7 | 23.5 | 23.4 | 23.2 |
| -3  | 26.6 | 26.4 | 26.2 | 26.1 | 25.9 | 25.7 | 25.5 | 25.3 | 25.2 | 25.0 |
| -2  | 28.4 | 28.2 | 28.0 | 27.9 | 27.7 | 27.5 | 27.3 | 27.1 | 27.0 | 26.8 |
| -1  | 30.2 | 30.0 | 29.8 | 29.7 | 29.5 | 29.3 | 29.1 | 28.9 | 28.8 | 28.6 |
| **-0** | **32.0** | **31.8** | **31.6** | **31.5** | **31.3** | **31.1** | **30.9** | **30.7** | **30.6** | **30.4** |
| **+0** | **32.0** | **32.2** | **32.4** | **32.5** | **32.7** | **32.9** | **33.1** | **33.3** | **33.4** | **33.6** |
| 1   | 33.8 | 34.0 | 34.2 | 34.3 | 34.5 | 34.7 | 34.9 | 35.1 | 35.2 | 35.4 |
| 2   | 35.6 | 35.8 | 36.0 | 36.1 | 36.3 | 36.5 | 36.7 | 36.9 | 37.0 | 37.2 |
| 3   | 37.5 | 37.6 | 37.8 | 37.9 | 38.1 | 38.3 | 38.5 | 38.7 | 38.9 | 39.0 |
| 4   | 39.2 | 39.4 | 39.6 | 39.7 | 39.9 | 40.1 | 40.3 | 40.5 | 40.6 | 40.8 |
| **5** | **41.0** | **41.2** | **41.4** | **41.5** | **41.7** | **41.9** | **42.1** | **42.3** | **42.4** | **42.6** |
| 6   | 42.8 | 43.0 | 43.2 | 43.3 | 43.5 | 43.7 | 43.9 | 44.1 | 44.2 | 44.4 |

| °C | .0 | .1 | .2 | .3 | .4 | .5 | .6 | .7 | .8 | .9 |
|---|---|---|---|---|---|---|---|---|---|---|
| 7 | 44.6 | 44.8 | 45.0 | 45.1 | 45.3 | 45.4 | 45.7 | 45.9 | 46.0 | 46.2 |
| 8 | 46.4 | 46.6 | 46.8 | 46.9 | 47.1 | 47.3 | 47.5 | 47.7 | 47.8 | 48.0 |
| 9 | 48.2 | 48.4 | 48.6 | 48.7 | 48.9 | 49.1 | 49.3 | 49.5 | 49.6 | 49.8 |
| **10** | **50.0** | **50.2** | **50.4** | **50.5** | **50.7** | **50.9** | **51.1** | **51.3** | **51.4** | **51.6** |
| 11 | 51.8 | 52.0 | 52.2 | 52.3 | 52.5 | 52.7 | 52.9 | 53.1 | 53.2 | 53.4 |
| 12 | 53.6 | 53.8 | 54.0 | 54.1 | 54.3 | 54.5 | 54.7 | 54.9 | 55.0 | 55.2 |
| 13 | 55.4 | 55.4 | 55.8 | 55.9 | 56.1 | 56.3 | 56.5 | 56.7 | 56.8 | 57.0 |
| 14 | 57.2 | 57.4 | 57.6 | 57.7 | 57.9 | 58.1 | 58.3 | 58.5 | 58.6 | 58.8 |
| **15** | **59.0** | **59.2** | **59.4** | **59.5** | **59.7** | **59.9** | **60.1** | **60.3** | **60.4** | **60.6** |
| 16 | 60.8 | 61.0 | 61.2 | 61.3 | 61.5 | 61.7 | 61.9 | 62.1 | 62.2 | 62.4 |
| 17 | 62.6 | 62.8 | 63.0 | 63.1 | 63.3 | 63.5 | 63.7 | 63.8 | 64.0 | 64.2 |
| 18 | 64.4 | 64.6 | 64.8 | 64.9 | 65.1 | 65.3 | 65.5 | 65.7 | 65.8 | 66.0 |
| 19 | 66.2 | 66.4 | 66.6 | 66.7 | 66.9 | 67.1 | 67.3 | 67.5 | 67.6 | 67.8 |
| **20** | **68.0** | **68.2** | **68.4** | **68.5** | **68.7** | **68.9** | **69.1** | **69.3** | **69.4** | **69.6** |
| 21 | 69.8 | 70.0 | 70.2 | 70.3 | 70.5 | 70.7 | 70.9 | 71.1 | 71.2 | 71.4 |
| 22 | 71.6 | 71.8 | 72.0 | 72.1 | 72.3 | 72.5 | 72.7 | 72.9 | 73.0 | 73.2 |
| 23 | 73.4 | 73.6 | 73.8 | 73.9 | 74.1 | 74.3 | 74.5 | 74.7 | 74.8 | 75.0 |
| 24 | 75.2 | 75.4 | 75.6 | 75.7 | 75.9 | 76.1 | 76.3 | 76.5 | 76.6 | 76.8 |
| **25** | **77.0** | **77.2** | **77.4** | **77.5** | **77.7** | **77.9** | **78.3** | **78.3** | **78.4** | **78.6** |

| °C | .0 | .1 | .2 | .3 | .4 | .5 | .6 | .7 | .8 | .9 |
|---|---|---|---|---|---|---|---|---|---|---|
| 26 | 78.8 | 79.0 | 79.2 | 79.3 | 79.5 | 79.7 | 79.9 | 80.1 | 80.2 | 80.4 |
| 27 | 80.6 | 80.8 | 81.0 | 81.1 | 81.3 | 81.5 | 81.7 | 81.9 | 82.0 | 82.2 |
| 28 | 82.4 | 82.6 | 82.8 | 82.9 | 83.1 | 83.3 | 83.5 | 83.7 | 83.8 | 84.0 |
| 29 | 84.2 | 84.4 | 84.6 | 84.7 | 84.9 | 85.1 | 85.3 | 85.5 | 85.6 | 85.8 |
| **30** | **86.0** | **86.2** | **86.4** | **86.5** | **86.7** | **86.9** | **87.1** | **87.3** | **87.4** | **87.6** |
| 31 | 87.8 | 88.0 | 88.2 | 88.3 | 88.5 | 88.7 | 88.9 | 89.1 | 89.2 | 89.4 |
| 32 | 89.6 | 89.8 | 90.0 | 90.1 | 90.3 | 90.5 | 90.7 | 90.9 | 91.0 | 91.2 |
| 33 | 91.4 | 91.6 | 91.8 | 91.9 | 92.1 | 92.3 | 92.5 | 92.7 | 92.8 | 93.0 |
| 34 | 93.2 | 93.4 | 93.6 | 93.7 | 93.9 | 94.1 | 94.3 | 94.5 | 94.6 | 94.8 |
| **35** | **95.0** | **95.2** | **95.4** | **95.5** | **95.7** | **95.9** | **96.1** | **96.3** | **96.4** | **96.6** |
| 36 | 96.8 | 97.0 | 97.2 | 97.3 | 97.5 | 97.7 | 97.9 | 98.1 | 98.2 | 98.3 |
| 37 | 98.6 | 98.8 | 99.0 | 99.1 | 99.3 | 99.5 | 99.7 | 99.9 | 100.0 | 100.2 |
| 38 | 100.4 | 100.6 | 100.8 | 100.9 | 101.1 | 101.3 | 101.5 | 101.7 | 101.8 | 102.0 |
| 39 | 102.2 | 102.4 | 102.6 | 102.7 | 102.9 | 103.1 | 103.3 | 103.5 | 103.6 | 103.8 |
| **40** | **104.0** | **104.2** | **104.4** | **104.5** | **104.7** | **104.9** | **105.1** | **105.3** | **105.4** | **105.6** |
| 41 | 105.8 | 106.0 | 106.2 | 106.3 | 106.5 | 106.7 | 106.9 | 107.1 | 107.2 | 107.4 |
| 42 | 107.6 | 107.8 | 108.0 | 108.1 | 108.3 | 108.5 | 108.7 | 108.9 | 109.0 | 109.2 |
| 43 | 109.4 | 109.6 | 109.8 | 109.9 | 110.1 | 110.3 | 110.5 | 110.7 | 110.8 | 111.0 |
| 44 | 111.2 | 111.4 | 111.6 | 111.7 | 111.9 | 112.1 | 112.3 | 112.5 | 112.6 | 112.8 |

| °C | .0 | .1 | .2 | .3 | .4 | .5 | .6 | .7 | .8 | .9 |
|---|---|---|---|---|---|---|---|---|---|---|
| **45** | **113.0** | **113.2** | **113.4** | **113.5** | **113.7** | **113.9** | **114.1** | **114.3** | **114.4** | **114.6** |
| 46 | 114.8 | 115.0 | 115.2 | 115.3 | 115.5 | 115.7 | 115.9 | 116.1 | 116.2 | 116.4 |
| 47 | 116.6 | 116.8 | 117.0 | 117.1 | 117.3 | 117.5 | 117.7 | 117.9 | 118.0 | 118.2 |
| 48 | 118.4 | 118.6 | 118.8 | 118.9 | 119.1 | 119.3 | 119.5 | 119.7 | 119.8 | 120.0 |
| 49 | 120.2 | 120.4 | 120.6 | 120.7 | 120.9 | 121.1 | 121.3 | 121.5 | 121.6 | 121.8 |
| **50** | **122.0** | **122.2** | **122.4** | **122.5** | **122.7** | **122.9** | **123.1** | **123.3** | **123.4** | **12.6** |
| 51 | 123.8 | 124.0 | 124.2 | 124.3 | 124.5 | 124.7 | 124.9 | 125.1 | 125.2 | 125.4 |
| 52 | 125.6 | 125.8 | 126.0 | 126.1 | 126.3 | 126.5 | 126.7 | 126.9 | 127.0 | 127.2 |
| 53 | 127.4 | 127.6 | 127.8 | 127.9 | 128.1 | 128.3 | 128.5 | 128.7 | 128.8 | 129.0 |
| 54 | 129.2 | 129.4 | 129.6 | 129.7 | 129.9 | 130.1 | 130.3 | 130.5 | 130.6 | 130.8 |
| **55** | **131.0** | **131.2** | **131.4** | **131.5** | **131.7** | **131.9** | **132.1** | **132.3** | **132.4** | **132.6** |
| 56 | 132.8 | 133.0 | 133.2 | 133.3 | 133.5 | 133.7 | 133.9 | 134.1 | 134.2 | 134.4 |
| 57 | 134.6 | 134.8 | 135.0 | 135.1 | 135.3 | 135.5 | 135.7 | 135.9 | 136.0 | 136.2 |
| 58 | 136.4 | 136.6 | 136.8 | 136.9 | 137.1 | 137.3 | 137.5 | 137.7 | 137.8 | 138.0 |
| 59 | 138.2 | 138.4 | 138.6 | 138.7 | 138.9 | 139.1 | 139.3 | 139.5 | 139.6 | 139.8 |
| **60** | **140.0** | **140.2** | **140.4** | **140.5** | **140.7** | **140.9** | **141.1** | **141.3** | **141.4** | **141.6** |

## II. Relationship of Air Pressure in Millibars with the Height of Mercury Column in a Barometer

Air pressure in the atmosphere is expressed in meteorlogical terms in millibars. Actually air pressure is measured by the height of the mercury column in a barometer. The heights are measured either in millimeters or inches. These heights are then converted to millibars which indicate the force (air pressure). In the following table the relationship between the heights (millimeters) and millibars are shown.

### *Millimeters of Mercury to Millibars*

| mmHg | 0 | 1 | 2 | 3 | 4 | 5 | 6 | 7 | 8 | 9 |
|------|-----|-----|-----|-----|-----|-----|-----|-----|-----|-----|
| 600 | 799.9 | 801.3 | 802.6 | 803.9 | 805.3 | 806.6 | 807.9 | 809.3 | 810.6 | 811.9 |
| 610 | 813.3 | 814.6 | 815.9 | 817.3 | 818.6 | 819.9 | 821.3 | 822.6 | 823.9 | 825.3 |
| 620 | 826.6 | 827.9 | 829.3 | 830.6 | 831.9 | 833.3 | 834.6 | 835.9 | 837.3 | 838.6 |
| 630 | 839.9 | 841.3 | 842.6 | 843.9 | 845.3 | 846.6 | 847.9 | 849.3 | 850.6 | 851.9 |
| 640 | 853.3 | 854.6 | 855.9 | 857.3 | 858.6 | 859.9 | 861.3 | 862.6 | 863.9 | 865.3 |
| 650 | 866.6 | 867.9 | 869.3 | 870.6 | 871.9 | 873.3 | 874.6 | 875.9 | 877.3 | 878.6 |
| 660 | 879.9 | 881.3 | 882.6 | 883.9 | 885.3 | 886.6 | 887.9 | 889.3 | 890.6 | 891.9 |
| 670 | 893.3 | 894.6 | 895.9 | 897.3 | 898.6 | 899.9 | 901.3 | 902.3 | 903.9 | 905.3 |
| 680 | 906.6 | 907.9 | 909.3 | 910.6 | 911.9 | 913.3 | 914.6 | 915.9 | 917.3 | 918.6 |
| 690 | 919.9 | 921.3 | 922.6 | 923.9 | 925.3 | 926.6 | 927.9 | 929.3 | 930.6 | 931.9 |
| 700 | 933.3 | 934.6 | 935.9 | 937.3 | 938.6 | 939.9 | 941.3 | 942.6 | 943.9 | 945.3 |
| 710 | 946.6 | 947.9 | 949.3 | 950.6 | 951.9 | 953.3 | 954.6 | 955.9 | 957.3 | 958.6 |
| 720 | 959.9 | 961.3 | 962.6 | 963.9 | 965.3 | 966.6 | 967.9 | 969.3 | 970.6 | 971.9 |

| mmHg | 0 | 1 | 2 | 3 | 4 | 5 | 6 | 7 | 8 | 9 |
|---|---|---|---|---|---|---|---|---|---|---|
| 730 | 973.3 | 974.6 | 975.9 | 977.3 | 978.6 | 979.9 | 981.3 | 982.6 | 983.9 | 985.3 |
| 740 | 986.6 | 987.9 | 989.3 | 990.6 | 991.9 | 993.3 | 994.6 | 995.9 | 997.3 | 998.6 |
| 750 | 999.9 | 1001.3 | 1002.6 | 1003.9 | 1005.3 | 1006.6 | 1007.9 | 1009.3 | 1010.6 | 1011.9 |
| 760 | 1013.3 | 1014.6 | 1015.9 | 1017.2 | 1018.6 | 1019.9 | 1021.2 | 1022.6 | 1023.9 | 1025.2 |
| 770 | 1026.6 | 1027.9 | 1029.2 | 1030.6 | 1031.9 | 1033.2 | 1034.6 | 1035.9 | 1037.2 | 1038.6 |
| 780 | 1039.9 | 1041.2 | 1042.6 | 1043.9 | 1045.2 | 1046.6 | 1047.9 | 1049.2 | 1050.6 | 1051.9 |
| 790 | 1053.2 | 1054.6 | 1055.9 | 1057.2 | 1058.6 | 1059.9 | 1061.2 | 1062.6 | 1063.9 | 1065.2 |
| 800 | 1066.6 | 1067.9 | 1069.2 | 1070.6 | 1071.9 | 1073.2 | 1074.6 | 1075.9 | 1077.2 | 1078.6 |
| 810 | 1079.9 | 1081.2 | 1082.6 | 1083.9 | 1085.2 | 1086.6 | 1087.9 | 1089.2 | 1090.6 | 1091.9 |
| 820 | 1093.2 | 1094.6 | 1095.9 | 1097.2 | 1098.6 | 1099.9 | 1101.2 | 1102.6 | 1103.9 | 1105.2 |
| 830 | 1106.6 | 1107.9 | 1109.2 | 1110.6 | 1111.9 | 1113.2 | 1114.6 | 1115.9 | 1117.2 | 1118.6 |
| 840 | 1119.9 | 1121.2 | 1122.6 | 1123.9 | 1125.2 | 1126.6 | 1127.9 | 1129.2 | 1130.61 | 1131.9 |

# Glossary

*Abyssal zone* – The deep ocean is known as the *abyssal zone*. The water in this region is very cold (around 3°C), heavy and rich in oxygen content, but low in nutritional status. The abyssal zone supports many species of invertebrates and fishes.

*Absolute humidity* – *Absolute humidity* is the weight of water vapour per unit volume of air, usually measured in units of grams of water vapour per cubic metre of air.

*Adiabatic expansion / adiabatic cooling* – As the air moves upward the pressure becomes lower with ascending heights. So there is loss of weight of the atmosphere upon it and the parcel of air expands in volume. The rising air is thus subject to expansion by which some energy is expended through work. This is known as *adiabatic expansion*. Due to expense of energy the temperature drops and air becomes cooler. This is known as *adiabatic cooling*.

*Adiabatic heating* – The increase in molecular movement causes an increase in the temperature of the air parcel. This process is known as *adiabatic heating*.

*Adiabatic temperature change* – When temperature of air changes without the addition or removal of energy, it is called the *adiabatic temperature change*. So there is no exchange of heat with the surrounding environment to cause the cooling or heating of the air.

*Adiabatic expansion* – When a parcel of air rises upwards it expands in volume due to drop of pressure with increasing height. This is called the *adiabatic expansion* (without addition of heat).

*Adiabatic cooling* – When the air expands, the molecules spread out and ultimately collide less with one another. Due to expenses of energy for expansion, the temperature of the air drops instantly. This is known as *adiabatic cooling*.

*Air mass* – Air spreads horizontally hundreds of kilometres and maintains uniform physical characteristics, particularly temperature, moisture and lapse rate. Such vast expanse of air over a place, whether land and seas is called the air mass.

*Albedo* – Albedo is known as the surface reflectivity of solar radiation. The term has its origins from a Latin word *albus*, meaning 'white'.

*Alize* – The *alizé* is the most well known trade winds which blows across Central Africa and the Caribbean as a steady, mild northeasterly wind.

*Alto clouds* – *Alto* means high. Hence Alto clouds are high clouds.

*Ammonification* – Ammonification is the process in which nitrogen in organic matter of dead plants and animals is converted to ammonia and amino acids.

*Applied classification systems* – *Applied classification systems* are those approached for a purpose, or as an outgrowth of a particular climate-associated problem.

*Aurora Borealis and Aurora Australis* – Above 100 km. ionisation process starts due to absorption of X-Rays and UV rays. Oxygen atoms and nitrogen molecules are converted to electrons by this process. Sometimes these electrons penetrate lower to an altitude from 300 km. to 80 km. As a result *Aurora Borealis and Aurora Australis* could be observed over high latitudes from the earth's surface. These are unique phenomena observed in this layer.

*Baroclinic air mass* – In a region where the isobars and isotherms cross each other baroclinic conditions prevail. The air mass lying over this region is called the *baroclinic air mass*. Baroclinic conditions denote unstable conditions where the mixing of two barotropic air masses takes place.

*Barotropic air mass* – In a region where the isobars and isotherms lie parallel the air mass over that region is called the *barotropic air mass*. Barotropic air masses are stable and extend over wide areas. They possess uniform characteristics.

*Beaufort scale* – Beaufort scale was devised by British Rear-Admiral Sir Francis Beaufort in 1805 to indicate the wind speed by natural observations. No instrument is required. The scale ranges from 0 to 12.

*Biome* – A biome is a climatically and geographically defined area of ecologically similar climatic conditions such as communities of plants, animals, and soil organisms. Biomes are often referred to as ecosystems and defined based on factors such as plant structures (such as trees, shrubs, and grasses), leaf types (such as broadleaf and needle-leaf), plant spacing (forest, woodland, savanna), and climate.

*Bond events* – The climate cycle during the Holocene period is known as Bond events. The North Atlantic climate fluctuations occurred every $H \approx 1470$ years throughout the Holocene period as the *Bond events*.

*Cirrus clouds* – *Cirrus* clouds are very high clouds and are formed of ice crystals. They are thin, feathery clouds observed on a fair weather day.

*(Climate) anomaly* – A climate *anomaly* is an event in which the magnitude of the deviation from normal conditions is unusually large, occurring infrequently in the historical record. The unusual conditions are regional in extent, involving multi-month periods of extremely high or low temperatures and/or wet or dry conditions.

*Climatic region* – A region having uniform characteristics of weather and climate sustainable for a long period is termed a *climatic region.*

*Coefficient of temperature efficiency* – Following Livingstone, Thornthwaite devised such a *Thermal Index* which can relate the growth of vegetation. He called this thermal index as the *Coefficient of temperature efficiency.*

*Cold front* – A *cold front* is defined as the advancing edge of a cooler and drier mass of air. Cold air mass pushes the warm air mass to move further up. A front formed along the cold air mass has a shape like nose penetrating the warm air mass. This is known as *cold front.*

*Condensation* – *Condensation* is the phase of change of water vapour into a liquid.

*Condensation level* – The bottom of the cloud is the place where the air temperature is the same as the dew-point temperature. This point is referred as the *condensation level.*

*Contact cooling, diabatic cooling* – When air comes in contact with a cooler surface and conduction takes away heat out of the air, it is known as *Contact cooling*, cooling by contact. It is also called *diabatic cooling.*

*Convective uplift* – A parcel of air is uplifted when it initially gains heat from the surface. This process of uplift is known as *convective uplift.*

*Convergent uplift* – *Convergent uplift* occurs when air enters a centre of low pressure. As air converges into the centre of a cyclone it is forced to rise off the surface.

*Coriolis force* – The Coriolis force is an apparent force that is caused due to the rotation of the earth.

*Cryogenian Ice Age* – The earliest well-documented ice age, and probably the most severe of the last 1 billion years, occurred from 850 to 630 million years ago during the Cryogenian period. During this period permanent ice covered the entire globe and a Snowball Earth was produced. This period was terminated very fast as water vapour returned to the earth's atmosphere.

*Cumulus clouds* – *Cumulus* clouds are the puffy, cotton-like clouds formed by upward rising of air.

*Cyclogenesis* – *Cyclogenesis* is the development or strengthening of cyclonic circulation in the atmosphere (a low pressure area). Cyclogenesis is a general term for several different processes, all of which result in the development of some sort of cyclone.

*Denitrification* – Denitrification is the process in which ammonia and oxides of nitrogen are reverted back to nitrogen by different forms of bacteria, namely *bacillium cereus, B. licheniformis* and *P. denitrificans*. The gaseous nitrogen is released to the atmosphere and the cycle continues.

*Diabatic temperature change* – Temperature change caused by an exchange of heat between two bodies is called *diabatic temperature change.*

*Dissolved Oxygen* – *Dissolved Oxygen* indicates the amount of oxygen dissolved in the water and states the environmental quality favoured for the aquatic animals.

*Depletion of Ozone* – Worldwide monitoring has indicated that stratospheric ozone layer declined for at least two decades, with losses of about 10 percent in winter and 5 percent in summer and autumn in various locations of the earth's surface. The problem is becoming more and more critical each year. From the satellite imageries taken recently the depletion of ozone over the poles has been confirmed. The major culprit for the depletion of ozone layer happens to be chloroflurocarbon (CFC11 and CFC12). F. Sherwood Rowland and Mario Molina who shared the Nobel Prize in chemistry in 1995 for their discovery of the role of CFCs in the destruction of ozone suggested that most of the chlorine atoms would combine with ozone and initiate a chain reaction. As a result of chain reaction, a single chlorine atom can remove as many as 100,000 molecules of ozone.

*Dew point temperature* – *Dew point temperature* is the temperature at which condensation takes place.

*Doldrums* – Doldrums happen to be a calm zone, where no surface wind is cognisable.

*Dry Adiabatic Lapse Rate* – The cooling rate for the unsaturated air is called the *Dry Adiabatic Lapse Rate (DALR).*

*Dry line* – The dry line is the boundary between air masses with significant moisture differences. The dry line may occur in regions between the arid lands and warm seas.

*Effective Solar Income* – Budyko estimated the solar income of the earth per year at 250 kilo calories per square centimetre. Budyko accepted that an amount of 14 percent for the earth's surface and 26 percent for cloud cover as albedo (nearly 40 percent of the income) is lost and the earth's *effective solar income* appears to be 150 kilo calories / square centimetre / year.

*Effective solar radiation* – Only 66 percent of solar energy is available for heating the earth-atmosphere system. This 66 percent is called the *effective solar radiation.*

*Ekman spiral* – The *Ekman spiral* is the vector form, how wind vector velocities change with altitude, flowing along the isobars above the boundary layer, then decreasing in speed and veering as height decreases and friction effects increase, until the surface vector lies across the isobars at low speed.

*Electromagnetic Wave* – Any body above the absolute zero can emit radiant energy. A hot body like sun emits heat in the form of radiant energy that can be absorbed by a colder body, such as the earth, thus raising its temperature. Radiant energy has a wave like form. Maxwell defined this as electromagnetic waves.

*El Nino* – In Spanish El Nino means 'child' or specifically 'child Christ', as the local people call the phenomenon. El Nino is actually identified with the warm episodes over the tropical eastern Pacific every two to seven years, when much than normal warm condition prevails.

*Empirical systems of classification* – *Empirical systems* of classification are those which are based on observable features.

*Environmental Lapse Rate (ELR)* – The rate of change in temperature with altitude is called the *environmental lapse rate of temperature (ELR)*. ELR varies from day-to-day at a place, and from place to place on any given day. Normally a decrease of 6.5°C temperature is recorded with a rise of 1000 metre altitude, which is the average value of ELR.

*Evaporation* – Free water molecules in air are vapourised and transformed into water vapour. This process is known as *evaporation*.

*Evapotranspiration* – *Evapotranspiration* (ET) is defined as the sum of evaporation and plant transpiration from the earth's land surface to air.

*Exobase and Critical level* – The lowest altitude of the exosphere is known as *Exobase*. It is also called the *Critical level* as the height above which there are negligible atomic collisions between the particles and the constituent atoms are on purely ballistic trajectories. It is also the designated layer where the space shuttle orbits.

*Eye (of tropical cyclone)* – A strong tropical cyclone has the anchorage in an area of sinking air at the centre of circulation. If this area is strong enough, it can develop into an eye. Weather conditions in the eye remain normally calm and free of clouds, although the sea conditions may be extremely rough.

*Faint young sun paradox* – It is supposed that, early in earth's history, the sun was too cold to support liquid water at the Earth's surface, leading to what is known as the *Faint young sun paradox*.

*Ferrel cell* – Ferrel conceived the idea that in the circulation model there exists a secondary circulation feature, dependent for its existence upon the Hadley cell and the Polar cell. It behaves much as an atmospheric ball bearing between the Hadley cell and the Polar cell, and appears to be the eddy circulations (the high and low pressure areas) of the mid-latitudes.

*Front* – The boundary of separation between two contrasting air masses is called a *front*. Fronts are often characterised by unstable and inclement weather. Fronts are usually associated with areas of low atmospheric pressure called frontal systems.

*Frontal uplift* – *Frontal uplift* occurs when greatly contrasting air masses meet along a weather front. For instance, when warm air collides with cool air along a warm front, the warm air is forced to rise up and over the cool air.

*Fujiwhara effect* – The two vortices will be attracted to each other, and eventually spiral into the centre point and merge. If the two vortices are of unequal size, the larger vortex will tend to dominate, and the smaller vortex will orbit around it. Dr. Sakuhei Fujiwhara observed this phenomenon, so it is called the *Fujiwhara effect*.

*Genetic classification systems* – *Genetic classification systems* are those which are based on the cause of the climate.

*Geostrophic wind* – If the *horizontal pressure gradient force* is exactly balanced in magnitude by Coriolis effect, accelerations of the air will be relatively small and wind will flow horizontally at a constant speed proportional to the isobaric spacing gradient, perpendicular to the two opposing forces and parallel to straight isobars. This is known as *geostrophic wind (Gk. geo = earth, strophe = turning)*.

*Glacial cycle* – The glacial period in the pliestocene age was not continuous. There had been breaks of intervening warm periods resulting in the retreat of glaciers on a large scale. Again with the advent of cold conditions glaciers reemerged. A cycle could be evidenced in the glacial period, cold – warm – cold. This is called the glacial cycle.

*Glacial periods / Inter-glacial periods* – Ice age conditions were not uniform. Within the ice ages (or at least within the last one), more temperate and more severe conditions occurred. The colder periods are called *glacial periods* and the warmer are *interglacial periods*.

*Global cooling* – *Global cooling* in general refers to an overall cooling of the earth's surface. It refers primarily to a conjecture during the 1970s of imminent cooling of the earth's surface and atmosphere along with a posited commencement of glaciations.

*Global warming* – *Global warming* means the rise in the average temperature of the near-surface air and oceans since the mid-twentieth century and its projected continuation.

The average global air temperature close to the earth's surface increased $0.74 \pm 0.18°C$ ($1.33 \pm 0.32°F$) since 1905.

*Gradient wind* – The *gradient wind* is the equilibrium wind for the three forces – centripetal acceleration, pressure gradient and Coriolis force (or geostrophic).

*Green House Gases (GHG)* – Several gases in the atmosphere absorb infrared radiation causing warming of the earth's atmosphere. These gases, which absorb infrared radiation, are known as Green House Gases. These gases are Carbon Dioxide, Methane, Chloro-Fluro Carbon and Nitrous Oxide.

*Ground fog* – Moist air enveloping the land becomes cooler due to contact cooling. The condensation process is initiated and tiny water droplets form. As this type of fog lies near the ground surface it is also called the *ground fog.*

*Hadley cell* – Tropical convection cell where the winds converge is named after Hadley and known as Hadley cell.

*Holocene Climate Optimum* – The *Holocene Climate Optimum* was a warm period that occurred roughly during the interval 9,000 to 5,000 years ago.

*Holridge Life-zones system* – The *Holdridge Life-zones system* is a global bioclimatic scheme for the classification of land areas.

*Homosphere and Heterosphere* – It has been found that the composition of different gases in the lower reach of the atmosphere is nearly constant. Hence the zone (up to 100 km.) of this constant composition of constituent gases is termed *Homosphere.* Beyond this limit the composition of gases differs, and is called the *Heterosphere.*

*Humidity* – *Humidity* is a measure of the water vapour content of the air.

*Huronian Ice Age* – The earliest ice age, called the Huronian, probably prevailed around 2.7 to 2.3 billion years ago during the early Proterozoic Eon.

*Hydrologic cycle* – The *hydrologic cycle* is a conceptual model that states the storage and movement of water between the *biosphere, atmosphere, lithosphere,* and the *hydrosphere.* Water on this planet can be stored in any one of the following reservoirs: *atmosphere, oceans, lakes, rivers, soils, glaciers, snowfields,* and *groundwater.*

*Hygroscopic nuclei* – In the free atmosphere condensation is initiated around complex particles of dust, smoke, sulphur dioxide, salts (NaCl) or similar microscopic substances which possess the property of *wettability.* These are called the *Hygroscopic Nuclei.*

*Insolation* – Incoming solar energy that reaches the earth's surface is known as insolation.

*Inter-Tidal zone* – The *inter-tidal zone* is the meeting point of land with ocean. It is alternately submerged and exposed, as waves and tides come in and out.

*Inter-Tropical Convergence Zone (ITCZ)* – As the trade winds converge over the equatorial low, vertical movement predominates than the horizontal movement. The area of convergence is known as Inter Tropical Convergence Zone (ITCZ).

*Ionosphere* – Active process of ionisation is evidenced in the layer above 100 km. This layer is known as *Ionosphere*. In this layer electrons are heavily concentrated at altitude 100 – 300 km.

*IPCC (Inter-Governmental Panel of Climate Change)* – *IPCC* is a body to study different aspects of global warming, constituted with the participatory states of the world in terms of the United Nations Charter in 1992.

*Jet streams* – It was observed that strong wind with wind speed of 100-300 knots prevails above the subtropical highs at an altitude of 9-12 km. blowing both equator-ward and pole-ward. This strong wind in the upper air along a narrow corridor is known as *Jet Streams*.

*Kalt* – In German language *Kalt* means cold.

*Klima* – *Klima* means the slope of the earth. The Greek scholars considered the variation of regional climate due to variation of slope of the earth's surface. It resembles the present understanding of *latitudes*.

*La Nina* – *La Nina* in Spanish means young girl. It represents cold surface temperature of the water in the eastern Pacific Ocean.

*Law of Conservation of Angular Momentum* – An important physical law states that the angular momentum of an object does not change unless acted upon by a *torque* (force times perpendicular distance). This is known as the *Law of Conservation of Angular Momentum*.

*Law of Inertia* – Newton's first law of motion is known as Law of Inertia. It is stated as – 'Every body perseveres in its state of being at rest or of moving uniformly straight forward, except insofar as it is compelled to change its state by force impressed'.

*Lightning* – *Lightning* is an electrical discharge always associated with a thunderstorm. It can be seen in the form of a bright streak (or bolt) from the sky. Lightning occurs when an electrical charge is built up within a cloud.

*Limnetic zone* – The upper layers of water bodies where the sunlight can penetrate are known as *'limnetic zone'*. In this zone active photosynthesis and growth occur resulting in abundance of oxygen and rapid consumption of nutrients. So the limnetic zone is usually rich in aquatic species.

*Little Ice Age* – Even in the recent past a few centuries back in the early 15th century (1400 A.D.) to mid-19th century (1850 A.D.) Europe experienced a very cold climate. This period is known as *Little Ice Age*.

*Littoral zone* – The topmost zone near the shore of a lake or pond is the *littoral zone*. This zone happens to be the warmest since it is shallow and can absorb more of the insolation.

*Medieval Climate Optimum* – *Medieval Climate Optimum* was the warmer period before the advent of *Little Ice Age* when the cooling started.

*Milky Way* – The galaxy we live in is known as the *Milky Way*. It has a disc like shape and the earth is located halfway out from the centre. As we look toward or away from the centre or in any direction in the plane of that disc we find many stars and much dust and gas.

*Mixing ratio* – *Mixing ratio* is the weight of water vapour per unit weight of dry air.

*Neutron star* – Stars having 1.4-4 times mass of the sun collapse until they reduce to a quite small size. Most of the stars at that point contain gas of neutrons only, or of smaller particles called *Quarks* that make up neutron. The star is then known as *Neutron Star*.

*Nimbus clouds* – *Nimbus* indicates rain. So *Nimbus clouds* are rain clouds.

*Nitrification* – Nitrification is a process in which ammonia is first converted into nitrites by a group of microorganisms like *nitrosomonas, nitrospira, nitrosogloea* and *nitrococcus*. Nitrites are then converted to nitrates mostly by *nitrobacter* and *nitrocystis*.

*Nitrogen fixing bacteria* – A few bacteria like *Azotobacter, Clostridium, Derxia, Rhizobium* etc., can utilise nitrogen directly from air. These are called the nitrogen fixing bacteria.

*Noctilucent clouds* – In the mesosphere as the temperature drops with height and due to presence of water vapour, though negligible, the clouds are formed in this sphere. In high latitudes, particularly in summer, cloud formations are evidenced and known as *Noctilucent clouds*. Dust particles from the comets and upper convective current help the formation of noctilucent clouds. These noctilucent clouds appear white or pearly in colour and can have a wavy, web-like structure.

*Normal lapse rate, Environmental Lapse Rate (ELR)* – The decrease in temperature with elevation is called the *normal lapse rate of temperature, also known as environmental lapse rate (ELR)*. The *normal lapse rate of temperature* is the average lapse rate of temperature of 6.5° C / 1000 metres.

*North Atlantic Oscillation* – The *North Atlantic Oscillation* (NAO) is a climatic phenomenon in the North Atlantic Ocean. The fluctuations in the difference of sea-level pressure are observed between the Icelandic Low and the Azores high. The seasonal shifting of the Icelandic Low and the Azores high, results in modifying the strength and direction of westerly winds and storm tracks across the North Atlantic. It is considered to be a part of the Arctic oscillation.

*Occluded front* – An *occluded front* is formed during the process of cyclogenesis when a cold front overtakes a warm front. When this occurs, the warm air is separated (occluded) from the centre of low of the mid-latitude cyclone.

*Orographic uplift* – *Orographic uplift* is the forced ascent of air when it collides with a mountain. As air strikes the windward side, it is uplifted and cooled.

*Ozone hole* – Some gaps are found in the tropopause, through which there is exchange of matter between the troposphere and stratosphere. Water vapour can penetrate to the stratosphere and the ozone gas from the stratosphere subsides into the troposphere. The gap is called the *ozone hole*.

*Pacific Decadal Oscillation* – *Pacific Decadal Oscillation*, or PDO, is often described as a long-lived El Niño-like pattern of Pacific climate variability.

*Pelagic zone* – The *pelagic zone* is basically the open ocean. In the pelagic zone constant mixing of warm and cold ocean currents occur.

*Permo-carboniferous Ice Age* – The *Permo-Carboniferous* period includes the latter parts of the Carboniferous and early part of the Permian period. During the Permo-Carboniferous period, about 300 million years ago, great glaciations occurred on the earth's surface.

*Photosphere* – The solar surface is called the *Photosphere*. The term has been derived from the Greek word *photo,* meaning light.

*Pliestocene Ice Age* – The present ice age, known as *pliestocene or recent ice age,* started 40 million years ago with the growth of an ice sheet in Antarctica. It intensified during the late Pliocene, about 3 million years ago.

*Polar cell* – The area of heat sink is known as *polar cell*. Sinking air in the polar areas (above 60° latitude) settles and results in the formation of high pressure over the poles.

*Potential evapotranspiration* – *Potential evapotranspiration* (PE), was defined by Thornthwaite as the evapotranspiration that would occur from a vegetation-covered surface if the soil moisture conditions were adequate for unrestricted transpiration.

*Pressure Gradient Force* – A pressure gradient is the difference in pressure divided by the distance measured in the direction from high to low pressure. Therefore, the pressure gradient is responsible for exerting force acting on the air in the direction from high to low pressure. This is sometimes called the *pressure gradient force*.

*Profundal and benthic zones* – Next to limnetic zone are the *'profundal zone'* and *'benthic zone'*. These zones are dark, where decomposition, mineralisation and nutrient accumulation occur. In these dark zones dissolved oxygen is deficient and anaerobic microbial activity predominates. Salinity of sea water appears to be high under arid condition.

*Quasi-Bienneial Oscillation* – A stratospheric circulation is evidenced over the tropical latitudes, and is driven by gravity waves. This is known as Quasi-Biennial Oscillation (QBO).The gravity waves appear to be convectively generated in the troposphere. The QBO induces a secondary circulation which is important for the global stratospheric transport of tracers such as ozone or water vapour.

*Radiation Budget* – *Radiation Budget* is the balance between incoming energy from the Sun and outgoing thermal (longwave) and reflected (shortwave) energy from the earth.

*Radiational fog* – Radiational fog occurs due to cooling of the land surface in the night hours through radiation of the earth's surface.

*Red Giants* – When the stars spend all the hydrogen in their interiors, they can no longer fuse hydrogen into helium. As a result they cannot offset gravity and their interiors begin to contract. During the contraction energy is released and outer layers are pushed out. These outer layers become larger and cooler, and the stars become *Red Giants.*

*Relative humidity* – *Relative humidity* is defined as the ratio of the amount of water vapour in the air to its saturation point.

*Residence Time* – *Residence Time* of a reservoir within the hydrologic cycle is the average time a water molecule will be spent in that reservoir. It is a measure of the average age of the water in that reservoir, though some water will be spent much less time than average, and some much more.

*Rime* – Water droplets in the fog may be deposited on the edge of a sharp body and form frosts. These are known as *Rime.*

*Roaring Forties* – The strongest westerly winds which blow in the middle latitudes of southern hemisphere between 40 and 50 degrees latitude are known as the *Roaring Forties.*

*Rodent-borne diseases* – Rodents cause a number of diseases whether as intermediate infected hosts or as hosts for arthropod vectors such as ticks. Certain rodent-borne diseases like leptospirosis, tularaemia and viral haemorrhagic diseases are caused by flooding. So above-average precipitation may lead to occurrence of these diseases. Plague, Lyme disease, tick borne encephalitis (TBE) and hanta-virus pulmonary syndrome (HPS) are other diseases associated with rodents and ticks.

*Rossby waves – Rossby* (or *planetary) waves* are large meanders in high-altitude winds that have a major influence on weather condition. They are caused due to the variation in *Coriolis Effect* related to latitude.

*Saturated Adiabatic Lapse Rate –* The cooling rate in the saturated air is known as *Saturated Adiabatic Lapse Rate (SALR).*

*Saturation Vapour Pressure – Saturation Vapour Pressure* is simply the pressure that water vapour exerts when the air is fully saturated.

*Shear line –* A stationary front becomes a *shear line* when the density contrast across the frontal boundary vanishes. Usually it results due to equalisation of temperature. A narrow zone of wind shift persists for a time.

*Specific humidity – Specific humidity* is a measure of the weight of water vapour in the air per unit weight of air, which includes the weight of water vapour. The unit of measurement is grams of water vapour per kilogram of air.

*Stadium effect –* An inward curving of the eye-wall's top, resembling a football stadium is observed in the intense and mature tropical cyclones. This phenomenon is thus sometimes referred to as the *Stadium effect.*

*Stationary front – Stationary front* is a boundary between two different air masses, neither of which is strong enough to replace the other. So they remain almost in the same area for extended periods of time, and waves sometimes propagate along the frontal boundary.

*Stratus clouds – Stratus* clouds are the layered, sheet-like clouds. They are found at lower altitudes.

*Sub-geostrophic wind –* As the magnitude of the Coriolis force is directly dependent on wind speed it follows that the wind speed around a low is less than would be expected from the pressure gradient force and the gradient wind is called the *sub-geostrophic* wind.

*Sunspots and Sunspots Cycle –* The surface of the sun has, in general, uniform brightness, being slightly darker along its edges. Even there are few darker areas on its surface. These dark areas are relatively cooler regions of the solar surface and are called *Sunspots.* The Sunspots cycle significantly influences the global climate.

*Super-geostrophic wind –* For the three forces to be in equilibrium the Coriolis force must exceed the pressure gradient force and consequently the gradient wind speed must be greater than would be expected from the pressure gradient force and is thus called the *super-geostrophic* wind.

*Supernova* – After a star becomes a Red Giant, it grows larger and turns to be super-giant. At that point it may explode totally, giving birth to a *Supernova*. A Supernova is the brightest object in the galaxy. The Supernova completely destroys the star, forming heavy elements and then ejecting these elements out into space.

*Sweeping theory* – The larger droplets lead to coalescence of the smaller droplets. This is caused just like sweeping and is known as *sweeping theory* of rainmaking by Langmuir.

*Terrestrial Heat Balance* – It is assumed that the 66 percent of solar radiation gained is exactly balanced by an equal amount of energy radiated back to space as long wave terrestrial radiation. This is known as *Terrestrial Heat Balance*.

*Tethys* – The shallow seas extensive over the earth's surface during the cretaceous period were known as *Tethys*.

*Thunderstorm* – *Thunderstorm* is characterised by lightning and thunder and results due to intense convectional instability. Fundamentally the thunderstorm is a thermo-dynamic machine in which the potential energy of latent heat of condensation and fusion in moist conditionally and convectively unstable air is rapidly converted into kinetic energy of violent rising air associated with torrential rain, hail, gusty surface squall winds, lightning, thunder etc.

*Thunderstorm - Cumulus stage* – The first stage of a thunderstorm is the cumulus stage, or developing stage. In this stage, masses of moisture are lifted upwards into the atmosphere.

*Thunderstorm - Mature stage* – In the mature stage of a thunderstorm, the warm air continues to rise until it reaches existing air which is warmer than it. The air then can not rise further.

*Thunderstorm - Dissipating stage* – In the dissipating stage, the thunderstorm is dominated by the downdrafts only. If there be no super cellular development, this stage occurs rather quickly, some 20-30 minutes within the life of a thunderstorm.

*Thunderstorm - Single cell* - This refers to a single thunderstorm with one main updraft.

*Thunderstorm - Multi-cell cluster* – Multi cell storms form as clusters of storms but may then evolve into an organised line or lines of storms.

*Thunderstorm - Multi-cell line / Squall lines* – Multi-cell line storms are commonly known as *squall lines*. These occur when multi cellular storms form in a line rather than clusters.

*Thunderstorm - Super-cell* – Super-cell storms are large, severe quasi-steady-state storms with characteristic wind speed and direction varying with height ("wind shear"), separate downdrafts and updrafts (i.e., precipitation is not falling through the updraft) and a strong, rotating updraft (a meso-cyclone).

*Tornado* – The term tornado has been derived from the Spanish word *tronada,* which means thunderstorm. Actually the tornado is associated with the development of an intense thunderstorm.

*Total mass of the atmosphere* – The total mass of the atmosphere is approximately $5.2 \times 10^{15}$ tons and nearly 50 percent of the total mass of the atmosphere (air) lie within 5500 metres (5.5 km.) and about 90 percent lie within 30 kilometres from the surface of earth.

*Tropical cyclone* – A *tropical cyclone* is a storm characterised by a centre of low pressure and numerous thunderstorms that generate strong winds and heavy rainfall causing widespread floods. A tropical cyclone feeds on the heat released when moist air rises and the water vapour condenses to release the latent heat.

*Tropo and Troposphere* – *Tropo* means the mixing or turning. So troposphere is a layer where there is mixing or turning of air.

*Tropopause* – The isomorphic temperature condition prevails at the end of troposphere. This isomorphic layer determines the limit of troposphere and is known as *Tropopause.* So tropopause is a transition zone between the troposphere and the stratosphere.

*Vapour partial pressure* – The balance between condensation and evaporation gives the quantity called *Vapour partial pressure.* The maximum partial pressure (*saturation pressure*) of water vapour in air varies with temperature of the air and water vapour mixture.

*Vapour pressure* – *Vapour pressure* is the partial pressure exerted by water vapour.

*Vector-borne diseases* – A vector-borne disease is caused when the pathogenic microorganism is transmitted from an infected individual to another individual by an arthropod or other agent, sometimes with other animals serving as intermediary hosts.

*Veering and backing* – The terms *veering* and *backing* originally are referred to the shift of surface wind direction with time. Meteorologists now use the term when referring to the shift in wind direction with height. Winds shifting anti-clockwise around the compass are *'backing'*, those shifting clockwise are *'veering'*.

*VIVGYOR* – Human eye can detect the emission of radiant energy in the wavelengths ranging between 0.4 to 0.7 microns. This is known as *VIBGYOR* (violet, indigo, blue, green, yellow, orange and red). Red is the longest and violet is the shortest in this range.

*Warm fronts* – *Warm fronts* are the leading edge of a homogeneous warm air mass, located on the equator-ward edge of the gradient in isotherms, and lie within broader troughs of low pressure than cold fronts. A warm front travels more slowly than the cold front.

*Weather front* – A *Weather front* is a boundary separating two air masses of different densities, and causes typical weather phenomena, particularly in the mid-latitude region.

*Western disturbances* – *Western Disturbances* (WD) are considered to be the eastward-moving extra-tropical upper air trough in the subtropical westerlies, often extending down to the lower atmospheric level.

*White Dwarf* – A few nebulae represent shells of gas ejected by dying stars. The escaped nebula appears blue as it shows the inner layers of the star and is hot. Say after 50,000 years or so, the planetary nebula drifts away, and the central star cools. As a result the central star contracts until it is about the size of earth. If a mass less than 1.4 times the mass of the sun is left, the star continues at this stage indefinitely. This stage is called *White Dwarf.*

*Wurst* – In German language *Wurst* means hot.

*Younger Dryas (stadial)* – The term *Younger Dryas* stadial was coined after the alpine / tundra wildflower *Dryas octopetala.* It was a brief (approximately 1300 ± 70 years) period of cold climate, also known as the *Big Freeze.* It occurred at the end of the Pleistocene between approximately 12,800 to 11,500 years ago following the Bölling/Allerød interstadial and preceding the Preboreal of the early Holocene.

www.ingramcontent.com/pod-product-compliance
Lightning Source LLC
Chambersburg PA
CBHW082132210326
41599CB00031B/5950